How to Implement Lean Manufacturing

About the Author

Lonnie Wilson has been teaching and implementing Lean techniques for more than 39 years. His experience spans 20 years with an international oil company where he held a number of management positions. In 1990 he founded Quality Consultants which teaches and applies Lean techniques to small entrepreneurs and Fortune 500 firms, principally in the United States, Mexico, and Canada.

How to Implement Lean Manufacturing

Lonnie Wilson

New York Chicago San Francisco
Lisbon London Madrid Mexico City
Milan New Delhi San Juan
Seoul Singapore Sydney Toronto

The McGraw·Hill Companies

Cataloging-in-Publication Data is on file with the Library of Congress.

McGraw-Hill books are available at special quantity discounts to use as premiums and sales promotions, or for use in corporate training programs. To contact a representative please e-mail us at bulksales@mcgraw-hill.com.

How to Implement Lean Manufacturing

1 2 3 4 5 6 7 8 9 0   DOC/DOC   0 1 4 3 2 1 0 9

ISBN 978-0-07-162507-4
MHID 0-07-162507-0

Sponsoring Editor
Judy Bass

Acquisitions Coordinator
Michael Mulcahy

Editorial Supervisor
David E. Fogarty

Project Manager
Madhu Bhardwaj, International Typesetting and Composition

Copy Editor
Michael McGee

Proofreader
Sonal Chandel, International Typesetting and Composition

Indexer
Steve Ingle

Production Supervisor
Pamela Pelton

Composition
International Typesetting and Composition

Art Director, Cover
Jeff Weeks

Contents

Preface

Why I Am Writing This Book

I am writing this book for several reasons.

First, I have been asked to. On numerous occasions, clients have asked me to write a book. The first time was on the subject of measurement system analysis (MSA); the second on statistical process control (SPC); the third time on statistical problem solving—and in the past five years I have been asked on no less than three occasions to condense my thoughts on Lean Manufacturing into book form. Writing a book is also something *I* have wanted to do, but just never had the time to do so. However, in the end, it was other reasons (which I later list in this chapter) that drove me to the effort.

Second, I am very tired of seeing managers everywhere looking for this "silver bullet" called *Lean Manufacturing*. They see it as a catch-all for attacking all their business woes—including poor profitability and low levels of competitiveness—and transforming their business into the pinnacle of profitability. I want to stand up and yell at the top of my lungs and make it very clear that *there is no silver bullet*. In this regard, I now offer up three quotes here.

Third, as a consultant I frequently find myself quite frustrated in being unable to sell Lean Manufacturing to a facility that desperately needs it. Maybe this book can put my thoughts into a clear and convincing format that I am unable to otherwise convey.

Perhaps I get particularly frustrated because I embody a tough combination of characteristics. For instance, I am reasonably talented in what I do, which sadly helps me less than you would think when it comes to selling the concept of Lean Manufacturing. In addition, I am burdened with a high degree of frankness. And finally, I am one of the world's worst salesmen. I laughingly tell people I could not sell free water to millionaires dying of thirst in the desert. Maybe I'm not that bad—but when it comes to selling things, both physical and ideological, I have a lot of

> **"In** the choice between changing one's mind and proving there's no reason to do so, most people get busy on the proof."
>
> John Kenneth Galbraith

> **"Men** stumble over the truth from time to time, but most pick themselves up and hurry off as if nothing happened."
>
> Sir Winston Churchill

> **"Opportunity** is missed by most people because it is dressed in overalls and looks like work."
>
> Thomas Edison

room for growth. Thus, this combination of skill, frankness, and lack of sales ability puts me in the position of having to tell plant managers (PMs) and general managers (GMs) that their plants are very sick, when, indeed, they are very sick. What they really would like to hear is that their plant only needs a little tweaking, and that this tweaking can be done in a few weeks. Furthermore, they often want to hear that *I* am the person who can do it, do it after only a few visits to their plant, and that I can imbue a person on the grounds with such fantastic skills that they will never have to hire another consultant again. Unfortunately, to their dismay, I usually have to tell them what they don't want to hear. I tell them that their plant is really sick (meaning their present as well as future business is in danger, and that the jobs of hundreds of their employees are also in danger); that their problems are fixable (if they *are* fixable, and they usually are), but that it will take months to fix them, not weeks; and finally I tell them that the specific skillset they wish to bring to particular employees at their facility will not come in only a few weeks either.

Thus, my approach is to tell them what *can* be done, as I see it. I have been in this business for a long time and have developed the ability to quickly size up problems like this and make reasonable value judgments about the effort and time it will take to reach certain goals. However, when the typical manager hears this, he is disappointed and all too frequently will politely say, "Thank you. Let's keep in touch," which translates to "*Adios amigo,*" and so off I go.

Losing a good job is bad enough, but my frustration comes from a deeper part of me that wants to make their facility a better money-making machine and a securer work environment for their employees. So I am doubly frustrated. Therefore, in this book, I will offer up numerous examples of these kinds of situations, with which I'm sure you will be utterly amazed. You may even mutter, "This can't possibly be true." But *au contraire…* true it is.

I have often thought about this problem—that is, the one about the plant manager who wants to solve his problem without taking the bull by the horns. I have discussed it with others, including my wife, psychiatrists, psychologists, my pastor, as well as dozens of top-level managers. As might be expected, the root cause of this problem has myriad descriptions—some call it denial, others laziness, and still others say it results from a "quick-fix mentality." But the one I keep coming back to is simplistic thinking.

So just what is simplistic thinking? I like to say that there are two groups of people who can solve any problem—those who know nothing about the problem but its name, and those who clearly know what is happening.

- For those who know nothing, everything is simple. If you have teenagers, all too often they fall into this category. Unfortunately, many managers also reside there. Fueled by success stories that are simplified by TV and literature in which—in a world of instant gratification—television detectives can solve any problem in under an hour, and despite facing persistent problems in meeting ever-tougher business objectives, executives continually want to believe that life is simple. Well, it is not. Life is not only complicated, it is difficult. It's easy to focus on the success stories, simplify them, and ignore many of the bumps and bruises so inherent in any change initiative. Unfortunately, many believe lock, stock, and barrel in these fairy tale stories. Even worse, they believe that such stories detail all the effort involved. This is never the case. Any journey always involves complications, confrontations, disagreements, and wrong-paths-taken. To ignore such things makes a story simpler and more publishable, but it does not make it any more true.

- On the other hand, a select few who have been blessed with both clarity of thought and the ability to take seemingly complicated situations and reduce them to a simplicity that is not only amazing in its clarity but also revealing in its truth. Justice Oliver Wendell Holmes is credited with the saying, "I don't give a fig for the simplicity on this side of complexity, but I would die for the simplicity on the other side." I find his statement extremely profound.

Really, It's Not That Complicated, so Let's Get Started

Now for the fourth reason I'm writing this book. Once understood, Lean Manufacturing—its technical engineering aspects and basic concepts—is really quite simple. However, in my dealings with clients, most do not really understand Lean and many try to make it much more complicated than it is. Consequently, throughout this book two concepts shall appear.

- First, "Points of Clarity" will be scattered through the pages, where seemingly complicated concepts will be reduced to their simplest form—usually just one pithy sentence. After all, if things truly are simple, they can typically be expressed in only a few words.
- Second, the application of Lean concepts has been reduced to a simple prescription—a prescription that, once the concepts are understood, can be readily applied to a wide variety of situations.

Huge Gains Can Be Made

Fifth, I have found from experience that if Lean Manufacturing is implemented, it is highly possible to derive huge early gains from the effort. I will give several examples of this throughout the book. These large early gains, which I call "low-hanging fruit," are sometimes the fuel used to catalyze a truly deep and profound plunge into the heart of Lean, including the cultural change that is both so necessary and so beneficial. Only when the cultural change is completed will the benefits be realized and fully exploited.

Unfortunately, harvesting this low-hanging fruit frequently feeds the bias of management, since they now have living proof that there really are "quick fixes," and so while the Lean implementation is really in its infancy the focus is sometimes changed to something else causing the Lean implementation, and its benefits, to predictably regress. The message is this: **Often there are huge early gains to be made, but if these benefits are to be sustained, a cultural change must occur**. This change does not come easily, nor quickly. All too often, in manufacturing as in life, things that come very easily often disappear just as quickly. This can also be true of gains made from the implementation of Lean.

Separating Out the Intimidating Cultural Aspects

Sixth, many of those who seek to begin a journey into Lean start by reading a book or two on the topic. Unfortunately, some find the literature too complicated, inapplicable, or its cultural aspects too intimidating. They become discouraged by the published materials and do not proceed. Many find the answer to "What is Lean Manufacturing?"—indeed, some books are very good in describing the "what is" aspect. However, very few address the issue of "How can I implement Lean Manufacturing in my facilities?" I like to say they are long on knowledge but short on applications.

Regarding the books on Lean, I find the materials to be in two categories.

First, some literature is just way too complicated for many to deal with, at least regarding the initial phases of implementation. Although there is a reasonable amount of good literature on Lean, some of it does not really reach the audience I am trying to reach, such as those starving for information on "Just how do I implement this thing called Lean?" For example, the best source of Lean information, for me, is Ohno's book (Ohno, 1988). I find it rich in both information and insights. But his book is a hard read. It's sometimes difficult to follow Ohno's train of thought. In addition, he has the ability to say some very powerful things in very simple sentences. As a result, the depth of his statements often goes unnoticed by those just beginning their study of Lean or the Toyota Production System (TPS). Thus, Ohno is often misunderstood. Worse, some of what he says goes completely unnoticed by the novice. I am not sure who Ohno characterizes as his audience, but he takes many things for granted that the typical reader does not fully comprehend, especially if they are just beginning to research Lean. For example, Ohno's explanation of how quality was under control when they first launched the TPS* is not fleshed out fully enough for the novice. Such things make his book, as well as Shingo's, a bit ineffective with some readers (Shingo, 1989).

Still other books have been written by ex-Toyota personnel. Some of these books are quite good yet many readers are unable to identify with them. I found this curious, since usually books written by insiders are quite good. Nonetheless, I would get comments from readers like, "Sounds good, but it's not something we can do here," or other remarks about the ideas' inapplicability, despite the fact that the system could indeed be applied to their situation. Still, it prevented these people from using these resources as they embarked on their own Lean journey—or worse yet, it prevented them from even starting a Lean implementation. Hence, I have written this as a simplified version of "How to Implement Lean," complete with a methodology for assessing a company's needs, as well as a prescription for implementation.

The second intimidating factor about some of the literature has to do with the deep cultural changes needed to fully implement Lean. For example, some people embarking on a potential Lean journey will read about how the line operator is empowered—actually *required*—to shut down production when problems occur. They then contrast this to their facility where they see this outcome as a practical impossibility. The gap is just too intimidating for them. Or perhaps they read about the belief that defects and failures are seen as opportunities to improve production rather than as obstacles to overcome, and they just cannot envision this idea as part of their culture. These are not technical issues, they are cultural issues, and in some cases they will require huge changes—changes larger than they can imagine. And so they become discouraged.

Make no mistake about it, these are powerful aspects of Lean, aspects that separate it from other manufacturing philosophies, and aspects that should be sought after.

Again, make no mistake. Toyota, as well as others that developed these cultural changes, spent a significant amount of effort and time on the process, and it came with many bumps, bruises, and wrong-paths-taken. But these cultural changes were *exactly* what was required—in fact, they are the reasons Ohno was so successful. He guided his culture through the murky waters of change, and did so with skill.

These required cultural changes should be the reason people adopt, not avoid, a Lean implementation effort.

---

*See the section "It Is Not a Complete Manufacturing System" in Chap. 2.

Later in this book, in Chap. 11, cultural change will be discussed. However, in this book, as much as practical, we will separate the behavioral level skills from the deep cultural changes required and try to highlight them when they appear. Ohno did not develop the culture of the TPS, nor did he get it fully implemented in just a few years. He spent over 30 years in this effort, and had built on what others accomplished before him. So to expect that in a year or two you can achieve what Toyota and others now have in place is not reasonable. The development of a culture is truly a journey, not an event.

To Enhance Your Understanding

Finally, I hope this book will enhance your understanding of Lean and its many applications, as well as its limitations. Lean, they say, is a journey without any destination. T. S. Eliot expressed this idea quite elegantly: "We must not cease from exploration. And the end of all exploration will be to arrive where we began and to know the place for the first time." My hope is that this book will enhance your own exploration.

Lonnie Wilson

Acknowledgments

I owe a debt of gratitude and wish to thank many folks for making this book possible. First, thanks go to the entire McGraw-Hill team with special thanks to Judy for giving me the opportunity and Madhu for guiding me through the process. Next, this would not have been possible without the assistance of my two sisters, Jani and Brenda, who are accomplished authors in their own right. Third I wish to thank my peers and technical review team of Randy Kooiman, Kermit Kirby, Lynn Torbeck, Don Bartlett, and David Hoffman who gave me invaluable advice. Most importantly, I want to thank the love of my life, my wife Roxana, who lovingly assisted me through this entire adventure.

How to Implement
Lean Manufacturing

CHAPTER 1

What Is the Perspective of This Book?

The perspective of this book is fourfold. It is written by a consultant who is focused on engineering principles and very practical applications so progress can be made in short order. It is filled with theory but has a "how to" approach for easy application. Finally, it is definitely geared to the manufacturing world and focuses on applications of Lean principles so that large early financial gains can be made.

From a Practical Perspective

First, this book is written from the perspective of a consultant. That means it offers a great deal of practical advice, is filled with examples, and is geared toward making progress—in some cases very rapid progress. Every effort is made to keep the principles and applications as simple as they can be and, where applicable, these principles are reduced to *Points of Clarity*.

From an Engineering Viewpoint without Much Cultural Advice

Second, this book is written primarily from an engineering viewpoint. Lean Manufacturing, also known as the Toyota Production System (TPS), is part of a *business* system, so there clearly are business aspects to it. However, an engineer should easily understand this book, a production supervisor should also, and almost anyone involved in processes will identify with these materials. Most of the principles invoked here are engineering principles that support the business. We have steered away from covering many of the large and deep cultural changes needed to fully implement Lean (TPS) for three reasons. But before getting into the specific reasons for this, let me tell you a story I have seen repeated many times.

An All Too Typical Story

Arthur, the plant manager, has just returned from the annual division planning meeting, where a great deal of attention was given to the need for manufacturing cost reduction. All plant managers were advised to look into the Lean Manufacturing system. Pumped with enthusiasm and a touch of fear, Arthur appointed John, his young, bright, energetic production manager to research the topic and make a proposal that they could discuss by the end of the quarter. So John searched the Internet and found a three-day Lean seminar that he could attend to broaden his understanding of Lean Manufacturing.

It looked good. The seminar leader was an ex-Toyota supplier support engineer with 25 years experience, and the course syllabus included some of the tools John was already familiar with, as well as some that were new to him. He was particularly interested in this seminar since it dedicated a full day to value stream mapping. The location of the seminar was nearby and the cost was reasonable, so off to the seminar John went.

During the seminar, John called Arthur to tell him it was a good investment; the speaker knew his stuff and John was able to meet other manufacturing people who seemed to have problems similar to theirs. All in all, he thought it would be a valuable experience.

Upon returning to work Monday morning, John was to meet with Arthur for a quick discussion—just to "calibrate their thinking," as Arthur said. When they met, John was enthusiastic as usual, but even more so, having been energized by the new-found hope for manufacturing progress. John said he could envision literally hundreds of improvements they could make.

Arthur, a somewhat impatient man, said, "John, I know you want to digest this further before you make your final recommendations, but why don't you give me the *Reader's Digest* version of this Lean thing."

So John began… "Lean is a system of cost reduction focusing on the elimination of waste. It is built upon two pillars: JIT and *Jidoka*."

Arthur said he was a little familiar with JIT but had not heard of *Jidoka*.

John explained *Jidoka* "as a way to use machines and manpower more effectively and to show respect for people. One principle of *Jidoka* is that no defective parts should advance in the production system. You see, one of the *kanban* rules is that no defective parts are sent forward, because it ruins the JIT concept since now you do not have the right quantity of good parts where they need to be."

"Ok," said Arthur, a bit impatient and confused, "but just what will we need to do differently in the plant?"

"Well, to implement *Jidoka*," John said, "we will need to make greater use of *Poka-yoke* technology; get everyone involved in problem solving. And when we do have a defect, we will need to shut down the line until the problem is solved and we're assured no defective product can go forward."

Arthur paused, took a deep breath, and asked about the Delta line, their newest production line, where they had thought about beginning their Lean initiative. "John," he said, "Just what is our dropout rate on the Delta Line and how frequently do we have a failure?"

"Production losses due to scrap are about 2 percent," John replied, "and during the day we'll have about 70 failed units."

"John, let me see if I understand this," Arthur said. "To do Lean, we need JIT and *Jidoka* concepts, and even JIT will not work without *Jidoka*. Plus, to make *Jidoka* work we will need to begin the practice of shutting down the line each time we have a failure, which occurs maybe 70 times per day. If that's the case, just how do we meet our daily production numbers, not to mention solve all these problems? Was it not just last week that you brought me a personnel requisition for a new production engineer to resolve the many problems on the Delta line which today is only making production demand because we are working 22 percent overtime?"

Arthur finally exhaled, allowing John to sheepishly reply. "Yes, that *is* what I understand about the system."

At this point in similar scenarios at many companies, the initiative simply loses momentum or dies completely.

Of course, the mistakes made by both John and Arthur are huge. The plant manager should not delegate this task, and if he does anyway, he should give John more time to organize his thoughts. Also, John should make sure he fully understands *Jidoka* before passing on this information; the decision should not be made in one sitting; and the list goes on.

Cultural Change Sometimes Seems Impossible

The preceding scenario, or some variation of it, I have seen dozens of times and so watched Lean initiatives die as a result. The primary point I wish to make here is that the decision not to undertake a Lean initiative was based solely on this large cultural change that a company envisions must be made initially at the implementation of Lean. It is typically their belief that allowing line personnel to shut down the production line is an impossible undertaking—one that will require major cultural changes that they feel is beyond their reach. Consequently, they decide not to even start a Lean initiative.

It is true that the shutting down of lines to remedy defects requires a deep cultural change within the organization. It is also true that this is one of the techniques that makes Lean Manufacturing so powerful—and it is, at some point in time, necessary. However, for Toyota, the transformation that was required to make *Jidoka* a powerful tool did not happen quickly, it did not happen easily, nor did it happen early on in the development of the TPS. Likewise, you need not—in fact, you should not, and more specifically you will *not* be able to—implement this technique early on in your implementation efforts. The order and the rate at which Lean techniques are applied is a very complicated topic. At times, the strategies, tactics, and skills employed are very obvious, at other times are not so obvious, and at still other times the technique needed seems completely counterintuitive. I am sure this is a topic you will need to discuss with your *sensei* (teacher of Lean) on numerous occasions.

Yet controlling the order of implementing the various Lean strategies, tactics, and skills is absolutely necessary. To implement a technique such as line shutdowns by operating personnel on Day One of a Lean initiative is as ludicrous as trying to teach your two-month-old son to play soccer. First, he must crawl, then he must walk, then when he is into running, there is some hope of teaching him soccer—before that, it is not a reasonable goal. Nor is undertaking the huge cultural change of implementing line shutdowns by operating personnel, at day one, a reasonable undertaking as part of a Lean initiative. The production plant must first learn to crawl—it must have a product to sell, must have good leadership, good problem solvers, and a willingness to change. Only then can it walk and develop a solid quality system with high levels of product stability, good process stability, and a deep understanding of variation. Building on this foundation, it can now try to run by learning about **quantity** control and such techniques as JIT and *Jidoka*. Afterward, it can endeavor to play on the field—and be a part of the game—which is the cultural change needed to solidify and perpetuate the system.

To try to dramatically change this natural progression is to invite, if not guarantee, failure. A failure just as large as if you actually expected your two-month-old to venture onto the soccer field and compete—something that is patently ludicrous. And if you have seen the things I have, you would agree that both are ludicrous.

Some Difficult Aspects of Cultural Change

As far as deep cultural changes are concerned (something typical of the TPS), I have used the example of line shutdowns here as a quality control technique. Many more exist. Another example of this deep cultural difference is the Toyota practice of not laying off employees. To support this practice, Toyota has developed specific and mature business skills designed to not only handle their employees but their business as well. Compare their system to the system within your company. Are you willing to hire the way you hire, train the way you train, acquire business the way you acquire business, and THEN promise your employees that you will never lay them off?

Other activities also require strong cultural development before they can be implemented. These include the ability to properly develop and promote from within, as well as conduct effective group problem solving, to name but a few. However, the biggest of all is the ability to mold your company's culture to not only accept but *invite* and even *encourage* change—as the TPS does.

The TPS is an odd culture indeed. While all cultures seek stability, the TPS seeks out change—and though it sounds as if it is striving for chaos, the system is anything but chaotic. The change it desires is driven by a culture of continuous improvement with proven problem-solving techniques, as well as proven processes that strive to reach mutually understood goals. Although the Toyota culture is forever changing, it is not the kind of energy-draining change that occurs in many cultures—specifically those chaotic cultures that go about changing deadlines and altering their basic production schedule almost daily, if not hourly, and then still using tools like overtime and freight expediting to meet shipping targets when nothing out of the ordinary really happened. It all seems directionless and pointless. Far too frequently, there are no good reasons for these changes and they suck up both your physical and emotional energy.

Thus, we need to review our paradigms and rethink what chaos really is. However, back to the need for change.

A Difficult Cultural Change for This Implementation Initiative

Let's examine for a moment the typical culture and how the implementation of our new Lean initiative might play out. This typical culture is one in which change is not embraced but is instead fought at all levels. Our typical culture seeks stability. As soon as we announce that we have a new "improvement initiative," we will face a series of questions from the workforce. Two questions will be raised by the majority of cultures, as they always are, because they have seen "these initiatives" before. The two questions are:

1. How do I know this is not just another "program of the month (POM)"?

2. Does this mean my job is in jeopardy? (From those workers fearing "improvement" means "efficiency improvement.")

Why do employees think these two things? Well, because that's what has happened in the past. If that is the history, there is no dodging this issue. It must be dealt with squarely and up-front.

So our Lean implementation coordinator (LIC) tries to allay their fears, telling them that Lean is a much broader initiative and that it's the real deal. It is not just another POM. He explains in painstaking detail how a Lean system works, with its emphasis on pull systems, operating at *takt* by producing just what the customer needs, reducing inventory, and attacking waste of all kinds. Furthermore, he displays myriad charts, tables, and graphs that make the TPS truly sound like the formula for reaching the promised land of manufacturing. He speaks of management commitment, employee involvement, employee empowerment, and quality of work life, including a renewed emphasis on training and leadership, and goes on and on and on. Everyone is enthralled not only by the presentation but also by the energy with which it is delivered.

And then one salty old engineer in the back row raises his hand and asks a question. "I have read a little about Lean, and as I see it, the effort is about making better products using fewer resources. Is that it?"

The moderator is pleased and responds with an energetic "Yes!"

The salty old engineer then says, "Well, the thought of being more involved and more empowered is quite appealing. However, I have a problem with this effort since you are asking me to be more empowered and more involved in reducing the company's resources—of which I am one. So I am not sure why I should be too excited."

This is usually followed by some clearing of the throats and some really good dancing around the subject by the LIC. But once on the table, the topic sits there like the Cheshire cat in *Alice in Wonderland*, until it gets resolved. And if your history includes prior "improvement programs" that led to layoffs, there is some serious explaining to do. Quite frankly, there are ways to deal with this issue, but they are not easy or kind. They are just less mean than continuing to misrepresent the truth.

So now the problem is out in the open and the question is, "How do you get the support of these people to make changes when the changes will involve some of them losing their jobs?" The answer is, if that is all you do, you cannot get their support and you are now in real trouble.

Contrast this typical culture to the culture of an already Lean facility like Toyota. They simply do not have this problem because they have worked on their culture for as long as they have existed as a company. It was part of the genius of the Toyoda family (founders of the Toyota Industries Corporation) that they understood the importance of the culture and worked to mold it from the beginning. In addition, they recognized the need to provide job security and almost guarantee employees a job for life. With this history and the emphasis on job security, no one in Toyota need ask the question posed by the salty old engineer. So, for the Toyota manager, creating change is not such an antagonistic issue. To reach this point in your own Lean facility is possible, but the road is neither easy nor short. However, beyond a doubt, developing a culture like Toyota's is a worthwhile task.

Nevertheless, do not underestimate the need to manage these issues in a typical culture here. When it comes to cultural change, reshaping the culture to not only tolerate but support and promote change is the biggest issue of all in the area of cultural change and is often the place to start if you want to begin making conscious changes to the culture.

All of these cultural skills (which are apparent within the TPS) are, in the final analysis, the aspects that separate the TPS from other manufacturing systems. All these skills require a strong, healthy, and well-developed culture. This development takes a great deal of time, effort, and management skill, but it does not come quickly or easily. It is these cultural aspects that make the TPS almost unique.

> **P**oint of Clarity The things which make the TPS unique among Lean Companies, are not technical in nature, rather it is how Toyota has been able to manage the culture.

Three Reasons Why Cultural Change Is Not the Central Theme of This Book

Quite frankly, it is not too difficult to design and implement a pull production system, that is balanced, operating at *takt*, and that has short lead times and high levels of quality. But to sustain it year after year, and ever-improve the system year after year, requires a very special culture that fully, and *I mean fully*, embraces the concept of continuous improvement. In this book we will not embark in detail on these deep cultural changes. These must be addressed in the future. So for now, let's return to the three reasons why I have not made these cultural changes the central theme of this book.

1. Many of these aspects of cultural development take years to develop and some businesses will not use them for quite some time, if at all. For example, I can't give you a single example of any company that effectively implemented line stoppages for quality problems in the first three years of Lean implementation. Nor can I name but a few who changed the thinking in the facility about how to put problems in the proper perspective. In most facilities, problems are to be avoided and even hidden. In Lean thinking, however, problems are that "avalanche of diamonds" that allow us to improve our processes and systems. Changes of behavior and changes of thinking of this magnitude do not happen quickly. Often, even with great effort, they do not happen easily. So my message is, "Don't start there!"

2. The culture of the TPS, although strong and worthy of emulating, was not designed out of whole cloth, so to speak. It was both created (designed) and developed as a result of a continual effort to consciously manage it. In a word, it also evolved. In your efforts, the same process is required—both creating the culture and developing it through conscious, consistent, continuous effort. Although your Lean culture may be similar, it will not be the same as the TPS. However, I would be remiss if I did not say that much of the culture of the TPS can be copied and implemented directly into your culture, *but it must be done at the right time, and in the right way*.

3. While implementing the technical aspects of Lean Manufacturing, you will, without really knowing it, begin to make large changes in the culture. Guided by your *sensei*, he/she will be able to assist in standardizing these changes in your culture and also help decide precisely when and exactly how to approach the deeper cultural issues.

In this book, this lack of focus on the culture is quite frankly a bit ironic. Our firm is expert in cultural change, so you might think that would be the major contribution of this book. However, our focus here is more on early improvement and finding the "low-hanging fruit." It is precisely because we know how much effort it takes to make major modifications in a culture and how long these issues take to mature that we stay away from that topic in the early stages of a Lean initiative. We have found that it is better to focus attention on the needed behaviors within the organization and then take on the deeper cultural topics as they surface—and surface they will.

To Begin to Change Your Culture, Start Small and Choose Wisely

Despite this advice, if you still wish to dabble in your culture, I will not discourage you—rather, I will act as a guide. In this area, the most meaningful early effort should be focused on one topic, which should be "How do we modify the culture so it not only embraces change but encourages it as well?" The logic employed is the following:

- If we want to survive, we must improve.
- If we want to improve, we must change.
- If we want to change, we need a culture that not only:
  - Accepts change, but...
  - Embraces and encourages change as well.

As far as making improvements in the business, that shift of paradigms is the single most powerful cultural shift I can think of. If this paradigm shift is achieved, it will accelerate all progress beyond your most optimistic goals.

This Book Has a "How to" Perspective

This topic of "How To" is something that many will advise you to avoid. They say flowery things like "The TPS is a roadmap, not a prescription," or "Lean is a journey in which everyone must find their own way." Well, flowery those phrases are; helpful they are not. Even if it is an accurate metaphor, every roadmap has a starting place and a destination even if there are many routes to your destination. I just find that many people are afraid to generalize, even when generalization is helpful. I wonder if they do not use this as an excuse, maybe because they do not really deeply understand the concept and the applications of Lean Manufacturing well enough. At any rate, I will describe several "How To" prescriptive methods coupled with an occasional word of caution that should easily carry your effort through the first three years of implementation.

The First Prescription—Four Strategies to Become Lean

The first prescription is "How to apply the four strategies to become Lean." It is a process that can be used for *kaizen*, a process of continual improvement. The prescription can be used for *point kaizen*—that is, some local situation such as might be used to make a work station or a cell leaner. Or this prescription can be used in *value stream kaizen* (sometimes called *flow kaizen*), where an entire value stream is addressed. This covers the diagnostic tools used and how they can assist you in your efforts to eliminate the seven wastes in the production process. Once the four strategies have been addressed, there is a large "to-do" list of *kaizen* activities designed to make your system Lean. This prescription is described in detail in Chap. 7, and the application of this prescription is further clarified in three case studies in Chap. 16.

The Second Prescription—How to Implement Lean... The Prescription for the Lean Project

The second prescription addresses how to take all the information and resources and apply them in a project format so your Lean design will become a reality. This prescription starts with a complex evaluation of the overall manufacturing system, goes into the specifics of the value stream improvements and turns it all into an action plan. In fact, this prescription is more than an action plan.

> **P**oint of Clarity This book is about the application of Lean principles.

It is an eight-step process that results in a leaner process. This second prescription is a proven technique (described fully in Chap. 8) and its application is shown in three case studies in Chap. 16.

"How to" Become More Secure and a Better Money-Making Machine

Make no mistake, this is not a primer on "What is Lean" or "What is the TPS," although much of that is contained herein. It is not a book about theory, it is a book to guide actions, and it is a book about applications. It is about how you can apply the principles of Lean and lead your facility through the first three years of your effort, making it:

> **P**oint of Clarity This book is about making your business a more secure money-making machine.

- A more secure work environment, and
- A better money-making machine

And to Those in Manufacturing Who Seek Huge Gains

Finally, this book was written for those in manufacturing who wish to excel in making their business healthier and more robust, as well as for those who are struggling to survive. It is one of the themes of this book that you can make huge early gains—for those in survival mode, this should be encouraging. As you read, pay particular attention to how quickly many of these companies achieved gains—and how large those gains were—by applying the prescriptions outlined here for you.

Although manufacturing is the focus, the Lean principles, once understood, have a very broad application. For example, we have applied these principles to staff functions within manufacturing such as training, procurement, engineering, and testing labs to name just a few. In addition, we have applied these principles to Doctors' offices, hospitals, hotels, private schools, restaurants, and other service functions. They have a broad, not infinitely broad, but very broad application to a wide variety of functions and industries.

Chapter Summary

Four themes run throughout this book. First, this is a practical book about Lean Manufacturing that includes many illuminating examples. Second, the book has an engineering perspective, but we have steered away, as much as is practical, from the deep cultural aspects that are more appropriately addressed later in the initiative. Third, this book takes a definite "how to" approach so as to show the applications of principles. Finally, this book is written to those in manufacturing who hope to make their business more secure and more profitable. For these individuals, there are often huge early gains to be made. (For a better understanding of cultures, including elements of the TPS culture, refer to Chap. 11.)

CHAPTER 2

Lean Manufacturing and the Toyota Production System

In this chapter, we explore what Lean Manufacturing is and what it is not. In so doing, we will define Lean Manufacturing from several different perspectives and compare it to other philosophies, including how Lean Manufacturing and the Toyota Production System (TPS) compare. The creator of the TPS, Taiichi Ohno, wrote a great deal about the TPS, and we will explore his thoughts on its uniqueness. Finally, we will discuss the limitations of Lean Manufacturing and its applicability outside the "Lean Stereotype."

The Popular Definition of Lean

The popular definition of Lean Manufacturing and the Toyota Production System usually consists of the following:

- It is a comprehensive set of techniques that, when combined and matured, will allow you to reduce and then eliminate the seven wastes. This system not only will make your company Leaner, but subsequently more flexible and more responsive by reducing waste.
- Wikipedia says "Lean is the set of 'tools' that assist in the identification and steady elimination of waste (*muda*), the improvement of quality, and production time and cost reduction. The Japanese terms from Toyota are quite strongly represented in 'Lean.' To solve the problem of waste, Lean Manufacturing has several 'tools' at its disposal. These include continuous process improvement (*kaizen*), the '5 Whys' and mistake-proofing (*poka-yoke*). In this way it can be seen as taking a very similar approach to other improvement methodologies."

What Is Lean?

The TPS is often used interchangeably with the terms Lean Manufacturing and Lean Production. Regarding the technical issues of TPS and Lean, I will frequently use these terms interchangeably. It is called Lean because, in the end, the process can run:

- Using less material
- Requiring less investment

- Using less inventory
- Consuming less space and
- Using less people

Even more importantly, a Lean process, be it the TPS or another, is characterized by a flow and predictability that severely reduces the uncertainties and chaos of typical manufacturing plants. It is not only financially and physically Leaner, it is **emotionally much Leaner** than non-Lean facilities. People work with a greater confidence, with greater ease, and with greater peace than the typical chaotic, reactionary—change-the-plan-hourly-and-then-still-work-overtime-and-then-still-expedite-it-all manufacturing facility.

Lean and the Toyota Production System

To further explore the depth of what a Lean Manufacturing system really is, we will look deeply at the TPS. Not because the TPS is the best Lean system around, although it may be. I can say it is the best *I* have seen. Rather, we will look at the TPS because it is the best documented system and it has proven itself over a very long time. It has not only proven itself but stands as the example of "Lean done extremely well."

What Did Ohno Say about the Toyota Production System?

If we really want to understand what the TPS is, we certainly must listen to what its creator has to say. In any discussion about the TPS, Ohno will be my arbiter. A great deal has been written about Lean and the TPS, but some of it misses the point. If there is any question about what Lean or the TPS is, we will use Ohno's thoughts as the final word on the topic. His most notable writing on the subject is his book, *The Toyota Production System, Beyond Large-Scale Production*. In it, Ohno makes three key statements, which when taken together define his TPS.

- "The basis of the Toyota Production system is the absolute elimination of waste." (pg. 4)
- "Cost reduction is the goal." (pg. 8)
- "After World War II, our main concern was how to produce high-quality goods. After 1955, however, the question became how to make the exact quantity needed." (pg. 33)

Taken together, we could then write a definition such as, "the TPS is a production system which is a quantity control system, based on a foundation of quality, whose goal is cost reductions, and the means to reduce cost is the absolute elimination of waste."

The TPS and Lean Manufacturing Defined

I find that all of these definitions miss part of the essence of Lean Manufacturing. I do not think Ohno defined it more carefully because oftentimes we do not feel the need to define those things that are very near and very obvious to us. I believe others may not define it better because they simply miss the point, while others may understand it but not be able to articulate it. My definition of the TPS is a manufacturing system that:

1. Has a focus on **quantity** control to reduce cost by eliminating waste

2. Is built on a strong foundation of process and product quality

3. Is fully integrated

4. Is continually evolving

5. Is perpetuated by a strong healthy culture that is managed *consciously*, *continuously*, and *consistently*

I refer to this as the five Fold Definition of the TPS. It is a much better characterization of the TPS, but unfortunately to those within the TPS this is old news, while to those outside the TPS this definition may not be very meaningful. So, as we go further into this chapter, I will try to put these five aspects of the TPS in context.

Who Developed the TPS?

The accepted architect of the TPS is Taiichi Ohno, the Chief Engineer of Toyota for many years. Others contributed greatly, including Shigeo Shingo and the members of the Toyoda family, but Ohno gets most of the credit for its creation, development, and implementation. Ohno may also get most of the credit because he wrote the most about it, or maybe just did the most about it. Neither is important. What *is* important is that the TPS specifically, and Lean Manufacturing in general, is a tremendous contribution to society and manufacturing in particular and we owe a huge debt of gratitude to those individuals who created it and caused it to mature even further.

The Two Pillars of the TPS

Ohno describes the TPS as consisting of many techniques that are designed to reduce the cost of manufacturing. His method of reducing cost is to remove waste. This waste elimination system, the TPS, is built on two pillars.

Just In Time

The first pillar is *Just In Time* (JIT). This is the technique of supplying exactly the right quantity, at exactly the right time, and at exactly the correct location. *It is quantity control.* It literally is at the technical heart of the TPS. Most people envision this pillar as inventory control, and this is a part of it. However, JIT is much more than a simple inventory control system. What's surprising to a large number of practitioners is that at the heart of quantity control—at the heart of JIT—is a deep understanding and control of variation.

> "**A**fter 1955, however, the question became how to make the exact quantity needed..."
> Taiichi Ohno

Jidoka

The second pillar is *jidoka*. This is a series of cultural and technical issues regar-ding the use of machines and manpower together, utilizing people for the unique tasks they are able to perform and allowing the machines to self-regulate the quality. Technically, *jidoka* uses tactics such as *poka-yoke*, (methods of fool proofing the process) *andons* (visual displays such as lights to indicate process status especially process abnormalities), and 100 percent inspection by machines. It is the concept that no bad parts are allowed to progress down the production line. This not only is needed to protect the customer and reduce scrap costs, it is a continuous improvement tool and is a key element in making *kanban* work. It is a violation of *kanban* rules to allow bad parts to be transported.

> **P**oint of Clarity   The TPS is a quantity control system.

What Is Really Different about the TPS?

Technical Issues

So just what is different about the TPS? What makes it so revolutionary? The answer to that is not so simple. First, let's look at the technical aspects, particularly some of the industrial engineering aspects. These technical skills and tactics are the foundation for the quality and quantity control aspects (items 1 and 2 from our Five-Fold Definition of the TPS). Some very old engineering techniques are used within the TPS. In addition, some old techniques with new twists, as well as some totally new techniques, are also included. Later, we will address the deeper issues of the integration (item 3), the evolution of the system (item 4), and the cultural differences (item 5). But as I said, let's first discuss some of these technical issues. We will compare several aspects of a manufacturing system as either Normal Mass Production Model (MassProd) or as Lean. And where the TPS stands out among Lean facilities, we will highlight that as well. These manufacturing aspects include:

- The makeup of the typical production cell/line and how quality is handled
- Handling multiple models of a product
- The use of "pull" versus "push" technology
- The issue of changeover times
- How parts and subassemblies are transported in the plant
- How finished product demand and supply variations are handled
- How quality is managed
- How cycle time variations are managed
- How line availability is managed

In the next subsection, we shall review them in order.

The Typical MassProd Production Cell/Line

The makeup of the typical MassProd production cell/line is normally a flow line. The typical flow line has work stations lined up in a row. It has inventory in front of and after each station. These work stations operate more like production islands rather than a connected process. The production of each station is maximized to improve equipment utilization, they call this "optimization" and product is pushed to the next station. This station-by-station "optimization" and "push production" are extremely key issues. The production rate is determined by the plan and all too frequently each island has its own plan. Normally product is produced, one model at a time, in large batches. There is an attempt to finish each batch before the appointed delivery time. The typical quality system is inspection-based, is normally done by humans, and rework is not only common it is necessary and designed into the process flow. This differs greatly from the Lean solution, which is typically a *cell, with a pull system, balanced at takt, flowing one piece at a time, with a system of jidoka.*

We will explain each of these more clearly later, but the key differences are that each work station has the same cycle time, they are balanced, and all operations are synchronized. All process steps are placed close together—in a cell, for example—so one-piece flow is possible. There is no buildup of inventory between processing steps. This close connection and lack of inventory allow the product to "flow." Furthermore, all work stations are balanced, in that each station has the same cycle time. This allows synchronization of manufacturing. Furthermore, the balanced cycle time is designed to be at *takt*, which is the demand rate of the customer. Product is made at the same rate at which the customer wishes to withdraw it, and production only occurs when the customer removes the product. This is the "pull" concept, which is different from "push" production. Quality is managed by the *jidoka* concept, which not only finds and removes any defects, it initiates immediate root-cause corrective actions. Using the preceding reference techniques, the Lean process itself, rather than any of its individual steps, is optimized, forgoing the optimization of local work stations. Many Lean facilities have very good *jidoka* systems, but the TPS is second to none in the application of *jidoka* principles, largely due to the way Toyota has managed, and continues to manage, their culture.

Handling Multiple Models of a Product

The handling of multiple models of a product is a uniqueness of Lean. In the typical MassProd where we have multiple models, they are normally produced in large batches even if they use the same production facilities. Generally, the reason for the batch operation is because when switching models, changeovers are required. The Lean solution employs the concept of model mix leveling. Say, for example, we have three models of our product. They are A, B, and C, all produced on the same process facilities. The model mix is 50 percent A, and 25 percent of each B and C. We then would produce these three models simultaneously in our cell, producing in this order ABACABACABAC. Of course, to do this single minute exchange of dies (SMED), or quick changeovers, or more likely its refinement, OTS (one-touch setups) must be very mature in this cell. Surprisingly enough to those not acquainted with the TPS, this is not as difficult as it sounds. In addition to accomplishing the leveling and knowing which model to produce, a *heijunka* board is employed to schedule the leveling of the various models.

The *heijunka* board will accept the production *kanban*, which are removed form the product when the product is withdrawn. These *kanban* normally are circulated directly to the pacemaker step and the replenishment signal, the *kanban* itself goes directly to the production cell, normally bypassing planning altogether.

> **P**oint of Clarity   Lean Manufacturing is a batch destruction technique.

Pull Production Technology

Pull production is a confusing concept to many, but it is a critical aspect which makes Lean work. It is one of the key tools used to avoid overproduction, both local overproduction at a specific work station and overproduction at finished goods. Pull production is the concept that the production process is not initiated by orders or schedules.

What triggers production is customer consumption. To implement pull, the business must make the huge cultural change of responding to what the customer **does**, rather than what the customer **wants**. It is the "take one, make one" concept and this production mode is called "replenishment". It is virtually impossible to have a pure pull system in most manufacturing environments, but all pull system have three characteristics.

- First, no production is initiated until actual consumption has occurred.

- Second, since the production process sometimes has mismatched rates or machines which require changeovers, storehouses are often needed to maintain flow. Hence, the production process must feed the storehouse, at a rate greater than the consumption of the customer. However, to maintain control, all storehouses must have a maximum upper limit on inventory. In fact the entire production process must have an upper limit on inventory to make pull a reality.

- And third, the production is triggered only when the next downstream customer in the production process comes to pick up the product. This aspect has gotten lots of press from the GM supervisor who said, "Yeah, I understand pull, cuz in a pull system we ain't sendin nuttin nowhere, somebody comes to git it".

MassProd systems use "push" production systems. They are the opposite of pull systems. They rely on schedules and forecasts and try to "push" the product to the next work station. They do not have controls on inventory so it is possible to have work in process (WIP) explosions with highly variant lead times and inventory problems galore. But the real killer problem solved by pull production is variation caused by scheduling changes, often initiated by customer order changes.

The downside of a pure pull system, is that it must be a "make to stock" system. You must have inventory to make it work, and recall inventory is one of the named seven wastes. Sometimes, holding this inventory is not practical so the inventory buffer is handled with a time buffer or approximated with a FIFO (first in first out) lane.

Finally I only know of two pull systems that have proven themselves in industry. These two are *kanban* and CONWIP.

Changeover Times

The issue of changeover times is generally an item that is largely ignored in MassProd. Long changeover times are often just simply accepted. Consequently, you have two options when employing model changeovers. First, you can shut down the entire line when the machinery undergoes a changeover. Of course, this causes a loss of production. In this case, all equipment needs to be oversized to account for the downtime. This then requires a greater initial investment. Alternatively, what is most often done is to oversize only the equipment that needs the changeover, to account for the downtime, and design inventory buffers in front of and behind the machines so the rest of the line can continue producing during the changeover. This helps keep the investment in the rest of the line down, at the expense of work in process (WIP) inventory. In addition, the conventional wisdom is that if the changeovers are done less frequently, the time to do the changeover, which is a form of downtime, can be distributed over a larger volume of parts, thereby reducing the per-part cost. This conventional wisdom leads to long runs and extremely large inventories before and after the machinery requiring the changeover. The Lean solution is a large paradigm shift. That is, in Lean we do not accept that the changeovers will take a lot of time. The technique employed is SMED or quick changeovers. With SMED technology, the need to oversize the machinery is dramatically reduced. In addition, the need to hold inventory is also reduced. SMED technology was developed and refined by Toyota. Shigeo Shingo not only designed much of it, he published several books on the subject and is considered the architect of SMED technology. SMED is a staple in the Lean toolbox.

Transportation of Goods

How parts, finished goods, and subassemblies are transported in the plant in the Mass-Prod seldom looks like a Lean facility. Nearly all materials are "pushed" and huge volumes of WIP build up between processing steps, taking up space and greatly inflating both inventory and operating costs. More and more, even in a MassProd facility, raw materials are being handled like in a Lean facility—Lean in the sense that *kanban* cards are used for raw materials replenishment. However, there the similarities end. In the MassProd system, often *kanban* cards are used but the entire *kanban* system is not employed. Rather, *kanban* rules are seldom followed; the most egregious errors being:

- The failure to reduce the number of *kanban* to achieve process improvement
- The willingness to transport defective products

These two rules are always followed in a Lean plant, along with the other four rules of *kanban* (see Chap. 3 for a discussion of *kanban*). However, in MassProd, seldom is *kanban* used for subassemblies and for finished goods. Rather the demand is scheduled and the volume is pushed through the system until it shows up at the storehouse. Huge volumes of inventory accumulate in the process, and the lead time is both long and unknown. The Lean solution is again a huge paradigm shift. In Lean, a pull system is used as described above. and inventory is volume, time, and location controlled. In addition, *kanban* was an invention of the TPS and they use it not only for raw materials, finished goods, and subassemblies, but for such items as tools as well.

Handling Product Demand and Supply Variations

How finished product demand and supply variations are handled in a MassProd facility is a function of the scheduling. As demands change or production rates vary, the scheduler tries to manipulate the planning program to respond. This is a terribly ineffective solution that leads to not only a lot of overtime, a lot of expediting, a large number of late shipments but very large inventories as well. In addition, it leads to very high levels of uncertainty and stress. Adjusting to these demand and supply variations in this manner is not only ineffective, it is frequently impossible. The planning model is designed to be updated weekly or sometimes daily, for example, but the production volume is changing hourly or by the minute, thus there is no way the planning model can be responsive enough to the actual production line needs.

In the Lean solution, the effort to respond to demand and production variations is accounted for by two factors. First, the variation in the processes is considered a huge problem and so is attacked from that standpoint. Stable processes are a foundational issue in Lean, and are taken care of early in the initiative. As a result, variations are dramatically reduced through good process management. Second, the variations that remain are managed by segregating the inventory into logical categories and responding to inventory withdrawals with problem solving when appropriate. These inventory segregations are:

- Cycle stocks (normal pickup volumes)
- Safety stock (to account for internal variations in production)
- Buffer stocks (to account for external variations in demand)

So, for example, when the customer comes to pick up product in a quantity greater than the scheduled demand, some of the buffer stock is removed. Once removed, buffer stock removal rules require that immediate corrective action be implemented (likewise, corrective action is required for the removal of safety stocks). In addition, by using a *kanban* system, once the product is picked up by the customer, the *kanban* cards are circulated and show up at the *heijunka* board at the production cell. The cards for buffer and safety stocks are a different color, which signals there has been a change that requires some countermeasure by the production cell, such as scheduling some overtime to replenish the inventory. If the *kanban* cards are picked up each hour, then in two hours or less, the signal that something unusual has happened is sent to the pacemaker process. The cell can then implement some countermeasure for replenishment. If that information goes to the typical MRP system, a countermeasure may be a full week away. The Lean solution is both more responsive and relies less on centralized planning functions for daily operations. It should be noted that for production floor scheduling, the classic models of MRP, MRPII, SAP, ERP, BRP, SCM and any other generic models are simply inadequate. Many efforts have been made to make them responsive enough, but they have internal inadequacies when it comes to shop floor planning. In most cases, the best system is *kanban* for triggering production. The inadequacies of these scheduling tools such as MRP, is a topic outside the scope of this book but your experience has already told you that your scheduling model will never make you Lean. However, if you wish to read about this, it is covered well in *Factory Physics,* (McGraw-Hill, 2008) by Hopp and Spearman.

Managing Quality in MassProd

How quality is managed in MassProd is changing somewhat. A few years ago bad quality was only a volume and financial issue. If production rates could not be met due to quality drop out, it was a crisis, but if rates could be met, then the motivation to reduce defects was one of finances alone, using short-term economics and immediate effects only. Since the rework of product was commonplace, this system caused little improvement in quality.

However, more recently, driven by customer demands for a higher-quality product, all manufacturing facilities in the supply chain have been affected. Now almost all firms still in business, at least claim to have a continuous quality improvement philosophy. However, old habits change slowly and most do a better job of talking about quality than actually delivering it. Typical quality systems rely heavily on inspection done by humans and have a high dependence on attribute data. In the end, they generally just sort out the defective product. Frequently the only place good inspection is done is at final inspection and test.

These systems are not designed to be used for continuous improvement and the defect data are employed mainly for yield and maybe quality cost data. Sometimes problem solving is completely absent; most often it is superficial at best. These facilities are not staffed for problem solving, and so quality problems not only persist, but frequently even the ones that were "solved," rise again. Finally, the attitude toward problems, in the typical MassProd facility is one of trying to avoid dealing with them.

Compare all of these efforts to supply quality to the Lean facility whose primary job, like the MassProd model, is to protect the customer. However, quality management runs much deeper. First and foremost, rather than just inspect and sort the product, the process is scrutinized deeply so it will improve and the need for both process and final inspection will be dramatically reduced. The amount of emphasis placed on process control in the Lean facility is staggering when compared to a typical non-Lean plant. The result is that the Lean plant develops much more robust processes with more stable cycle times and improved process and product quality. This isn't rocket science, simply good old-fashioned hard work with a deep-seated belief in process management.

In the Lean plant, the primary purpose of inspection data, which is largely variables data rather than attribute data, is for problem solving. Real problem solving is deep and practiced by all. *Jidoka* is used in all steps of the manufacturing process, not just at final test. The underlying principle of *jidoka* is that no bad parts are allowed to progress in the production cycle; even if it means shutting down the production line and ceasing production until the root cause of the problem is found and removed. Using this technique, it is clear that quality is equal to production. The second underlying principle of *jidoka* is that it is a continuous improvement process. *Poka-yoke* are used extensively and quality problems become immediately obvious through the use of *andons* and other forms of operational transparency. The uniqueness of Lean, in the area of quality, lies in four areas. Note that the first three of these distinctions are cultural, rather than technical:

- First, a quality problem is not just a reject, it is failure of the system, which is owned by all.

- Second, the quality problem is not bad news, rather it is often good news, signaling a weakness that can now be understood and corrected, thus leading to a more robust system, rather than be ignored and forgotten only to reappear again.

- Third, everyone participates in the technical solutions to problem solving.

- Fourth, and finally, the system uses a system of tools such as *poka-yokes* to attain 100 percent inspection.

How Cycle Time Variations Are Managed

In MassProd, cycle time variations are not considered a problem. They are seldom quantified, hence they are largely unknown and ignored. Average cycle times are understood, but to maintain average rates, large volumes of inventory are held between stations. As long as average cycle time is maintained, the variations do not affect the overall production, but only at the cost of huge inventory volumes (for an example of this, fast-forward to Chap. 18 and perform the Dice Experiment). Frequently, without understanding the damaging effects of cycle time variations, efforts will be made to reduce inventories. This is almost always met with a significant drop in overall production rates.

It is extremely common in MassProd facilities that even when individual stations perform at design rates, on average, the overall lines do not perform at the design rate. This is caused by the interaction of variation and dependent events that occur on the line. (This is completely explained by the experiment in Chap. 18.) In the Lean facility there is always a significant effort to optimize the cycle times, including significant efforts to reduce the variation in the cycle times. In the Lean facility, both the average and the variation of the cycle times, are known, understood, and managed.

Frequently, there will be specific tools at the line to signal if cycle time is being met or not. Common among these tools are the *heijunka* board. Where *heijunka* boards are not used, it is common to see counters and/or clocks to assist the workers and advise them if the process has slowed for some reason. These tools are not supervisory in nature, they are supplied for diagnostic purposes so process deterioration can be found and corrected. In addition, the concept of transparency of Lean systems has the element known as Standard Work (SW). SW is a set of tools, one of which is a flow chart with cycle times for each process step, so a supervisor, engineer, or manager can evaluate how well the process is performing and assist in process improvement.

How Line Availability Is Managed

In MassProd, line and machine availability, like cycle time, is seldom measured, thus it is typically not known or managed. If production should fall behind schedule, the typical response is usually overtime and expediting. In the Lean facility, availability is known, understood, and managed. The key issues in availability problems usually center on two issues. The first issue is materials, either defective material not allowing the process to perform, or stock outs. The answer to both can be *kanban* or other pull system tools. The second problem that adversely affects availability is usually equipment downtime. Tools like *andons* exist to signal problems, but the key tool used in the improvement of availability is Total Productive Maintenance (TPM).

Summary of Technical Issues

The following essentially summarizes the majority of the technical issues in the TPS:

- Production cells flowing using pull production systems
- Balanced, so synchronized flow is achieved
- Producing at *takt* rate
- Using *kanban* to reduce inventory
- Rate and product mix leveled to minimize inventory
- Using cycle/buffer/safety stocks to handle internal and external rate fluctuations (while keeping cell production stable)

What we find by reviewing this list of tools and techniques in the Lean facility is a clear lack of uniqueness. The use of cells and line balancing to minimize labor costs are engineering techniques that have been around for a long time and clearly preceded the TPS. The use of strict inventory controls, operating at *takt*, SMED and *poka-yoke* techniques are not really new, but are applied with a rigor and a vigor in Lean not seen in MassProd. So they are different only in the way they are applied, but they certainly are not revolutionary. So technically there is a difference in the application, but compared to the last three aspects of the five-part definition of the TPS definition, the technical and operational differences are relatively minor.

The Integrity of the TPS

The third major difference is the integrity of the system. Most people think honesty is a synonym for integrity. It is not. Integrity has come to mean honesty because so many people misunderstand the word "integrity" and misuse it. This is not all that uncommon. Words will often change meaning due to misuse. Integrity, integration, and the word integer all have the common root in the Latin word, *integrare*, which means to make whole. The word integrity means "state or quality of being complete; undivided; unbroken," (from my *Webster's New Collegiate Dictionary*). The TPS is a highly integrated manufacturing system, and is distinct from most manufacturing systems. It is integrated internally as well as externally; horizontally as well as vertically.

How is it internally integrated? What are the characteristics it has that lead us to call it integrated? First, it operates as one system from beginning to end, which is the supply to the customer. It is a whole system, with the customer being the pacesetter of the process by consuming the product. A series of pull signals, starting with customer consumption, cause the system to operate in unison: a take-one-make-one system. Every technology is challenged to make the system "flow." Every technology is challenged to speed up the flow of the product through the system. Every technology is challenged to reduce the distance the product travels and the time it takes to deliver the product to the customer. Every effort is made to shrink the system in both time and space; this makes it more compact and smaller. The opposite of an integrated system would be a disintegrated system. A system that is scattered and a series of islands connected in some fashion. The TPS is designed to eliminate these islands of production whether they are batches or large "monument" machines requiring very long setup times.

Second, and most importantly, the TPS is integrated because there is a deep understanding of the concept of systems and that the system is the entity that requires optimization. They fully understand that "the system optimum is not necessarily the sum of the local optima."

Most manufacturing systems strive for local optimum conditions that are often at odds with the overall optimum. For example, they will strive to have high machine or manpower efficiencies, even if that means overproduction and finished goods sitting in the warehouse without sales to pay for them. The goals system and the accounting system in most plants are geared to drive toward these local optima. If managers proceed without an understanding of the local impacts on the system performance—that is, proceeding without an understanding of the integrity of the system—it is easy to push the entire system to a nonoptimum location: or simply put, to waste money. When a manager is striving to optimize the performance of his division amid the changing demands of the customer—a process that includes delivery, personnel, and raw materials problems—it is often difficult, if not impossible, to maintain focus on what is best for the system.

So internally the TPS is integrated as it operates as one entity—in a synchronous fashion—and always strives to optimize the system to supply maximum value to the customer.

However, the TPS was the first system to become externally integrated as well. It is externally connected to both the customer and the supply chain. It is connected to the customer because the customer wants value and that is the foundation upon which the TPS is built. Most production systems are built on the concept of providing volume at low cost, which does not connect them as directly to the customer.

The TPS is connected to the supplier chain as if the supply chain was an extension of the plant itself. Ohno started incorporating what he calls the "cooperating firms" in TPS concepts before the oil crisis of 1973. In his book, *The Toyota Production System, Beyond Large Scale Production*, (Productivity Press, 1988) he stated:

> **"T**hen after the oil crisis we started teaching outside firms how to produce goods using the *kanban* system. Prior to that, the Toyota Group guided cooperating firms ... in the Toyota system."
>
> T. Ohno

At this time, as early as 1955, the primary, if not sole, concern of all customers was the laid-in cost of an item, and yet Toyota was busy teaching their suppliers the TPS and techniques such as *kanban*. Today, it is commonplace for a supplier to seek advice and support from their customers. Prior to 1974, this was revolutionary. Nonetheless, it had the effect of better connecting suppliers to Toyota, which is integration.

Many companies have tried to mimic this concept of integrating suppliers. What most do is talk about a cooperative long-term relationship based on trust and mutual support. However, in the end, they just hammer the supplier to produce a lower-cost product without working with them on how to produce it less expensively. A few minutes after the discussion of trust and long-term relationships is finished, the topic of costs comes up. Altogether too often this is a discussion that ends with the open or veiled threat of "if you can't cut the costs, we'll be forced to find someone else who can." So much for mutual trust and that long-term relationship.

Also, having worked with many customers and suppliers alike, one thing I find very common is the case where customers will demand things from their suppliers that they themselves are not capable of doing. Just because they are not competent does not mean they cannot be demanding. This is not so with Toyota. Ohno refers to this as the "My plant first principle." Without exception, I have found that for any production technique they demand, they are also capable of assisting the supplier if not outright teaching them. I find this interesting that customers will require skills and standards from their suppliers, which they themselves do not have. I have often wondered how they can then evaluate if their suppliers are complying, much less be able to assist them.

Toyota has an integrated production system for many reasons, but at its deepest core are these concepts.

- They fully understand what the customer wants: *value*.
- They know how to provide value, using a system-optimizing production system, and forgoing local optima.
- They have readily available information that tells them how the system is producing.
- They are willing and able to respond to the system if it is not optimized.

A Philosophy of Continuous Improvement

Fourth among the differences is the reality that things can always be done better, faster, cheaper, and with less waste. In addition, Toyota has shown a great awareness of the world of manufacturing with their ability to be introspective and questioning. They clearly understand the concept of continuous improvement and recognize that their system is never fully optimized. If the TPS is to improve, it must change, it must evolve.

At the foundation of continuous improvement is the education process. Not only are workers and supervisors trained, but suppliers are also able to receive excellent training. They recognize that the minute you stop trying to learn is the minute you stop improving.

A Culture That Is Managed

Finally, and certainly the most important of all differences, is the culture. It is a culture of consciousness. They are aware of what is going on. It may sound odd, but most manufacturing plant managers are only modestly aware of how their plant is functioning. They are hampered by: little time on the floor; poor information systems; unclear goals and objectives; ever-changing philosophies; and myriad operational problems, including low line availability, poor quality, and delivery problems. Altogether, these issues then create a culture of chaos and firefighting. It is no wonder they are not on top of the action. The culture of Toyota is one of awareness, and this is no more obvious than in their characterization of such things as inventory and operational problems.

Ohno describes inventory as not only consuming space, raw materials, and time, but he explains that it hides the real problems of a system so they cannot be found and corrected. It is an oddity of the TPS that they seek out problems. The vast majority of managers yearn for the day when they will have no problems. Not so with Toyota. Awareness of problems, awareness of value supply, awareness of operational performance—pick your topic. Awareness is a cultural characteristic of the TPS.

There is a great deal more about the culture of Toyota (and cultures in general) in Chap. 11, but it too only scratches the surface of this topic. There is one cultural aspect, however, that deserves a little coverage here: *continuity*. W. Edwards Deming, in his book *Out of the Crisis*, (MIT, CAES; 1982) speaks about "constancy of purpose". Toyota is the poster child for this. They have exhibited the same principles for over 50 years. The principles apply to them and to all those who work with Toyota, especially their supply chain. The principles have been maintained through numerous management changes and through crisis after crisis, including those crises that have threatened the very existence of the company. Through all this, their principles have not changed. This type of continuity is almost unheard of in industry and it is the key reason why the culture has been so strong and why it has endured so well.

The Behavioral Definition

Reduction of waste and the specific definition of waste is almost a unique contribution of the TPS. Ohno defined waste in ways that no others had really thought. He described seven types of waste:

1. Transportation
2. Waiting
3. Overproduction
4. Defective parts
5. Inventory
6. Movement
7. Excess processing

Others have tried to remove these wastes, but the TPS has carefully defined them and made the continual reduction of waste an effort that is almost religious in its fervor. This is the definition of the TPS at the behavioral level; at the action level. However to really understand the TPS one must go deeper into what Ohno has written.

The Business Definition

In his book, *The Toyota Production System, Beyond Large-Scale Production*, (Productivity Press, 1988), when asked what Toyota was doing, Ohno is quoted as saying:

> **"A**ll we are doing is looking at the time line, from the moment the customer gives us an order to the point when we collect the cash. **"**
>
> T. Ohno

Beyond this discussion, there is little that would lead you to believe this is a business model. There are no discussions of markets and market share, no discussion of stocks and earnings per share, no discussions of paybacks and return on investment; only this discussion on the principle of cash flow, or more pointedly, how to use manufacturing to improve cash flow. He has no other financial discussions, not even the calculations to justify manufacturing projects. The TPS is clearly not a business system; it is a production system.

Several Revolutionary Concepts in the TPS

To appreciate the TPS and its genius, it is worthwhile to view some of its more revolutionary concepts.

- Supplying value to the customer
- Reducing lead times
- Focusing on the absolute elimination of waste; especially the waste of inventory

None of these concepts are new, but some aspects of the TPS are attacked with such a religious fervor that they seem almost a uniqueness of the TPS. Many businesses, long before the TPS became popular, would attack one or more of these issues, but none has packaged it in such an integrated way. Nor do they attack it with such single-mindedness as is seen in the TPS.

The Supply of Value to the Customer

This is a revolutionary concept as used in the TPS.

From the early years of mass production, the metrics of manufacturing focused on two parameters. Cost and production rate were the two items of major importance to manufacturing firms. Sometime later, in the 1960s or even the '70s, quality became a major issue. The typical plant manager was always working to meet the production schedule to reduce the costs of production and make sure the product met the quality standards.

Of course, today, these three factors are always on the mind of any businessman. However, Ohno began to think in other terms. He basically said, "I know what my plant needs from my perspective, but what does my plant need from my customer's perspective? His answer became his key metric, it was **value**. He described it as those things the customer was willing to pay for—which is what he called value.

> **Value-Added Work**
> To be defined as "value-added" work, the activity must meet these criteria:
>
> - It is something that adds to the form, fit, or function of the product.
> - It must be something that the customer is willing to pay for.

Suddenly, much of what a typical plant does is now questioned if this concept is truly employed. Of course, producing defective units is not a value-added activity, but consider packaging and transportation, for example. We go through serious design efforts to develop good packaging so we can transport the parts of an automobile, for example, to the final assembly plant. We utilize design tools such as Failure Mode Effect Analyses (FMEA) in the design of the packaging to assure we have no losses during transportation, and that the end packaging is suitable for our customer. Now Ohno calls not only the packaging we so carefully designed waste, but the transportation costs as well. The customer does not care that the steel came from Brazil. He does not care that the steel was packaged and transported to Mexico, where it was stamped into a wiper blade holder and then packaged and sent to Detroit to be assembled into a wiper blade assembly so it could be packaged and transported to the automobile assembly plant in Tennessee, where it could be installed on a car, which was then prepped and transported to Seattle, Washington for sale to some customer.

The customer does not care that this wiper blade traveled 25,000 miles and went through four packagings and four unpackagings, hundreds of handlings, and four tiers of suppliers with all the associated costs, before it was even attached to his car. His primary concern is that he gets good value for his expenditure.

This understanding and application of value is truly revolutionary.

Another way to look at this concept is by comparing it to what is called the Golden Rule: "Do unto others as you would have them do unto you." This is great advice and quite frankly a real stretch for me and others to aspire to. It requires that you think carefully about a situation, and think about exactly what you would like them to do to you. The question is, "How would you like them to act upon you?"—and then proceed to do the same with them. This level of introspection and detachment is extremely difficult, yet when followed it will lead to a higher level of awareness and a higher moral awareness as well. It is then hoped that this awareness will lead one to a more appropriate action on their part.

But I maintain that there is a flaw in this logic. It requires that you act on them *as you would like to be acted upon*. Well, what about their wishes? Sometimes these things can be reduced to simplicity for understanding. Apply this to something simple, such as buying a present for someone. If you follow this maxim, you will end up getting them what *you* want. Well, *maybe* that would be a good gift, but should we not get people what *they* want? I think so. And that is what makes getting gifts so difficult at times: It requires a high level of empathy—a quantity that is in increasingly shorter supply in our narcissistic world. Therefore, I believe the Golden Rule should instead be, "Do unto others as they would wish to have done unto themselves." Then they get what *they* want, not what *you* want.

This is what Ohno did, he put himself in the shoes of the customer and looked at value. Well, you might argue that this is what most typical plants do. On this point, I doubt this is the case, however. Plants look first to survive and second to prosper financially. Concepts to the contrary make for good discussions, but in the end if the place makes money, it stays in business. If it does not, it disappears. Not too complicated. So the typical plant, while looking at production rates, cost, and quality—and seemingly looking to the customer—are actually only looking internally, in order to survive and prosper.

Don't get me wrong, there is nothing wrong with surviving and prospering. After all, the customer counts on and needs the product; the people need the jobs created by the plant; and in the end the plant is providing several societal needs. Ohno simply took a huge leap beyond the normal thinking at a plant and in the end really tied his manufacturing system to the customer, letting the customer decide how he should redesign his system. He truly was connected to the customer, and that connection helped provide what the customer really wanted: *value.*

Reduction of Lead Times

In his book, *The Toyota Production System, Beyond Large-Scale Production*, (Productivity Press, 1988) Ohno commented on what Toyota was doing, writing that:

> **"A**ll we are doing is looking at the time line from the moment the customer gives us an order to the point when we collect the cash .... *And we are reducing that time line by removing the non-value-added wastes.* **"**
>
> T. Ohno

> **P**oint of Clarity   A key metric in the TPS is lead time; the key goal is lead time reduction.

The point here is that the TPS is clearly a system whose function is the reduction of lead times. In Ohno's writings, he does not really stress this point beyond what has been quoted. Instead, he focuses on the means to achieve reduced lead times, which are, of course, waste reductions.

The means of reducing lead time is through waste reduction, but the benefits of reduced lead time go well beyond the obvious savings regarding the waste that was eliminated. The beauty of reducing lead times can be seen in a variety of activities in the typical business. Take, for example, when a specification needs to be changed. This always raises the question of product obsolescence and the scrapping of those obsolete units. However, the bigger business question is, "Who pays for the obsolescence?" As you reduce lead times, the impact of this obsolescence is reduced. This fact is not missed by the typical production manager.

However, lead time reduction goes way beyond that and can be seen in two parameters that all managers want, but few know how to obtain. A plant with shorter lead times is both more *responsive* and more *flexible*. It is more responsive in terms of being able to change when the customer's schedule changes. It does not matter if the change is volume or model mix or both. With reduced lead times, a plant is better positioned for both types of changes.

There is yet another benefit of reduced lead time that is not discussed at all in the TPS literature: future business. It is not part of the TPS literature because the TPS is designed for a business that has a secured customer and some sense of future commitments—hence, a relatively stable demand. The concept of *takt*, for example, states that there is a commitment of product demand by the customer. This is not true in all businesses. Take a typical job shop in which each job is unique—possibly a cabinet maker or your standard air conditioner supplier. It is not easy to calculate *takt* for them, yet unbeknownst to most of the mass production world, lead time is THE key metric for them. Having a short lead time not only improves their quality responsiveness and cash flow, it dramatically increases their possibility of getting future work. If the salesman can quote short lead times and deliver, he will get a lot of business. He will literally steal business from long-lead-time suppliers even if he does not have the lowest cost. Interestingly enough, in my experience I have found that those with the shortest lead times are often also the suppliers with the lowest cost.

In fact in our rapidly changing world, although short lead times do not guarantee success for this type of business, long lead times will almost certainly guarantee failure. So in a nutshell, short lead times for them are equal to future business. (Chap. 5 has more information about lead times.)

> **P**oint of Clarity The way to become more responsive and more flexible in a manufacturing business is by reducing lead times.

Initially, most engineers and manufacturing managers do not see the power of lead-time reductions. It is no underestimation to say that at the very heart of a plant's flexibility and responsiveness is the topic of lead times but Ohno sought to reduce lead times so he could get paid sooner.

However, once lead-time reductions are achieved, a number of other equally powerful manufacturing qualities are unleashed. First, the plant becomes both more responsive and more flexible. These are both obvious and powerful manufacturing skills to have. However, the fact that you have a shorter lead time has a significant impact on variation reduction, most notably the variation in the production schedule. Think about it, how accurate is your production schedule today or one day out? Generally, the schedule is pretty good. What about one week out or even one month out? As you go to one week out, or one month out, you can count on some demand variation. But what about six months out? How much do you think the production demand will change? Thus, it's simple: In order to reduce this demand variation, reduce the lead time.

Through the Absolute Elimination of Waste

In his book, Ohno states "The TPS, with its two pillars, advocating the absolute elimination of waste, was born in Japan out of necessity." Think about that: the "absolute elimination of waste." Not the reduction of waste, but its *elimination*.

Ohno categorized wastes into seven principle types. They are:

- **Overproduction**. This is the most egregious of all the wastes since it not only is a waste itself but aggravates the other six wastes. For example, the overproduced volume must be transported, stored, inspected, and probably has some defective material as well. Overproduction is not only the production of product you cannot sell, it is also making the product too early. An interesting note about overproduction is that, in my experience, I have found that nearly all of the overproduction is planned overproduction. It is planned, and often for a variety of good-sounding reasons. However, upon scrutiny, I find that nearly all planned overproduction should be eliminated. For example, to assure they have sufficient finished goods, many companies plan for extra production and purchase extra raw materials because they will have quality fall out during the process. This planning process is really just guesswork and adds considerably to the variation in the process. Even worse, many companies work hard to fine-tune this planning process so as to minimize the waste of planned-overproduction. *Thus, we have the already scarce supply of technical manpower working to remove the planned-overproduction, which is caused really by the planning process, which saw a need because there is a quality problem which affects production quantities.* So why not attack the quality problem and get rid of all this waste, including the waste of the lost technical manpower? Sounds simple, but it is often overlooked.

> **P**oint of Clarity   Don't work at getting good at something which should not be done at all!

- **Waiting**. This is simply workers not working for whatever reason. It could be short-term waiting, such as what occurs in an unbalanced line (see the story of the Bravo Line in Chap. 15), or longer waits, such as for stock outs or machinery failure.

- **Transportation**. This is the waste of moving parts around. It occurs between processing steps, between processing lines, and happens when product is shipped to the customer.

- **Overprocessing**. This is the waste of processing a product beyond what the customer wants. Engineers who make specifications that are beyond the needs of the customer often create this waste in the design stage. Choosing poor processing equipment or inefficient processing equipment increase this waste also.

- **Movement**. This is the unnecessary movement of people—such as operators and mechanics walking around, looking for tools or materials. All too often, this is frequently overlooked as a waste. After all, the people are active; they are moving; they look busy. The criterion is not whether they are moving, it is: Are they adding value or not? I can't think of any example of people movement that is value added. Work design and workstation design is a key factor here.

> **P**oint of Clarity   The TPS is a batch destruction technique.

- **Inventory**. This is the classic waste. All inventories are waste unless the inventory translates directly into sales. It makes no difference whether the inventory is raw materials, WIP, or finished goods. It is waste if it does not directly protect sales.

- **Making defective parts**. This waste is usually called scrap. But the phrase Ohno uses, "making defective parts" is classic Ohno. Most people use the term "scrap," so they view the defective part as waste. Ohno moves far beyond this. He not only categorizes the part as scrap, but the effort and materials to make it. Ohno was a natural process thinker. In this case, he not only lamented the loss of a production unit but the fact that people spent valuable time, effort, and energy to make the unit—all of which was lost, not just the production unit.

The TPS Is Not a Complete Manufacturing System

The TPS is not a complete manufacturing system. In fact, it is only a part of a manufacturing system. To better understand what part of a manufacturing system it is, or rather what it is *not*, we need to return to Ohno's book for a moment. While discussing flow as the basic condition, he writes:

> **"A**fter World War II, our main concern was how to produce high quality goods and we helped the cooperating firms in this area. After 1955, however, the question became how to make the exact quantity needed. **"**
>
> T. Ohno

These two sentences are so simple that their significance is missed by almost everyone. However, the implication to these two sentences, especially to those wishing to undertake a TPS initiative, must be thoroughly understood. For example, let me paraphrase it a bit:

"From the end of WWII until 1955, we had focused our attention on improving the quality of our goods. By 1955, we thoroughly knew how to provide quality to our customers. We could discuss the key quality concerns with them, we could determine how to supply quality and we could provide it to a very high level. We used a long list of tools to achieve these communications skills, but the most important two were the simple customer quality questionnaire—which we statistically analyzed, of course—and we also became very proficient at Quality Function Deployment (QFD). Long ago, we ceased using inspection, especially visual inspections by humans, as a means of achieving quality. Instead, we moved to process control as a means to make the process more robust. To do this, we first became proficient in data gathering and analysis using such techniques as those that *Ishikawa* outlines in his writings. Now the vast majority of our data is used for process improvement rather than product evaluation. In addition, we became very proficient in a wide range of statistical techniques so we could analyze and make better decisions with our data. The four fundamental statistical techniques of Measurement System Analysis (MSA), Statistical Process Control (SPC), Designs of Experiments (DOE), plus Correlation and Regression are widely understood at even the supervisory level in our plants. We also made all levels of personnel responsible for root cause problem resolution, which means we trained them in various levels of problem solving—the "5 Whys" being the cornerstone technique. Another significant effort was the transition in our quality, process, and product data. Initially, the majority of our data was attribute data on the product. We moved from a high percentage of attribute quality characteristics of the product to variables characteristics of the process. We recognized early on that high levels of quality could not be achieved if we used attribute data, so this meant we needed to correlate the attribute defects to process parameters, and so we became skilled at this very early in our quality efforts. We had been very committed to providing quality products and had been working very hard on quality. With the help of Deming and a unified effort pushed by JUSE, we could supply excellent quality and our costs and losses associated with quality were very low. We had what most Westerners would call a very mature manufacturing system that could consistently produce high-quality goods and deliver them on time for a reasonable price. Quality was no longer a production problem. Now we needed to look at the losses caused by producing the wrong quantities—especially the wrong quantities produced and delivered to the wrong places at the wrong times."

If I were Ohno, that is what I would have written, because that is where they were as a manufacturing company. So what Ohno built, the TPS, had a foundation of quality, but his TPS is not a quality system. Yes, it had *jidoka*, but we will learn that *jidoka* is there to support JIT. In addition, and as an example, Ohno makes nearly no mention of Cp and Cpk which are the two accepted measures of process capability or "process goodness." Nearly every book you read on manufacturing and process quality reduces the concept of quality to Cp and Cpk for all measurement data, yet Ohno hardly mentions it. One has to ask why? Well, it is because of what he said—they simply could supply high-level quality, and quality improvements were not what they needed to focus on to reach a higher level of manufacturing excellence. Their focus, as he says, was on quantity. Make sure you spell that: q-u-a-n-t-i-t-y.

> **P**oint of Clarity The TPS was built on a foundation of quality and the focus is to control quantities.
> —(see House of Lean in Chap. 20).

The implication of this information, to some company that wishes to embark on the journey into Lean, is usually quite sobering. Ohno says they just spent seven years—seven very focused years—learning how to deliver *quality*, after which they embarked on the journey of *quantity* control. I can say with assurance that most companies who entertain the option of mimicking the TPS do not have the sound foundation that Ohno describes in his writings. After all, that is what made them Toyota and the TPS only took them farther along.

Herein lies an interesting aside. It is part of the genius of Ohno and others like him that they do not care if you try to copy them. They know that you can't unless you have undertaken and built the foundation that they had in place when they started their individual journey of quantity control.

Most think they can bypass this step and are always disappointed to find that there are no shortcuts. If you want the benefits of the TPS, the foundational issues must be addressed. It does not mean you should not embark on the journey. I am not saying that. In short, the foundational issues must be addressed, or your effort will be in vain. However, it is possible, with good guidance, to attack the foundational issues as well as the quantity control issues, simultaneously.

To Not Understand This Concept Is Dangerous

This concept alone, specifically that the TPS is built upon a strong foundation and that one huge element of that foundation is high levels of delivered quality, is the reason most companies fail while trying to implement a Lean initiative. I do not mean their efforts yielded less than they had hoped for, I mean some downright failed.

For example, I got a call from a potential client who was trying to mimic the TPS. They called me when their production rates flagged and on-time delivery had dropped below survival levels. They described their efforts as a JIT Implementation. Over the phone, it took me about two minutes to diagnose their problems. They had tried to install a JIT system without the help of an expert. They had plunged headlong into an inventory reduction effort to improve lead times but were burdened by two major flaws. First, they had only a superficial understanding of the TPS. Thus, they had, in effect, placed a high-powered rifle in the hands of a child. Second, even if they understood the basics of JIT, their system was not able to undergo inventory reductions without attacking the underlying and necessary foundational issues, principally the reduction of process variation. Consequently, the JIT system exacerbated, rather than solved, their problems. My advice to them was to hire an expert, immediately, to help them. They said they did not have the resources to do that. My next best recommendation was to undo what they had done and return to the method they previously had, so they would at least survive. I am not sure exactly what they did, but I later learned that the business had closed—and with it over 200 people lost their jobs.

Still others implement the TPS system and fail to achieve what the system is capable of delivering. Quite simply, there are no shortcuts, and to that end, shortcuts of understanding are the most devastating. So if you want to embark on any initiative, make sure you have both a thorough understanding of the initiative, as well as a commitment to implementing it. (See the Five Tests of Management Commitment in Chap. 19.)

A Critical and Comparative Analysis of Various Philosophies

A number of philosophies exist—some of them still popular—that are an effort to improve business. Most focus on manufacturing and are generally a response to the success achieved by the Japanese in the automobile industry. In no particular order, they are:

- Theory of Constraints (TOC)
- Deming's Management Technique
- Total Quality Management (TQM)
- Crosby Approach to Zero Defects
- Six Sigma

I will touch on each briefly since there are many books and articles available to those who wish to research this further.

The Theory of Constraints

The Theory of Constraints (TOC) is a concept developed by Eliyahu Goldratt while he was trying to create a planning program to make chicken coops for a friend. The TOC addresses three major concepts. First, it covers process bottlenecks, the logic of problem solving, and contains a touch of business theory that nicely simplifies the topic of money in a manufacturing business. His system is strong on inventory reductions, reduced lead time, and reduced batch sizes, all needed to accelerate cash flow—much as Ohno discusses. There the similarities end, however. His theory is very weak on quality and many other aspects of waste. I have found that learning and applying the TOC is often a solid place to start for many businesses before they embark on a journey into Lean Manufacturing. On the other hand, if you have a pure make-to-order system, with multiple routings and highly variant machine cycle times, many tools in the Lean tool kit, become less effective. Some of the tools and techniques of the TOC become more effective. Since almost no business is a pure make-to-stock system, it is a good idea to have an understanding of the TOC as you embark on your Lean journey.

Deming's Management Technique

Deming's Management Technique, along with his 14 Obligations of Management and 7 Deadly Sins, is pure gold. The wisdom contained therein is simply wonderful, but the problem is that few have found a way to turn it into a solid management or business practice. His writings contain a number of solid thoughts and principles, but it is not clear that they are woven into an overall philosophy—at least not one that many can apply. I have found plenty of companies who have embraced many of his teachings, but only a few who have been able to turn it into a clear business or manufacturing system. Some have tried to do so under the name of TQM. Most writings about TQM picture it as a comprehensive philosophy that supports the principle of continuous improvement in a business. In the design of the TPS, and in Lean, it is easy to see the extensive and profound influence of Deming, his teachings, and his 14 Obligations of Management.

Crosby's Approach to Zero Defects

Crosby's approach to zero defects is an idea that had a great deal of traction in the 1980s, and many companies made improvements based on the concept of quality cost reductions. At that time, companies could survive with quality levels that were measured by percents of defects, and rework was a way of life. Today, quality levels have improved dramatically and are measured in the parts per millions (PPM). Almost no one who is serious about quality embraces this philosophy today. First, it is based on a fallacy: that zero defects are achievable. Second, the key metric used to drive the effort is "quality costs." With quality costs, there are at least two major problems. The first issue is that many of the major quality costs are not quantifiable. For example, what is the true cost of a customer return? What is the cost of a field warrantee failure? What is the cost of the loss of future business? However, even worse than that, the American system of cost accounting does not capture quality costs very well. Our financial accounting systems are designed to take care of two business needs: the profit and loss statement and the issue of taxes. It is not designed to capture things like the cost of poor quality.

Consequently, a zero defects program finds itself unable to account for the costs, and then the engineers whose job it is to reduce these quality costs (as well as justify their own existence) find themselves buried in the cost system trying to extract the savings from their efforts. They then become experts in the cost system, and so quality issues suffer.

Six Sigma

The concept of Six Sigma has achieved a great deal of publicity and, quite frankly, a great deal of success lately. Most of the publicity is directly or indirectly related to the large effort put forth by GE, which was made very public by their now-retired CEO, Jack Welch. Six Sigma owes its roots to Motorola and the efforts of Mikel Harry. I first read about it as a design concept and tolerancing mechanism popularized in the publication *Six Sigma Producibility Analysis and Process Characterization*, put out by Motorola University. The purpose of this technique was to start at the design phase and try to produce a "Six Sigma" quality product. This was defined as one with less than 3.4 defects per million opportunities. Later, the concept of Six Sigma was broadened to become a problem-solving technique, and the Six Sigma curriculum has become standardized with Six Sigma Blackbelt and Greenbelt certifications now available.

Since the early days at Motorola, the Six Sigma concept has grown and has various degrees of success. Welch in his writings claimed that GE sent $10 to the bottom line for each dollar they spent on Six Sigma efforts. Most large GE facilities had a complement of Blackbelts and Greenbelts and actually set up these internal consultants as cost centers. Still others have tried to link Six Sigma with Lean Manufacturing under the name of Lean-Sigma or some such double-barreled effort to sell yet another product. Today, to most people: "*Six Sigma is a project-based, problem solving initiative which uses basic as well as powerful statistical methods to solve business problems and drive money to the bottom line of the business*" (from www.qc-ep.com).

This being the case, Six Sigma is neither a manufacturing system nor a manufacturing philosophy at all; rather, it is a fine set of tools that can enhance problem solving in any sort of business, manufacturing or otherwise. For example, when teaching and training Blackbelts, we always require they undertake a project during the four-month training program. We also keep track of the financial earnings that are driven to the bottom line; they must be identifiable on the balance sheet. A recent group of 14 Blackbelts, four months after their training, had booked $1,030,000. By the end of a year, their projects had driven $3,500,000 to the bottom line. Other groups perform similarly, so the Six Sigma problem solving concept is a sound one and helps make the company a more powerful money-making machine. But Six Sigma is not a manufacturing philosophy. It is a completely different animal than the TPS, yet it is totally compatible with the TPS.

So What Really Is the Defining Difference between Lean and the Toyota Production System?

When James and Womack published their landmark book, *The Machine That Changed the World*, (Rawson Associates, Macmillan, 1990) they either created or popularized the term "Lean Manufacturing." They called it Lean because it generated products using:

- Less material
- Less investment
- Less inventory
- Less space
- (And) less people

The term Lean Manufacturing has since become synonymous with the Toyota Production System, but there are at least two differences. The first is a rather subtle difference and has more to do with the implementation of Lean, while the second is the fundamental difference between Lean and the Toyota Production System.

The first difference is subtle and is lost on many people. It has to do with the starting point of the journey into Lean. Recall that quantity control is the defining characteristic of Lean. When Ohno started in 1955, he had in place an extremely sound quality control system. His foundation of quality control was more than sound, it was very mature. In fact, the first application of Toyota's *jidoka* system predated the Toyota Motor Company. It was done in the Toyoda Spinning and Weaving Company in 1902. Today, few companies have this same solid and mature foundation of quality when they embark on a Lean initiative. So they must simultaneously work themselves out of a serious quality problem while trying to implement quantity control measures. Consequently, to implement a Lean initiative today, companies must embark on a renewed effort in quality control. Hence, Lean efforts today have become synonymous with not only quantity control but also *quality control*, which was never an issue with Ohno.

The second difference is more obvious. Many businesses can become Lean by simply following the outline in this book. They will achieve large gains in profits, be able to reduce lead times, become more flexible and responsive, and generally become a better business. Quite frankly, this is not too hard. What it takes is sound leadership, a decent plan, a motivating environment, a few problem solvers with a willingness to implement change, and good old-fashioned hard work. Couple those attributes with sufficient doses of both humility and introspection, and you have enough to make you Lean.

The difficulty is not getting there, but staying there. Here is where the Lean facilities, which have sustained their effort, stand out—and, of course, the granddaddy and greatest of them all is the Toyota Production System. Toyota has not only been an innovator in improving manufacturing techniques (and that may be the understatement of the century), but they have sustained this excellence for over 50 years. They have done this by not only implementing Lean techniques, but also by managing the culture in such a way as to sustain these gains through every kind of change and challenge imaginable. They manage their culture consciously, continuously, and consistently.

Ohno was a master at changing the culture and then creating the type of environment that would sustain those cultural changes. Herein is the main difference between Toyota and many other firms—some of which are very Lean. Toyota has been able to manage their culture in such a way that the gains are sustained. Sounds simple, but simple it is not. (We will touch on the subject of cultures in Chap. 11, but you will find that Chap. 11 is only a brief introduction to the topic of cultures.)

In simple terms, the TPS is a production system, focusing on quantity, and was built on a sound foundation of quality control. Lean is also a quantity control system but nearly always, in the Lean application, the quality control system must also be developed. Second, the TPS is a manufacturing system that is driven and supported by the Toyota culture. Other Lean firms, at least in the first several years of implementation, seldom have the strong, focused, mature culture like Toyota. However, with serious work, specifically on the culture, these Lean firms can have a manufacturing system that approaches the TPS in excellence. That is why we can say that although the TPS is Lean, not all Lean Manufacturing is done to the standards of the TPS.

Do not let that be discouraging. Ohno and Toyoda embarked on the development of the TPS over 60 years ago and they built on what others had done before them, particularly those in the Toyoda Spinning and Weaving Company. They did not create the Toyota culture in just a few years. It took decades of hard work, decades of dedication, and literally decades of trial and error to create the culture they wanted. But the key is that they figured out what they needed to do, what Ohno calls "out of necessity" and then not only did it, but managed it with a long-term philosophy of growth and integration. You can do the same, out of your necessity, but only if you're willing to make the short-, medium-, and long-term commitments and sacrifices that Toyoda and Ohno made.

Where Lean Will Not Work... or Not Work Quite so Well

Imitations and Lack of Understanding

Imitation is the sincerest form of flattery... This is a natural truth, and it is no more apparent than in the number of ways in which Lean is trying to be imitated. Everywhere you look you find Lean this and Lean that. Do an Internet search on Lean and it is amazing what you will find. There is Lean management, Lean education, and Lean healthcare to name just a few. Then, some practitioners use Lean as a new lead-in for some double-barreled title to spiff up or differentiate some already mature field such as Lean Six Sigma and Lean Software Development to name a few. Lean is "in" and all the salesmen know it!

It is interesting to me that I find many people who just can't quite see how Lean principles apply to their business when it is a natural fit for their situation. And yet I also see some people stretching the principles of Lean so far out of shape as to make them fit what they would like Lean to be. For example, I was assisting a small group trying to use Lean principles to guide their improvement efforts in education. They had spent many hours developing the "7 Wastes of Education." I asked how they decided on seven. Their reply was "That's what the TPS has." They had spent a tremendous effort to catalogue the wastes and then force-fit them into seven categories. I found it interesting to say the least, and nonproductive to say the worst.

Some Questions

However, this brings up some questions that some of us should address:

- What are the limits to Lean?
- Where will it not work so well?

- Where will it simply not work at all?
- Does it apply to the production line only?
- Does it apply to the staff functions in the manufacturing plant as well?
- Does it apply equally well to all aspects of manufacturing, regardless of product or customer?
- Can it be applied to all businesses regardless of product or customer?
- Does it only work in manufacturing and not in the service sector?
- Are there applications for it in the nonprofit sector as well?

A Two-Part Discussion: The Enterprise Level and the Product Level

The Toyota Production System and, hence, Lean Manufacturing was designed based on a certain set of business conditions. It would seem to make sense that the uses of Lean Manufacturing might have applications beyond the production of automobiles. Likewise, it seems reasonable to assume that if you diverge significantly from these basic business conditions, the applicability of Lean Manufacturing concepts and tools might wane as well. So it would make sense that the application of lean concepts and tools may not have infinitely wide applications.

In fact, after some review it seems that at the enterprise level—or some may call it the business level—the enterprise must be driven by four basic concepts. Lacking any of these, the enterprise is poorly suited to use Lean as a primary business philosophy. These basic concepts are:

- The enterprise must be in a competitive free-market environment. For those entities that are not struggling for profits and or survival, there is simply insufficient motivation to undergo the discomfort of the huge cultural changes it takes to implement a Lean initiative.
- There must be a clear customer focus. The enterprise must know who the customers are, what they need, and what they want. The enterprise must continually work to supply their needs and work to be ever improving in both finding and meeting the needs of the customer. In Leanspeak, these needs and wants of the customer are value.
- In supplying value to the customer, a key strategy must be the elimination of waste.
- The business must have a long-term focus, even at the expense of short-term gains.

Hence, entities that do not have a strong customer focus, that are not interested in survival and growth, and that are not willing to drive out waste over the long term would not be good candidates for Lean as a guiding business philosophy. In fact, in almost all cases such as this, Lean simply will not work.

Some examples of this are covered in the following sections.

Sports Teams

Professional sports teams are very bad candidates since they are focused almost solely on short-term gains. They may look only to the end of the season for the Super Bowl, for example, but if that is out of reach, it is not uncommon for them to make a huge change in the middle of the season. They will fire the coach and release veteran players with their large salaries, with the hope of becoming more competitive next year.

Recall the Florida Marlins who, immediately after winning the World Series, completely liquidated their high-paid roster and sunk to the bottom of the league the next year. I am surprised some ticket holders did not start a class-action lawsuit against them. Whether to liquidate a roster like that is legal or not, I do not know, but it certainly is not aimed at providing value to the customer. Nor is it the sign of a business striving to compete. It is the height of arrogance toward ticket holders.

Sports teams may be the perfect place that Lean is doomed to failure. They have no survival issues at all. In fact, they are a monopoly with practically a guaranteed income via television. Second, they have no interest in waste reduction at all. Indeed, they intentionally increase wastes as they can pass the costs on with impunity. Third, there is no sense of long-term stability. In fact, their mantra is "What have you done for me recently?" Finally, all their protestations to the contrary, they no longer consider the individual fan to be their customer. With the price of tickets so high, the vast majority of tickets are bought by businesses. In addition, the majority of the income is from television so the networks are their customers, in reality.

Charities

What about charities? They have no profit motive and consequently there is insufficient motivation to make Lean work. In fact, I have worked with charities that have the end-of-the-budget-year problem of not spending all of their grant money. So fearing they will get less next year, they find ways to spend the money. Rather than reduce waste, they frequently create waste.

Not-for-Profits

What about other not-for-profits? A lot is said about Lean in the government—the entity that is supposed to be serving you and me. I think they have lost sight of who their customers really should be. To apply Lean as a guiding philosophy in the top levels of government management, I see no hope whatsoever. The top few are interested in survival, but the survival issue is not the survival of the business (government), rather it is their individual job survival that is of importance to them. Their primary focus is on the self-serving survival issue of reelection. Reelection efforts are fueled by money, which is received through such things as PACs, which are largely supported by businesses. The "customer" of the high-ranking government official is more likely to be a PAC or large donor than Joe citizen. The PACs and large businesses are not interested in Grandma Jones getting her Social Security check. They are interested in their own self-serving purposes. So in the application of Lean as a business philosophy, this is a complete misfit.

I do see *some* hope for the application of many Lean tools (refer to Chap. 20), but not from the top. Lean tools are exactly what is needed at the level of government with which you and I interact. For example, at the Social Security Office, or the Department of Motor Vehicles, it has tremendous applications. At the "service provider" level, far removed from the top-level politicians, all these agencies use processes that could easily benefit from applications of the tools of Lean Manufacturing. Since the Lean tools are so powerful at waste reduction, some clever politician who wants to make a name for himself has a powerful tool at his disposal. If he applied it at the right time in the right way, he could use it as a way to get reelected and promote Lean in government along the way. Waste reduction is so sorely needed in the U.S. government. Quite frankly, it is a Lean opportunity well past its time. The key will be how to apply it and by whom?

The Health Care Industry

Some possibilities in using Lean tools exist for health care companies, but not as many as I would like to see. In the small doctor's office, with one or two doctors and a small staff, Lean should work fairly well. These offices are usually customer sensitive, interested in making money, and focused on the long haul.

However, in the hospital, serious issues are present that prevent Lean applications. The first problem is… "Exactly who is the customer?" The hospital will tell you it is the patient. But that is only partially true—and is more untrue than not. So why is the patient not really the customer? Well, think of what makes a customer a customer. Generally, to be a customer you:

- Are courted or otherwise sought out by the provider
- Use whatever they are selling
- Pay for it
- Can complain about a problem and get action

Well, in the hospital situation, seldom do I hear about someone shopping around like you would for a present or a new car. If the typical patient needs an MRI or knee surgery, it is rare that they shop about to find a low-cost provider. Rather, most people just go where their doctor sends them, unless their insurance will not pay for that location, then they instead go where their insurance will cover the cost. So guess who the hospital's customer is by this measure? Here, they are courting the doctors and insurance companies, so by this measuring stick, *you* are not the customer. You use the service, but who pays for it? Well, here comes the insurance company. Guess who qualifies as the customer in that instance? In fact, I have never been to a health care facility where they didn't first check my insurance before they checked me—emergency room service included. So, just who is the customer by this measure? Well, I guess you see my point. The large providers are mostly disconnected from the patient as being the customer. If they claim the patient is the customer, they are confused, at the very least. Hence, there is little hope here to apply Lean as a business philosophy until some dramatic changes occur.

But Lean Can Apply ..... Just Less Broadly

In the cases just listed, the primary driving forces of the business are so distant from the driving forces behind Lean that I cannot envision Lean principles becoming the guiding philosophy of the business or entity.

However, do not lose hope. In every case I have mentioned here, there are a series of processes, within the business, where Lean process tools still have some limited application. In many processes internal to these entities, the customer is much better defined and quality characteristics can be determined, measured, managed, and improved.

For example, we see the best as well as the worst applications of Lean principles in professional sports. While the removal of waste is not important in most sports, watch the execution of a football team during their two-minute drill. That is a very special application of driving out the wastes in the "process to score a touchdown." Another outstanding example is a NASCAR team while it is undergoing a pit stop. Where else can you see four tires changed, a car fueled up, the windshield washed, and the driver get a drink in just seconds? Here, every wasted motion is eliminated in the drive to create the shortest pit stop possible. Or, just because a professional football team is not a Lean enterprise that doesn't mean you can't get a good quality hot dog, with a minimum wait at the concession stand when you attend the game, even if it costs you a 500 percent premium. Likewise, just because your hospital is more interested in your insurance company than they are in you does not mean they can't serve their patients a good meal, on time, or get you to your MRI on time.

Remember that "wherever there is a value stream, Lean principles will apply." Even if the top levels of the business of government are driven by large businesses and PACs, that doesn't mean you can't get your driver's license renewed using a Lean process. In government, as in all the other examples given, Lean principles may work at some level—these applications are just limited.

What I am saying, and will repeat to avoid confusion, is that those entities that are not customer-focused, without a survival motive and without a concerted effort to reduce waste and provide value to the customer—those entities that do not embrace a long-term philosophy of growth and service can't become Lean enterprises, and cannot become candidates for Lean as a guiding business philosophy until they change. They, however, can still utilize some of the process management tools of Lean for some of their internal, particularly lower-level, processes.

However, if the business does meet these criteria, is competitive with a clear customer focus on supplying value by driving out waste, and is in it for the long term, does that mean Lean as a business philosophy will necessarily work for it?

The answer to that question is yes, but to a varying degree due to three basic conditions of the product. Those conditions, when combined, are known as the *Lean Stereotype*. Specifically, the more the business fits the Lean Stereotype, the more it will be able to utilize the Lean tools in its battle to survive and become more profitable. The Lean Stereotype is the specific type of business for which the strategies, tactics, and skills of Lean Manufacturing were developed. Consequently, the more a business approaches this stereotype, the more the business will be able to directly apply the strategy, tactics, and skills of the House of Lean.

The Lean Stereotype is a business in:

- Manufacturing
- Discrete parts
- Stable product demand

Lean Applicability: Continuous Process Industries

The first and smallest negative effect on the applicability of Lean is the shift from discrete parts manufacturing to the continuous process industry. This would include industries such as petroleum and chemicals manufacturing, food processing, and pharmaceuticals to name a few. My background was in petroleum refining and we were able to make almost total applicability of the earlier mentioned principles, with one major exception: destroying the batch. We were able to see and work at waste reduction, we were able to produce to *takt*, create flow, and use pull systems. There was one large problem with refining: The unit you deal with is the batch. And the batch is sometimes, out of necessity, large. Other times, it was possible to reduce the batch dramatically with resultant improvements in lead time. As you might expect, this batch reduction could translate into improvement in those two wonderful business weapons: flexibility and responsiveness. In most cases, reduction of batch size could be easily done. However, because the history of the petroleum business is to make batches, that paradigm was very hard to change. The momentum was very much against batch size reduction. Another factor in these businesses that must be understood is the continuous process industries, such as refining and chemicals, where the capital investment per employee is much larger. For example, in refining, it is not uncommon to have $4,000,000 of capital investment per employee. By contrast, many of my current clients who are typically tier 1 automobile suppliers, have $3000 to $50,000 of capital investment per employee. This high capital investment will cause those businesses to view their wastes somewhat differently.

Lean Applicability, Unstable Demands

The next largest negative effect that makes the operational techniques of Lean less applicable is unstable customer demand. When you compare the three-year contractual demand that most tier 1 automobile suppliers are blessed with to the come-and-go demand that many job shops face, it is easy to visualize that this is a huge obstacle to implementation of Lean principles. For example, there is no *takt* to calculate, so synchronizing with the customer is difficult, and since the life of a job is often very short, continuous improvement requires a completely different philosophy. However, it is still possible to create a pseudo-*takt* to use in synchronized supply with the customer and synchronized production flow. Often flow can still be balanced and, most importantly, the flow velocity can be accelerated by reducing the batch size. In the typical job shop, lot size reduction is a powerful tool (see the Story of Excalibur Manufacturing in Chap. 5 and the Story of the Bravo line in Chap. 15, for specific examples). To keep lead times reasonable, quick changeover (or in Leanspeak, Single Minute Exchange of Dies, [SMED]) technology must be very strong in this type of business. In most job shop applications, to create a "flow" for each job is not too difficult, and all the basic principles of Lean apply.

The complexity is borne in the concept that lots of products, or jobs, exist. These jobs have different routings with variant cycle times and complex interactions of people and machinery. Yet examples of businesses that have made huge improvements by applying the principles of Lean abound. Two principles that seem to repeat time and again are the use of SMED technologies to reduce setup times, and the use of small batch sizes to reduce lead times. The efforts in job shops to reduce lead times pay triple dividends.

- First, with short first piece lead times, rework is reduced dramatically. This is an extremely powerful quality weapon that should not be overlooked by these businesses.

- Second, short lot lead times translate into quicker deliveries with improved cash flow.

- Third, the ability to quote shorter lead times is a power weapon to acquire future business.

Lean Applicability: The Service Sector

The third and even larger negative effect that reduces the applicability of Lean is seen when the business is not manufacturing, but instead is typical service sector work such as that of a hotel, restaurant, or hospital. Service sector work has at least two very large problems.

- First, but less important, is the demand instability a typical service sector business must deal with. Seldom do they have the stability of the tier 1 automobile supplier. Sometimes doctor's offices and dentist's offices have a fairly stable demand rate, but they are the exception rather than the rule in the service sector. Compare them to the comings and goings of customers at the typical hotel or restaurant.

- Second, and more importantly, unlike manufacturing, the service sector has a paucity of specifications. The time it takes to check in at a hotel, get served at a restaurant, and get your oil changed have no real specification. Consequently, the metrics of the service sector are often up for grabs, and it becomes very difficult to measure the quality. Do not forget that the foundation of Lean Manufacturing is good quality. It is hard to build on a foundation that barely exists.

So How Should We Proceed When We Have to Deal with These Effects?

All of this is interesting, at best, and possibly contrary to progress, at worst. Those who ask about the applicability of Lean are often looking for the "correct formula" to remedy their ills. All too often, they are trying to find a simple, proven, ready-made solution to what are often complicated problems. It's sad to say, but there is no ready-made formula; there is no "silver bullet" to solve these ills.

By developing the TPS, Ohno obviously found his remedy, and using the same logic, you too will need to find your solution—and do this by using his logic, which may or may not mean you will end up using his tools of improvement.

Ohno said the "TPS was developed out of necessity." My advice to you is this: Find your own necessity and then develop what you need to, for your unique circumstances. And keep in mind that although the TPS may not apply totally to your situation, I am equally sure that some of it will.

However, if you, for the moment, ignore the TPS itself and instead focus on the logic and method that Ohno used to create the TPS, you will find how to apply his method to your situation. In other words, you will find "your necessity" and then with some of the following:

- Good old-fashioned hard work
- Sound logic

- Good problem-solving skills
- A determination to get through resistance and failures
- Truly motivated introspection
- A touch of humility
- Sufficient courage

> **P**oint of Clarity Do not let what you can not do, prevent you from accomplishing what you can do.

… you will see that Ohno's method applies totally to your situation as well.

Chapter Summary

The Toyota Production System (TPS) is a quantity control system built on a solid foundation of quality, and was the manufacturing system perpetuated by the strong culture of Toyota. Herein are the two differences between Lean Manufacturing and the TPS: the strong Toyota culture and the solid quality foundation. Both are strong manufacturing philosophies designed to make your business more secure and a better money-making machine through the total elimination of waste, thereby supplying what the customer wants: *value*. It needs to be understood that Lean is primarily a manufacturing philosophy and is not a business philosophy. Finally, the tools of Lean were designed for—and work best in—what I call the Lean Stereotype, although you are only a little hard work and imagination away from applying these tools very broadly.

CHAPTER 3

Inventory and Variation

Many early efforts at imitating Lean production focused on the Just In Time (JIT) concept of inventory reduction. Here, we will explore why many of these efforts failed and introduce why we have inventory, why we need inventory, and the two key business reasons why we strive to reduce inventory. We'll explore the dynamics of inventory creation and its relationship to variation and dependent events, including making the sample calculation for all three types of inventory. Finally, we will discuss the powerful tool designed by Ohno called *kanban*.

Background

In the 1970s, it became clear to a select few that the Japanese, most notably Toyota, had found a better way to manufacture cars, which caused a number of very interesting things to happen. First, and most notably, the majority of the manufacturing world went into a huge case of denial. This was heard as "that will work in Japan, but not here" and a variety of other statements that could politely be said to have lacked insight.

However, some with a little more insight, curiosity, and humility asked, "Could there be something to this?" Well, from that small group came a series of efforts to try to capture parts of the Toyota Production System that were serving Toyota so well. The piece that seemed the most appealing was the JIT concept. It was rapidly popularized as an inventory reduction effort, which in fact is only a part of what it really is.

JIT practitioners came out of the woodwork and many companies went about implementing *kanban* and slashing inventories to reduce the high cost of producing and managing the inventory. Some went about using the slogan of "Zero Inventory" and slashed inventory with such fervor it was as if they were pursuing the Holy Grail of manufacturing. Inventory had become a bad word, much like "scrap."

Unfortunately, many of these efforts were grossly misguided. Their only focus was on inventory reduction. They reduced inventories as if it were an independent entity that had no relationship to anything else. JIT implementation efforts became nothing more than aggressively slashing inventories. Those that had this approach often caused irreparable damage. They found they needed to expedite nearly everything, needed to work large amounts of overtime, and then still frequently missed delivery dates. Others found the worst of all scenarios. They not only missed shipments but as they cut inventories they found that production rates flagged significantly. Due to these misguided efforts, many companies ceased to be competitive and some even went out of business.

Just Why Do I Have and Why Do I Need the Inventory?

The smart ones, once they got in trouble said, "Wow, maybe there's more to this inventory reduction than just slashing our inventory volumes. What's going on here?" The really smart ones asked—and answered—two basic questions. These were questions that were overlooked by the vast majority of the JIT implementers of the past, and are frequently overlooked by the managers of the present. These questions are:

- What is the basic purpose of inventory? Meaning, if it's so bad, why do I have any at all?
- What is causing the need for the inventory? In other words, why can I not seem to operate without it?

What Is So Bad about Inventory?

So what's so bad about inventory? Simple: Inventory costs a lot of money. First, there are the raw materials and operating expense it costs to produce it. Next, we must handle it, which means we need more people and machines like forklifts. Then we find ourselves moving the material around, usually more than once before it gets to its desired location. This in turn requires space and transportation and neither are free. Next, we must keep track of it, which means people, computer programs, and reports galore—almost all of which are filled with errors. We then try to fix these errors. The way we try to fix the errors is to use things like cycle counts which then take more people, more time, more computers, and worst of all more reports and more meetings. In addition, we must care for this inventory to make sure it does not get damaged. And finally, we must ship it before it becomes obsolete.

I have dealt with several firms who say their cost of inventory, *obsolescence excluded*, exceed 25 percent/yr of the product value. That is 2 percent per month, and if your company is operating on 12 percent earnings on sales—well, the impact is huge and you can see why many firms wanted to get rid of it. What else can be done to make such a huge bottom-line improvement? It's no wonder many firms jumped on this bandwagon of inventory reduction in the name of JIT.

All of these liabilities of inventory are obvious bottom-line opportunities, and yet the greatest advantage of reduced inventory is not even mentioned here. In fact, it is often not even recognized. In just a minute, we will get to that crucial advantage which so few see and even fewer appreciate.

Question 1: "What Is the Basic Purpose of Inventory?"

What is the basic purpose of inventory? This is not a complicated question—indeed, it is a rather simple one actually, but it is asked by a scarce few, and answered by even fewer. First, let's make sure we are on the same sheet of music here. I am talking about the use of inventory in a classic for-profit business. The objective of those companies is generally to do just that: make a profit. In those businesses, the purpose of inventory is singular and simple: You should only hold the inventory you will need to protect your sales.

I see no other reason to hold inventory. Any amount beyond this is an expense that is not justified, yet to hold less undermines your ability to supply your customer—and nothing will hurt profits like failing to sell. It is a very simple concept. Well, it is pretty simple, anyway, and yet is still missed by many.

We only hold the inventory we need to protect sales. There is a relationship between the amount of inventory and the volume of sales. So in a nutshell, if we know the sales volume and can understand the relationship, we could calculate the inventory we need to protect those sales. The relationship is a simple mathematical calculation, but just how is that calculation made?

> **P**oint of Clarity The only economic purpose of inventory is to protect sales.

First, let's address finished goods inventory. This inventory calculation has three parameters—and not surprisingly, there are three types of inventory. The three parameters are:

- Stock replenishment volume (that is, the picked-up volume by the customer)
- External variations, usually demand fluctuations
- Internal variations, usually production issues

The total inventory is the sum of the three types of inventory, which are:

Cycle stocks

This is the volume you need on hand to take care of the normal demand pickups by your customer, often this is called stores.

- For example, if your customer picks up each Wednesday, you will need their ordered volume ready for pickup then—and not before. Unfortunately, the information handling system, production, and the delivery system are not instantaneous, so we need some volume in cycle stock that is above the bare minimum the customer will pick up. So, to make sure you can achieve the customer's needs, we will calculate the cycle stock's volume to be the production rate multiplied by the replenishment time plus some arbitrary safety factor we will call Alpha (see the section "Finished Goods Inventory Calculations" later in this chapter). The stock replenishment time (see Fig. 3-1) is the sum of four variables:

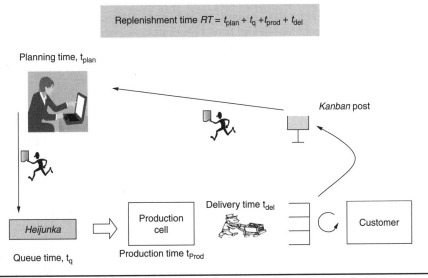

$$\text{Replenishment time } RT = t_{plan} + t_q + t_{prod} + t_{del}$$

Planning time, t_{plan}

Kanban post

Heijunka

Production cell

Delivery time t_{del}

Customer

Queue time, t_q

Production time t_{Prod}

FIGURE 3-1 Replenishment time.

Planning time   This is the time that the order takes to be processed and sent to the production line.

Waiting time   This is the time the order is waiting to be processed. This is sometimes referred to as queue time.

Production time   This is the time it takes to produce the desired quantity.

Delivery time   This is the time to get the lot from the production line to the storehouse.

Buffer stocks
This is the incremental volume of inventory, above the cycle stock's inventory volume, which is held to account for external variations, and is calculated based on historical data of the variation of these external causes.

Safety stocks
This is the incremental volume of inventory currently held that is above both the cycle stock and buffer stocks. It is held to account for internal variations in supply to the storehouse.

Question 2: "Just What *Is* Causing the Need for Inventory?"
The need for each of the three types of inventory is caused by different factors. These factors are:

- For the cycle stock, the need for the inventory is caused by the size of the picked-up shipment, which, for a constant demand product, is a function of how frequently the shipment is picked up. In addition, some inventory is needed to cover the time it takes to plan the shipment, make the shipment, and move the shipment within the plant. This is the replenishment time calculation.

- For both the buffer and safety stocks, the need for inventory is one of the world's best kept secrets. The need is caused by variation. When we have more variation in the system, we need more inventory. The buffer stock size is usually determined by two variables: changes in customer demand, and variations in delivery conditions. Often, this is due to weather, or in the case of products that cross an international border, customs can be an issue. So, the sources of variation for this volume of buffer inventory are somewhat outside of the control of the plant.

- Regarding safety stock in particular, the large sources of variation are issues of supply to the storehouse. These sources of variation include such items as line outages due to machinery failure or stock outs. Poor cycle time performance can cause production to fall short of goals, and of course quality problems can also be a major cause of variation. All three of these issues, which happen to be the three aspects of OEE (Overall Equipment Effectiveness), are largely under the control of the plant.

- In the case of the buffer and safety stock inventories, there is a simple way to calculate the volumes needed. If the variation of the volume swings is calculated over a reasonable time frame and stated as a standard deviation, your variation is now converted to numbers so we can have a common understanding of it. Now if you have a stable system and hold 2.33 standard deviations of inventory, you will have enough inventory to cover about 99 percent of these deviations, presuming your data are normally distributed. Since, in reality, there are many possible sources of variation, assuming the normal algorithm is reasonable. With weekly shipments, that would mean about one undersized or late shipment every two years.

What Creates the Need for Work In Process (WIP) Inventory?

The preceding discussion of inventory was focused on finished goods inventory, although the concepts are general and apply to all inventories. In many plants, the problematic large inventory is not in finished goods but in WIP. What causes the need for WIP inventory?

Take a simple cell, for example. Let's say we have a six station cell and all work stations have 60 seconds of work, which is also *takt*, and that there is no inventory between stations and we have one-piece flow. When station 1 finishes a piece, so do stations 2 thru 6—and in unison, all six pieces of in-process work are simultaneously pulled from the previous work station every 60 seconds. This is perfect synchronization of process flow, the ideal state.

But for the moment let's imagine that the cycle time for station 4, although it averages 60 seconds, varies from 50 to 70 seconds. When station 4 performs at 50 seconds, it finishes its process and then station 4 has a 10-second wait time before its product is pulled by station 5. There are 10 seconds of wait time, which is a waste for station 4. *But this is not a production rate problem.* The cell will still produce to *takt*. It is just that the operator at station 4 will sit around a while. On the other hand, when station 4 takes 70 seconds to produce its work, that subassembly is held up and station 5 is starved for work for 10 seconds. This delay passes through all the workstations of the cell in a wave, and that piece is produced on a 70-second cycle time.

So let's recap… If the station that varies—in this case, station 4—operates faster than *takt*, station 4 must wait for the subsequent station to pull the production. However, when station 4 just happens to operate slower than *takt*, station 4 will slow down the whole cell on that cycle and there will be no recovery. So even though the station may have a 60-second cycle time *on average*, any time the cycle time is above average, the production rate drops. This concept is known as the effect of variation and dependent events. (The dependency is that the "next step" depends on the "prior step" for supply.)

So the solution is, guess what?… You got it! *Add some inventory.* We will need to add inventory both before and after station 4, the one with the variation. We need the inventory in front of station 4 so when it produces faster than *takt*, say at 50 seconds, there is raw material available to keep it producing. We also need the inventory after station 4, so when it is operating slower than *takt*, say at 70 seconds, there is raw material to supply station 5. Then station 4 can have the variation AND maintain production at *takt* on average.

So the answer to the question is "The need for WIP inventory is caused by variation."

This destructive relationship of variation and dependent events interacting to cause a reduction in production rates is critically important. It is understood by a scarce few, yet you need to understand it if you wish to implement a Lean initiative. At this point, I suggest you go directly to Chap. 18 and perform the dice experiment to begin to understand this relationship. It may be the best 60 minutes you can invest in your understanding of Lean.

So this inventory—whether it is buffer and safety stocks in finished goods or WIP between work stations on the line—is nothing more than a response to variation so rate can be maintained. Once you understand this, it is easy to see that the answer to inventory reduction is to first reduce the variation, and then the reduction in inventory can be made with no loss in production.

> **P**oint of Clarity Inventory is a necessary response to system variation so rate can be maintained.

Although inventory is a waste, it is one of those necessary wastes. We wish to reduce it, but few businesses can survive without some inventory at some point in their process on the way to the customer. Remember that inventory is needed because of the variation that is present. Later in this chapter, you will learn that variation "is the inevitable differences…," hence it is unavoidable. So it is not possible to eliminate it totally, but we strive to do just that.

Just What Is This Key Advantage of Inventory Reduction That Was Alluded to Earlier?

To answer this question, you will need a lesson in physics, such as those in *Factory Physics* (McGraw-Hill, 2008), Hopp and Spearman's book. The particular law of factory physics to which I refer is known as Little's Law. In Leanspeak, it states that the WIP in any system is equal to the throughput rate multiplied by the lead time.

$$\text{Little's Law} \quad WIP = TH \times CT$$

Where:

WIP = the work in process … units of inventory between any two stations
TH = throughput rate … units produced/unit of time
CT = cycle time … which they define as the average time from when a job is released into a station to when it exits…*in Leanspeak, this is called lead time.*

If we replace the term CT with Lead Time (LT) and then rearrange them, we get:

$$LT = WIP/TH \text{ … which is Little's Law in Leanspeak}$$

As you can see, as WIP is reduced, lead time is reduced in direct proportion.

Consequently, we can double the throughput rate or cut the inventory in half. Both will reduce the lead time by 50 percent.

And as lead times shorten, flexibility in production will improve, as will the plant's responsiveness to changes—all this and cash flow will improve as well. *Lead time, more than any other metric, is the most descriptive measure of the health of a Lean manufacturing system.* And improved lead times come about largely by reducing inventories.

We have already discussed the importance of lead time in Chap. 2. It will be further amplified in Chap. 5 since it deserves a chapter of its own. In addition, it will be highlighted in several case studies throughout the text.

About Variation

Production process variation is everywhere. It affects every aspect of every step of your process and every specification of every part of your product. It is present in the materials, the manpower, the methods, the measurement, and the environment of all that we do to manufacture our products.

It is inevitable.

It is the enemy.

It is the enemy to not only good product quality, it is the enemy to rate, it is the enemy to rate stability, and consequently it is the enemy to operating costs—but most of all it is the enemy of bottom-line profits. Nothing is more basic to improving the manufacturing systems than the reduction of variation.

Variation in a production process is to be understood, sought out, and destroyed.

Variation is "the inevitable difference of the individual outputs of a system." I have taught that definition for years and am not sure of its origin. I believe I can thank either Walter Shewhart or Donald Wheeler for it, but I am not sure. But what is important is that it is a clear representation of variation. It is:

- Inevitable
- Applicable to all outputs—in fact, it is applicable to every characteristic of each output
- System generated—in other words, every part of everything that went into making the individual output varies

Sometimes, especially as it applies to attribute data, I use a definition I found taken from the writings of Walter Shewhart. Most people refer to this book as The Western Electric Handbook, but it's real name is *Statistical Quality Control Handbook,* (AT&T, 1956). It is the three-part definition of variation and it says:

- Everything varies.
- Individual items are not predictable.
- Groups of items, from a constant cause system, tend to be predictable.

It does not really matter which definition you use. In a process, they both converge and you find that variation is the enemy to both process stability and process capability. It is the enemy to the very foundation of a Lean effort and must be understood and aggressively reduced at all times, and in all processes.

So, to summarize… All systems have variation; hence, all systems will need some inventory to maintain rate. However, inventory is a waste, but at some level it is a necessary waste, so you will want to scientifically and economically minimize it. To minimize the inventory, you need to reduce the variation—there is no other productive way. In short, inventory reduction is reduction of variation by another name.

> **P**oint of Clarity   To reduce inventory levels, reduce variation.

Buffers

Whenever there is variation, we need inventory to compensate for the variation if we wish to maintain the production rate. This is not quite a true statement. Specifically, when we have variation, we need *something* to compensate for this variation, to maintain rate. We talk about inventory as being a countermeasure for variation, but in a more general sense we need a buffer. A buffer is some resource we have in excess that is designed to account for the fact that production cannot be in perfect lock-step with consumption.

Buffers come in three forms:

- Inventory
- Capacity
- Time

There are three types of buffers: inventory, capacity, and time.

- Finished goods inventory is a buffer because we must accumulate finished goods between customer pick ups. WIP inventory is a buffer. It is a natural response to variations in the production system, including scrap production, machine downtime, changeovers, and cycle time variations, to name just a few.

- Excess rate capacity in a machine that requires changeovers is a buffer, a capacity buffer.

- When we do not have a good understanding of our lead times—which for the sake of argument vary from three to five days—we may enter a time buffer of six days into our planning program to make sure that when we release an order, it will be completed on time. In addition, a typical lean strategy is to run a plant 2 to 10 hour shifts. This strategy coupled with a some overtime, provides both a time buffer and a capacity buffer.

Kanban

Basics

Kanban means sign board. A *kanban* can be a variety of things, most commonly it is a card, but sometimes it is a cart, while other times it is just a marked space. In all cases, its purpose is to facilitate flow, bring about pull, and limit inventory. It is one of the key tools in the battle to reduce overproduction. *Kanban* provides two major services to the Lean facility.

- It serves as the communication system.
- It is a continuous improvement tool.

Types of *Kanban*

Kanban provides two types of communication. In both cases, it gives the source, destination, part number, and quantity needed.

Rule No.	Rule	Function
1	Later process goes to earlier process and picks up the number of items indicated by the *kanban*	Creates pull, provides pick up or transportation information. The replenishment concept is formed here
2	Earlier processes produces items in a quantity and sequence indicated by the *kanban*	Provides production information and prevents overproduction
3	No items are made or transported without a *kanban*	Prevents overproduction and excessive transportation
4	Always attach a *kanban* to the goods	Serves as a work order
5	Defective products are not sent to the subsequent process	Prevents defective parts from advancing; identifies defective process
6	Reducing the number of *kanban* increases their sensitivity	Inventory reduction reduces waste and makes the system more sensitive

TABLE 3-1 The Six Rules of *Kanban*

- Parts movement information, the transportation *kanban*—this is like a shopping list.
- Production ordering information, the production *kanban*—among other things, this is primarily a production work order.

Kanban Rules

The Six Rules of *kanban* management provide several unique functions. The rules and functions are listed in Table 3-1.

> "The *kanban* method is the means by which the Toyota Production system moves smoothly."
>
> Taiichi Ohno
> From Toyota Production System

Kanban Calculations

Let's analyze a production *kanban* system. Recall that the *kanban* represents the entire inventory in the system. To assure delivery to the customer we will use a management policy with our finished goods inventory. It will involve the use of three types of finished goods inventory. To assure we have stock on hand for the normal pick ups by the customer we will carry cycle stock. In addition, in order to provide supply to the customer we will carry stocks to handle external demand variations and the internal supply variations of the finished goods. Hence we need a buffer and a safety stock volume, respectively. So our inventory management philosophy will include carrying three types of finished goods inventories and each one will be statistically calculated to minimize the total volume yet maintain a high level of customer service (in this case the level of customer service will be 99% on time delivery). Hence, the total number of finished goods *kanban* is the sum of these three stock volumes, divided by the container size.

No. of *Kanban* = (Cycle stock + Safety stock + Buffer stock) / Container size

Kanban Circulation

The kanban system is very flexible, and many types of *kanban* can be used. Likewise, as long as they follow the basic rules of *kanban*, they can be used in a large variety of ways. However, the majority of *kanban* follow a standard pattern. Let's follow a *kanban* as it is circulated (see Chap. 7, App. 5, which shows the *kanban* flow on the Value Stream Map for QED Motors). Since Lean thinking usually works best if we start at the customer and work backwards, let's do just that.

Since Rule 3 says that the product has *kanban* attached. When the customer comes for his pickup, the *kanban* are removed and placed in a *kanban* post. From here, the *kanban* are picked up, normally by a materials handler, and transported to Planning, or ideally they go directly to the *heijunka* box in front of the production line. If they go to Planning, they generally do little with the *kanban*, but they like to stay in the loop. From Planning, the *kanban* are sent to the front of the production line per the information on the kanban. The *kanban* are then placed in the *heijunka* box, a load leveling tool. From here, the production workers withdraw the *kanban* from the box in sequential order, and the process then produces the product in the quantity listed on the *kanban*. The *kanban* has just served to be a production work order and is infinitely superior to any MRP type system to trigger production. The worker then attaches the *kanban* to the products made and they are placed in the designated spot, ready for pickup. On his normal circulation, the materials handler picks up the products, with *kanban* attached. The *kanban* tell him exactly where to deliver the products—normally this is the storehouse, which completes the cycle. The *kanban* have moved a distance and have consumed time by:

1. Transportation to, and time in, planning
2. Transportation time to and time spent waiting in the queue, the *heijunka* box
3. Time spent in the production line
4. Time used to deliver the finished goods

The sum of these four times is the replenishment time.

How Do We Achieve Process Improvements in a *Kanban* System?

Process improvement in a *kanban* system is accomplished by the reduction of inventory. The reduction of inventory and hence the reduction of *kanban* can be achieved by:

- Reducing any of the four replenishment times or reducing the pickup volume by the customer, this is usually achieved by increasing the pickup frequency. Reductions in any of these items will reduce cycle stock inventory.
- Reducing the variation in the production rate, which allows safety stock reductions.
- Reducing the variation in the customer demand, which allows buffer stock reductions.

What Does the *Kanban* System Really Do?

Think for a moment about a perfect stockless (almost) manufacturing system. It would have a cell where all the necessary processing steps are connected with zero inventory between stations, one-piece flow, operating with 100 percent availability and 100 percent yield, and hence the steps would operate in total synchronization. We would simply tell the operators to keep one unit of production in the finished goods inventory and if the customer came and removed a unit, then and only then would we replace it. In this system, with 100 percent on-time delivery, once the customer withdrew an item, it would signal replenishment, and in total synchronization all stations would spring into action and another would be produced, almost instantaneously. The perfect pull production system. Once a customer arrived, product was ready; however, If the customer did not withdraw a product, no production would occur. One hundred percent on-time delivery, with no overproduction, a near perfect Lean system. This, of course, would only occur in a perfect—therefore, non-real—system. Unfortunately, we mortals need to deal with the realities of life.

These realities of life include several issues.

First and foremost is the issue of variability. Did we not say it was inevitable? Since perfect synchronization is not possible, 100 percent on time delivery and zero overproduction, are also not possible. Though these ideals might be ones to shoot for, they are typically impossible and many times impractical. Variations always exist in rate, quality, people, machine and environments. They are inevitable and omnipresent. All of this variation creates, guess what? You got it, inventory. So to compensate for the variation, we need some buffers. This causes our total inventory to rise and Little's Law tells us our lead time will increase, which likely will cause us to hold even more inventory as finished goods.

> **"D**on't be fooled. The system optimum is not necessarily the sum of the local optima. **"**
>
> Unknown

So how do we reduce the inventory—that is, avoid overproduction of both the local (WIP) and finished goods, bringing inventory to its minimum—and still supply the customer with high levels of on-time delivery?

Either task can be done simply, but doing both simultaneously—and well—is the trick of a good business system.

And that trick is *kanban*.

The essence of *kanban* is twofold.

First, it is direct communications to produce material—in other words, to supply the customer. It is the pull signal to produce. Once the product is withdrawn by the customer, at that moment the *kanban* tells us exactly what the customer is using, and hence what the customer will need later. This *kanban* is sent as fast as possible to the production line. In essence, the *kanban* system is doing the "talking" to the production system, telling it to produce because some product has been removed. This system easily bypasses all the accounting and planning systems that tend to not only delay this signal but also add variability along the way. The *kanban* system is dealing real-time with the realities of what is happening on the line. The planning systems deal with what the programmer believed should be happening. I can say with certainty, that when it comes to triggering production, with the minimum lead planning time, no planning system can come close to *kanban*. In this manner, the *kanban* system not only assures supply to the customer, but does so with the minimum planning time.

Second, *kanban* creates an absolute limit on total inventory. Since each *kanban* represents a certain amount of stock, and the number of *kanban* are strictly controlled and limited, this creates an upper limit on the inventory. We will show in Chap. 4 that this inventory limitation is a key factor in making a pull production system function. By utilizing pull production, we minimize overproduction. Furthermore, the continuous improvement aspect of *kanban* works to further reduce this overproduction.

What Is Value Added Time?

Total time—all the time it takes to produce the product, which is made up of:

- **Value added time**—the time the customer is willing to pay for
- **Non-value added time**—which is waste. It is comprised of:
  - Pure waste—Activities that can be eliminated or reduced immediately.
  - Necessary waste—Activities that cannot be reduced immediately due to the present work rules or technology.

Finished Goods Inventory Calculations

Cycle Stock Calculation

Cycle stock, you will recall, is the stock that is the volume you need on hand to take care of the normal demand pickups by your customer. Hence, we will calculate the cycle stock volume to be the production rate multiplied by the replenishment time plus some arbitrary safety factor we will call Alpha. A sample calculation of replenishment time is shown in Table 3-2, which quantifies the display shown in Fig. 3-1.

For example, if our typical daily shipment is 1400 units per day. Presume *takt* is 1 minute, so production time is then 23.3 hours. The time the *kanban* cards are in planning is 24 hours, and delivery time (due to the material handler's frequency) is two hours. In our typical queue, we have 16 hours of demand in front of this order and we use an Alpha of 0.05. With this, the replenishment time is as shown in Table 3-2.

Thus, the cycle stock volume is (65.3 h × 60 units/h)(1 + 0.05) = 4114 units. If, for example, there are 50 units to a box, the cycle stock inventory would be 83 containers, and if *kanban* were used, we would have 83 *kanban*. So we could have as much as 83 *kanban* of finished goods in the cycle stock inventory. This is very unlikely, but if the

Planning time	24 h
Queue time	16 h
Production time…1400/60	23.3 h
Delivery time	2 h
Total time	65.3 h

TABLE 3-2 Replenishment Time (RT)

customer failed to make a pickup for a few days, this would be the maximum volume stored on-site. At that time, since there are only 83 *kanban*, each *kanban* would be attached to a box that is held in cycle stock inventory. Since all 83 *kanban* are attached to boxes, none work their way through the system to the *heijunka* board where they would trigger production—hence, no more product would be made.

Buffer and Safety Stocks Calculations

For both buffer and safety stocks, the same logic and methodology is used. In both cases, you use historical information to calculate the variation. We then determine an acceptable level of on-time delivery, normally 99 percent, and to obtain this, we need a z score of 2.33 sigma for a one-sided test. Therefore, we need 2.33 sigma volume of stock to assure 99 percent on-time shipments in the case of buffer stocks. Now the required volume to protect the supply can be determined. The difference between the calculations is that for safety stocks you use the data, which depicts the internal variation, usually the production rate to the storehouse. On the other hand, for buffer stocks you use the external variations that are typically the effects of demand fluctuations, plus delivery variations.

Let's calculate the safety stock for the preceding case:

The production data for a 30-day period are listed in Table 3-3.

With this information, you can quantify the need for safety stock due to production variations. The standard deviation is 59 units. To cover ourselves for this variation to a 99 percent certainty, we can carry 2.33 Sigma of stock, or 138 units. Practically, we would

Day	Production	Day	Production
1	1460	17	1480
2	1410	18	1350
3	1390	19	1450
4	1300	20	1250
5	1390	21	1370
6	1450	22	1400
7	1400	23	1390
8	1410	24	1480
9	1420	25	1450
10	1460	26	1400
11	1410	27	1350
12	1380	28	1310
13	1370	29	1380
14	1400	30	1510
15	1390	Ave	1400
16	1420	Std Dev	59.0

TABLE 3-3   Production Data

Day	Demand	Day	Demand
1	1400	17	1200
2	1400	18	1600
3	1800	19	1400
4	1400	20	1000
5	1500	21	1400
6	1000	22	1400
7	1800	23	1400
8	1500	24	1000
9	1200	25	1600
10	1400	26	1400
11	1600	27	1400
12	1400	28	1000
13	1600	29	1400
14	1400	30	1600
15	1400	Ave	1400
16	1400	Std Dev	208.0

TABLE 3-4   Demand Data

carry three boxes; two boxes of which would only be 100 units. In the final result, that would mean we might short an order as much as 1 percent of the time. If we examine these data, the variation for this 30-day period is a high of 1510 and a low of 1250. If the pickup is 1400 units and that day we made only 1250, we would need an extra 150, to complete the order, this is exactly what we have, so our system worked nicely to assure on time delivery.

Buffer stock is calculated the same way, except that we use the external variations. These are normally caused by demand changes by the customer. The demand data for a 30-day period is listed in Table 3-4.

If we wanted to cover the demand variations to 99 percent certainty, we would carry 2.33 Sigma of stock, or about 485 units. That would be ten boxes. In these 30 days, we actually had a high demand of 1800 which then could be covered by the normal production of 1400 plus 400 of the 500 units in buffer stocks. So again inventory management philosophy assured that we could meet on time delivery for the customer even when they were the source of the variation.

In addition, many people like to mingle these stocks with no segregation of buffer from safety stocks. This is not a good idea for two reasons. Since the reasons for demand variations are independent of supply variations, it is normal to calculate the two inventories separately. Also, since the use of either buffer stock or safety stock inventory triggers immediate corrective action, it is worthwhile to keep them separate since the respective corrective actions typically solve dramatically different types of problems.

Stock Type	Theoretical Need (At 2.33 Sigma Coverage)	Practical Volume on Hand	# of Boxes; # of *Kanban*
Cycle	4114	4150	83
Safety	138	150	3
Buffer	485	500	10
Total	4737	4800	96

TABLE 3-5   Finished Goods Inventory

The Total Inventory Situation

The total inventory is summarized in Table 3-5.

Three other aspects of Lean inventory management make this a bit complicated.

- First, FIFO inventory management is a Lean tool. Stock rotation gets complicated with three types of stock for each part number.
- Second, each time product is withdrawn from the buffer or safety stock, it signifies an unusual event and triggers a formal corrective action.
- Third, it is very common for the *kanban* cards for safety and buffer stocks to be a different color than the cycle stock inventory. It is normal for buffer to be orange and safety stock *kanban* to be yellow. Red *kanban* are sometimes used as a signal of an emergency situation.

So How Good Is the System We Just Calculated?

First, with our inventory management philosophy, we saw that it operated at virtually 100% on-time delivery. Now we need a metric to work towards minimizing inventory Such a metric for inventory can be calculated in many ways, but inventory turns is a common one. For instance, if we have an inventory of 4800 units and make 1440 units/day, we have 3.33 days on hand. If we work 261 days per year (5 days per week plus holidays), we then have 78 annual turns—that's pretty good. But we can still improve it quite a bit.

Inventory Reduction Efforts

The responsibility for inventory reduction efforts normally falls to the planning and the purchasing organizations to execute. For finished goods inventory, planning is normally assigned the task to keep inventory turns up and on-time delivery high. In most cases, on-time delivery is the more important metric, and inventory turns of finished goods takes a back seat. The resultant effect is for finished goods inventory to rise to high levels since the planning department wants to make sure shipments are not missed. Furthermore, since inventory is usually needed to cover variation, for example, variation in production rates. And reduction of that variation is not within the control of the planning department, the response by planning is predictable, automatic, and almost justified. They simply add inventory until they feel comfortable that shipments will not be missed.

What is missed in the normal plant is the understanding of why the specific levels of inventory are held. In a phrase, there is no management philosophy on inventory management.

With this methodology of calculating cycle, buffer, and safety stocks, not only is the amount of inventory understood but the reasons why these volumes are needed are understood. Furthermore, it is much easier to assign responsibility of inventory reduction to the group that can actually make an impact on that specific inventory creating process.

In this particular case, if we wished to reduce inventories—and we do—it is easy to see that the largest contributor to overall inventory is the cycle stock inventory. Furthermore, the largest contributor to the cycle stock (review the replenishment time calculation) is the planning time. Why does it take 24 hours for the *kanban* to be massaged in planning? Very likely these cards are sitting in an in-box waiting to be processed. In most good systems, planning is completely bypassed and *kanban* go from the storehouse directly to the *heijunka* board. If we could do that here, we could eliminate 24 hours of replenishment time, cycle stock inventory would shrink to 2601 units (that's a reduction of 30 *kanban*) and total stock would now be 3200 units (or 2.22 days) and we would improve from a very good 78 turns to an outstanding 118 turns—all this at no cost and no risk.

Think of all the work that would be needed to reduce ALL the variation associated with the supply and ALL the variation associated with the demand, so we could eliminate ALL the buffer and safety stock—and still you would only reduce 13 *kanban* total. Yet by allowing the *kanban* system to do what it is designed to do, and bypass planning, we eliminated 30 *kanban*. Sometimes our own processes and procedures are the source of huge wastes and need to be addressed (see the discussion on Policy Constraints in Chap. 12).

However, do not lose sight of the goal: the total elimination of waste. Make no mistake about it, all of this inventory is waste. It is total non-value added work. But it is necessary for the time being. We would do away with even this minimum inventory if we could, but it is only the "least-worst" available option—for the time being.

Kanban Calculations

The basic *kanban* calculation for cycle stocks is:

No. of *Kanban* = (Replenishment Time × Production Rate) (1 + Alpha) / (Container Size)

Make-to-Stock versus Make-to-Order Production Systems

Background

Many Lean systems use a make-to-stock production system, at all steps in the process. The beauty of a make-to-stock system for finished goods is that it virtually assures 100% on time delivery as long as you have a good inventory management policy.

- In a make-to-order system, no finished goods inventory exists at all. The finished goods are made only after the receipt of an order. In a make-to-stock system, by definition there is inventory.

So When Does Make-to-Order Make Sense?

In a Lean system with a fairly stable demand, there are times when an entire family of parts is made in one production cell, for example. Let's say the family has 30 models, but only 5 models comprise 90 percent of the total production. We might call them A models or runners. The other 25 models we will call strangers. It may then be advantageous to produce the runners on a make-to-stock system and produce these strangers on a make-to-order basis., holding only the cycle stock on hand. In so doing, you forego much of the resources it would take to hold the buffer and safety stock inventory for these 25 strangers. To accommodate the variation you would then buffer these cycle stocks with a time-buffer strategy such as a plan to work a little overtime when it is needed. This is fairly common, easily calculable and very often a good, Lean, business decision.

The other time make-to-order makes sense is in the job shop—that is, the extremely high-mix production situation, which is usually low volume as well. The key problems with a make-to-order system are that very often you really do not know the demand volume or the due date, until the order is in your hands. Most of these orders are unique, so to be competitive most job shops either have a great deal of invested capital or long lead times—each of which creates a problem of its own.

What Is the End Result of It All?

The typical Lean manufacturing system is a make-to-stock system—that is, it normally has a finished goods inventory, with a sound policy for the management of the inventory, to assure supply to the customer. It is also a pull production system, that is, production is only triggered by customer consumption, to avoid the waste of overproduction. However, this finished goods inventory, although it is necessary, is still waste and, as such, we wish to eliminate it. As we get better at removing the variation and then reducing the inventory you can see that the optimum condition would be to have no inventory at all; which would be a make-to-order system. The logical extension of a fully matured make-to-stock system is therefore a make-to-order system. The catch is, to remove all the inventory, we need to remove all the variation, but since variation is "the inevitable differences….") we can not remove it all. Interestingly enough, the perfect system would be a make-to-order system with no inventory and the lead time would have to be zero. Of course, that is impossible, but it is interesting.

Chapter Summary

Becoming Lean is not synonymous with JIT, for JIT is only a part of becoming Lean. To affect inventory reductions, it is important to understand that inventory is created largely due to the variation that exists in the manufacturing system and that this variation, at some level, is inevitable. Consequently, we want to reduce the variation to a minimum, which will then allow us to reduce the inventory without hampering customer service. This inventory reduction lets us not only save money but also allows us to reduce our lead time and hence become more flexible and responsive as a business. Though a number of inventory reduction tools exist, *kanban* is one of the most powerful in the House of Lean and must be applied totally—following all six rules—to be really effective.

CHAPTER 4

Lean Manufacturing Simplified

Lean Manufacturing must be understood at three different levels. There is the philosophy, which drives the goals and culture; the foundational aspects of quality control on which it is built; and the strategy, tactics, and skills utilized in the quantity control to become Lean. The House of Lean (see Chap. 20) is a descriptive metaphor in graphic format that will assist you in understanding how all these aspects work together to describe the mature Lean Manufacturing system.

The Philosophy and Objectives

At the heart of Lean is its philosophy, which is a *long-term philosophy of growth by generating value for the customer, society, and the economy with the objectives of reducing costs, improving delivery times, and improving quality through the total elimination of waste.*

It is not practical to discuss all of these issues without discussing, in detail, the cultural aspects of Lean and of the Toyota Production System (TPS). We will briefly cover the topic of cultures in Chap. 11, but a comprehensive discussion of the Lean culture is beyond the scope of this book. The rest of the subject matter of this book will keep you busy for the next three years. That seems like an adequate first effort for the implementation of your Lean initiative.

If you need a quick refresher on the foundational aspects of Lean Manufacturing, and particularly Ohno's clear distinction between quality and quantity control, see the Ohno quotes and extracts in Chap. 2.

> **P**oint of Clarity   The TPS is a quantity control system.

The Foundation of Quality Control

Strategy

This foundation of high quality has two strategies. First is the training and development of the workforce. Second is the effort to make all processes stable and capable of meeting customer needs. It is a strategy designed to achieve high levels of delivered quality.

Tactics and Skills for Quality Control

People

People, and the proper handling of people—including training, career planning, and the commitment to a job—are forever are at the heart of the TPS. The culture of Toyota is built on the people, and the company makes few compromises in this area. Some of the basic needs to execute the TPS are covered in the following subsections.

Multiskilled Workers   Multiskilled workers are required to staff the production facilities, for two major reasons. First, to achieve process improvements it is often necessary to reduce or change the elements of the work. This in turn often requires a redistribution of the work. In addition, work cells are often designed so they can be operated by one, two, three, four, or five people, for example, depending on changes in demand. If the workers are not multiskilled, the dynamics of Lean are lost. Multiskilled workers are at the heart of flexibility in Lean Manufacturing.

Problem Solving by All   Problem solving by all has been a hallmark of the TPS since its inception. Workers are expected to solve simple problems, and the TPS incorporates a time trigger regarding the escalation of problems and the involvement of others. Here we have the very revolutionary concept of line shutdowns initiated by the line worker himself. To really give justice to allowing the operator to shut down the production line is a book in itself. But just for fun, let's touch on one of the topics here—that is, how problems are perceived within the TPS is much different than the typical attitude toward problems. In a normal Western plant, problems are seen as a nuisance and even a sign of failure of management, engineering, or even the worker himself. Hence, problems become a thing to hide and shrink away from. No one wants to accept the resultant blame handed out, and so many problems go unresolved even though they are obvious to many. This is commonplace, even today, in most facilities where we work. However, within the TPS, problems are viewed as a weakness in the system and an opportunity to improve the system and make it more robust. Guilt and finger-pointing are avoided and problems are addressed and solved.

Now, let's do a little exercise in imagination. First envision several problems within your organization. Think of a production problem that has persisted for a while, maybe one that people feel a little uncomfortable talking about. No one else is around, so be honest!

Now ask yourself, "What must we do as a company to remove the root cause of this problem?" Do not be surprised if myriad answers come to mind, few of which are really doable.

Next, ask yourself, "What must we change in our company so this problem will never appear again?" and you will get an idea of the deep cultural change needed to alter the attitude toward problems in your company.

Understanding of Variation   An understanding of variation is a topic almost skipped in Ohno's book. Yet this topic *is* the topic of problem solving, process improvement, and inventory reduction, to name just a few. So why is it missing form Ohno's writings? Well, after some thought I've concluded that he had both a deep understanding of, and an ability to manage, variation reduction to such a level that it was simply obvious, it was second nature, to him. And Ohno—if he has a weakness—sometimes does not state the obvious. Do not slight this topic. It is at the heart of your company's survival and nearly all the 20 Lean tools require an understanding of, and reduction of, variation to work.

Stability

OEE OEE stands for Overall Equipment Effectiveness and is the primary measure of production effectiveness. It can be used for value stream or individual work station performance evaluation. Good value stream OEE is one of the key precursors to the implementation of Lean and is the product of three important operational parameters. These are:

- Equipment availability
- Quality yield
- Cycle-time performance

To calculate OEE, you will need five parameters. First is the planned production time for the line. Second is the unplanned line downtime. Third is the line cycle time, or cycle time, of the bottleneck. Fourth is the total production including scrap, and fifth is the total amount of salable product. Let's say we have the following data:

- Planned production time is 20.5 hours. It is 24 hours less 1 hour per shift for lunch and breaks, and less one-half hour for planned preventive maintenance.
- Unscheduled downtime was 1.5 hours.
- Design cycle time is 30 seconds per piece.
- Actual total production was 2020 pieces with 50 rejects, yielding 1970 pieces of salable product. This then allows us to calculate:
  - *Availability* = Total uptime divided by total planned uptime = A = (20.5 – 1.5)/ 20.5 = 0.927
  - *Quality Yield* = The total of salable production units divided by the total production = Q = 1970/2020 = 0.975
  - *C/T Performance* = Total units produced, good and bad, divided by the volume, which should have been produced during the actual uptime at the design cycle time = P = 2020/[(20.5 – 1.5) × (3600/30)] = 0.886
  - *OEE* = A × Q × P = 0.801

This is a way to express how our production facility is performing, and then prioritize our problems and allocate our resources. In this example, OEE is 80 percent. The losses can be stated as about 2.5 percent due to quality issues, 7.3 percent losses due to availability issues, either materials delivery or machinery downtime, and we are losing 11.4 percent due to the line not performing at the design cycle time. It is a very good picture of plant performance and allows management to focus on the appropriate goals for improvement.

MSA MSA stands for measurement system analysis. It is the statistical calculation of the variation in the measurement system and applies to both attribute and variables data. MSA must be done on all measurement systems. The most common use of MSA is as a precursor to doing a capability study on a product characteristic or a process parameter. It is crucial to understand the variation in the measurement system since it detracts from the capability performance of the process. Frequently, process performance can be improved simply by working on the variation in the measurement system.

Cp and Cpk   Cp and Cpk are the industrially accepted measures of process performance. They are both called process capability indices. Several good books describe how to calculate Cp and Cpk, but one major point of understanding must be accepted—specifically, Cp and Cpk have no meaning if the process does not exhibit process stability—that is, process predictability. Process stability is best evaluated using a control chart and is absolutely necessary for Lean initiatives to be implemented. Nothing is more basic to successful Lean implementation than process stability.

Availability   Availability is the concept that the production process shall be capable to produce product, when it is scheduled to do so. High process availability is a necessary characteristic of a process ready to be Leaned out. Low process availability is almost always a sign of an unstable process. Usually, low availability is associated with machinery downtime or the inability to deliver on-spec raw materials to the production line.

Cycle-Time Reductions   Cycle-time reductions are very important to Lean implementations. It is best to work hard on cycle-time reductions prior to implementation of a Lean initiative. This helps stabilize the process and then the quantity control issues are more easily managed. However, often during a Lean implementation, cycle-time reductions will be found and they usually translate directly into higher production rates. These cycle-time reductions are truly the "low hanging fruit" of Lean implementations. Any time a cycle-time reduction can be achieved, the resultant extra production is the lowest cost product you can make. Basically, you are transforming the cost of raw materials into the value of the finished product.

Standard Work   Standard work, as defined by Ohno, has three elements:

- The cycle time
- The work sequence
- The standard inventory

However, it is a much misunderstood concept. In his book, Ohno says, "…I want to discuss the standard work sheet as a means of visual control, which is how the Toyota production system is managed."

Notice he uses two interesting terms. First, he uses the term "visual control," and second he says it is how the TPS is "managed." He does not say, "this is how the TPS is operated." He is very specific, so do not be confused. This explains why, when you enter a Toyota facility and see the standard work sheet at a work cell, it is not facing the operator. Rather, it is facing the aisle so it is available to the supervisor, the engineer, and the manager. Standard work is not used by the line operator but by the team leader, engineer, or manager so they can audit the work, understand the status of the process, and provide assistance if the process is not performing as designed. The standard work chart is part of the concept of transparency and is there for visual control by the management team. It is a myth that the Standard Work Chart is made for the operator.

Transparency *Transparency* is the concept that the performance of the process or the entire line is able to be "seen" simply by being on the floor. It is not generally a set of charts that will allow this—to the contrary, it is a set of visual controls such as *andons*, *heijunka* boards, and space markings that make the process performance "transparent." Where transparency is implemented properly, a manager can determine within one or two minutes if his process is performing as designed—and if the process is deficient, the manager can quickly discern the problem areas. For more on transparency, see Chap. 10.

5S *5S* is a set of techniques, all beginning with the letter "S." They are used to improve workplace practices that facilitate visual control and Lean implementation. The 5Ss in Japanese and English are:

- *Seiri* . . . . . .Separate
- *Seiton* . . . . .Set to order
- *Seiso* . . . . . .Shine
- *Seiketsu* . . .Standardize
- *Shitsuke* . . .Sustain

TPM *TPM* are the initials for Total Productive (*not preventive*) Maintenance. It is a revolutionary approach to the management of machinery. It consists of activities that are designed to prevent breakdowns, minimize equipment adjustments which cause lost production, and make the machinery safer, more easily operated, and run in a cost-effective manner. In most plants, wishing to implement a Lean Initiative, we find that equipment availability is a large source of the process losses. Frequently, the largest of the three losses in the OEE metric. TPM is therefore a powerful tool to improve overall performance of the plant. It is generally defined as having five pillars, which are:

- *Improvement activities*, designed to reduce the six equipment-related losses of:
  - Breakdown losses
  - Setup and adjustment losses
  - Minor stoppage losses
  - Speed losses
  - Quality defects and rework
  - Startup yield losses
- *Autonomous maintenance*, which is an effort to have many routine activities performed by the operator rather than the maintenance department.
- *A planned maintenance system*, which is based on failure history. This is not timed maintenance. Instead, it is based on historical evidence.
- *Training of operators and maintenance personnel* to improve operations and maintenance skills.
- *A system for early equipment maintenance* to avoid the loss that occurs upon new equipment startup.

Process Simplification   *Process simplification* is a basic concept, but is frequently overlooked by most. It is the idea of eliminating and simplifying steps in the production process. This is one of the most powerful variation reduction techniques you can employ.

Sustaining the Gains   Sustaining gains is the concept that once a process improvement is achieved, the next step is to standardize it. Thus, we want to institutionalize the gains so they will be there forever. We then want to build on this gain. It is curious that almost everyone knows this, but almost no one does it, not even modestly. In my work with over 200 companies, I can't give you one example of any company that does this well.

Quantity Control

Strategy

The quantity control strategy has two "pillars": *jidoka* and just in time (JIT).

Jidoka

Jidoka is a revolutionary 100 percent inspection technique, developed by Toyota. It is done by machines not men, using such techniques as *poka-yoke* (error proofing), which will prevent defects from advancing in the system by isolating bad materials and/or implementing line shutdowns.

It is also a continuous improvement tool because as soon as a defect is found, immediate problem solving is initiated, which is designed to find and remove the root cause of the problem. In the design case, the line does not return to normal operation until it has totally eliminated this defect-causing situation.

This powerful concept has been in place at Toyota since its inception as an automaker. In the Toyoda family, it was first implemented in 1902 when it was applied to looms to trigger shutdowns automatically when a thread snapped. Since then, *jidoka* has been continually evolving to higher levels of sensitivity. It is truly a revolutionary concept. (Read about the impact of *jidoka* in Chap. 15.)

A great deal has been written about *jidoka* in cultural terms, with such topics as the interworkings of men and machines, which allow the machines to do the repetitive simple checking and let men do the higher-value work, such as problem solving. Ohno called it "autonomation," and he speaks of it in terms of "respect for humanity." Here is also where the revolutionary concept of "shutting down the line by the operator for production problems" is also manifest.

JIT

Just in time (JIT), on the other hand, is designed to deliver the right quantity to the right place at precisely the right time.

Tactics, and Skills for Quantity Control

The *Jidoka* Pillar

Poka-yoke   *Poka-yoke* is a series of techniques, limited only by the engineer's imagination. The purpose of *poka-yokes* is to achieve error proofing of a process activity and thereby make the process more robust. *Poka-yokes* are also used in the inspection process to achieve 100 percent inspection. There are two types of inspection *poka-yokes*: those that control—that is, shut down—the process or isolate the product upon finding a defect; and those that warn the operator via an *andon*.

5 Whys 5 Whys is the cornerstone of the TPS problem solving effort. The "5 Why" technique is simple enough in concept. However, it will not work unless those using this technique have both expertise and experience in the problem area. They must fully understand the cause-effect relationships to utilize this seemingly simple technique. The check on the "5 Whys" is the "Therefore" technique.

Kaizen *Kaizen* is the concept of improving a process by a series of small continuous steps. Often times these improvements are small and hard to measure, however the accumulated effect is significant. Over the years, *kaizen* has evolved to mean improvement.

CIP CIP stands for Continuous Improvement Process or Philosophy. Many can talk about it, but few can show a process flow chart for their CIP, and even fewer can adequately measure it. An example of a Continuous Improvement flow chart is shown in Fig. 4-1. In addition, the Toyota Production System advocates the concept of *yokoten*, which concerns extending the process improvements to other locations, as well as other similar applications, as part of CIP.

The JIT Pillar

Takt *Takt* is the design process cycle time to match the customer's demand, normalized to your production schedule. It is the key calculation used when we synchronize supply to the customer. *Takt* is calculated by dividing the available work time by the product demand. The system is then designed to produce the product at this rate. If we produce at a cycle time higher than *takt* (hence, under-produce) we will not be able to supply the customer demand. If, however, we produce at a cycle time lower than *takt* (overproduce), we will either increase inventory or idle the line to stop the overproduction. Both of these are wastes—recall that the #1 waste is the waste of overproduction. For example, if we run our operation using two 10-hour shifts with a 30-minute lunch break and two 15-minute breaks each shift, and run five days per week, holidays included, and need to produce 500,000 units per year, our *takt* would be: $\{(365 - [52 \times 2]) \times (2 \times [10 - 1]) \times 60)\}/500,000 = 0.56$ minutes or about 34 seconds. That is, to stay in step with the customer's demand, considering our work schedule, we will need to produce one salable unit every 34 seconds. Consequently, since there are losses, our production cycle time will need to be shorter. For example, if OEE was 0.80, a production cycle time of about 27 seconds (0.8×34 seconds) would be required.

Balanced Operations Balanced operations are a simple industrial engineering technique to have all operation steps—of a cell, for example—operating with the same cycle time. It is the first step in synchronizing the internal production. This technique, not unique to Lean Manufacturing at all, is designed to avoid the waste of waiting. However, this technique places a large emphasis on the ability to standardize operations so we can avoid variation in the process. If any step in a process has high variation, that step will naturally unbalance the entire line or cell.

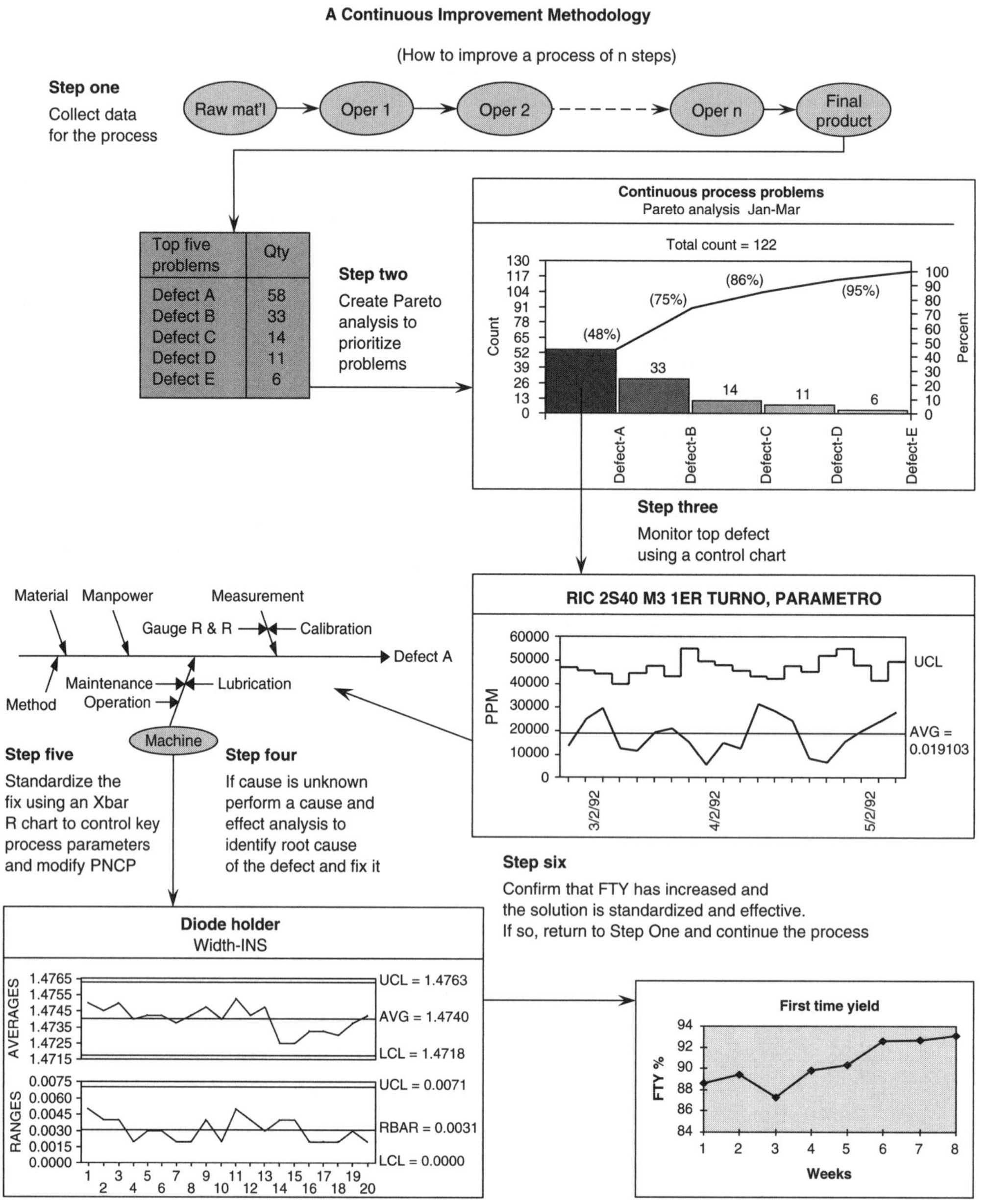
A Continuous Improvement Methodology
(How to improve a process of n steps)
Step one
Collect data for the process
Raw mat'l
Oper 1
Oper 2
Oper n
Final product
Top five problems
Qty
Defect A 58
Defect B 33
Defect C 14
Defect D 11
Defect E 6
Step two
Create Pareto analysis to prioritize problems
Continuous process problems
Pareto analysis Jan-Mar
Total count = 122
Count
Percent
(48%)
(75%)
(86%)
(95%)
Defect-A
Defect-B
Defect-C
Defect-D
Defect-E
Step three
Monitor top defect using a control chart
RIC 2S40 M3 1ER TURNO, PARAMETRO
PPM
UCL
AVG = 0.019103
3/2/92
4/2/92
5/2/92
Material
Manpower
Measurement
Gauge R & R
Calibration
Defect A
Maintenance
Lubrication
Method
Operation
Machine
Step four
If cause is unknown perform a cause and effect analysis to identify root cause of the defect and fix it
Step five
Standardize the fix using an Xbar R chart to control key process parameters and modify PNCP
Step six
Confirm that FTY has increased and the solution is standardized and effective.
If so, return to Step One and continue the process
Diode holder
Width-INS
AVERAGES
UCL = 1.4763
AVG = 1.4740
LCL = 1.4718
RANGES
UCL = 0.0071
RBAR = 0.0031
LCL = 0.0000
First time yield
FTY %
Weeks

Figure 4-1 Continuous improvement flow chart.

Pull   Pull systems are production systems that are designed to minimize overproduction, the most grievous of the wastes. Pull systems have two characteristics.

- They have a maximum inventory volume—for example, when using a *kanban* system.
- Production is initiated only by a signal from the customer, and that only occurs when some inventory has been consumed.

A pull system is one in which the customer, the next step in the process, removes some product that then is the signal for the upstream step to produce. For example, for some reason the finished goods inventory of our customer is full and the entire complement of *kanban* cards are attached to the finished goods in the storehouse. Since *kanban* cards are the signal to produce, and they are all attached to the finished goods, production has stopped. Hence, our overproduction is limited to whatever we have in finished goods for the cycle stock, buffer stock, and safety stock. However, when our customer arrives and withdraws product, then the *kanban* cards are removed and circulated back to the production cell, signaling that production is authorized to begin. Once it starts production, the cell will produce only that volume dictated by the *kanban*, and these finished goods will then be placed in inventory. This process of replacing the inventory that was withdrawn is specifically named replenishment.

A manufacturing system with a limit on the maximum inventory, and production based on replenishment, is the essence of a pull system. The opposite of a pull system is a push system. In a push system, there is no maximum inventory, the downstream process produces until it is told not to, usually by the scheduler. It then pushes that product onto the next step whether the next step needs the production or not. Hence, on the production floor, there is no maximum control on the WIP, so WIP can grow uncontrollably. With this uncontrolled growth of WIP, lead time will grow, with resultant quality, delivery, and cost problems escalating.

Minimum Lot Size   Minimum lot size is a means to reduce lead times. By reducing production lot size and transfer lot sizes, the process proceeds much faster. Two benefits are achieved. First, we reduce the lead time for the first piece through the process. This benefit is usually felt in quality responsiveness. If the first piece lead time is reduced, and there is a problem with the product, this information is fed back to the problematic station more quickly. The problem can be resolved more quickly, and if rework is required, fewer items will need to be reworked. (For a dramatic example of this effect, see Chap. 15's discussion of the Bravo Line.) The second benefit is that the overall product will be completed more quickly, reducing production lead time for the lot. Minimum lot size, with the ultimate being "one-piece flow," is the key to plant flexibility and product supply responsiveness.

Flow   Flow is the concept that parts and subassemblies do not stop except to be processed, and then only for value-added work. It is more of a concept to be attained than a reality. It is the primary tool used to reduce production lead time. The typical technique is to design the process so that as little inventory as possible exists at each work station, and the work stations are synchronized as close as practical. The design ideal is a multistation cell with no inventory between work stations. The ideal state we seek is one-piece flow with 100 percent value-added work only. This ideal state is frequently not possible, at least initially, because there are obstacles to flow. (The Seven Obstacles to Flow are detailed in Chap. 5, with a case study in Chap. 15.) For example, let's review a process running multiple products in which one of the steps is a large machine, say a press, that must undergo changeovers between production runs. To avoid stopping the process during a changeover, a buffer is built up both before and after the machine, so the rest of the production process can continue to run while the press is undergoing the changeover.

All the items in these buffers will arrive "too early" to be just in time. However, considering all the options, creating a buffer is the least-waste-generating choice for the process, so it was selected. This does not create the ideal system but it is the economically practical answer. In every case, if it is not currently possible to eliminate all inventory in a process, then the next best solution is to design a system with the minimum amount of inventory. The amount is calculated, and posted at the work station as a maximum. Whenever the upstream process has produced that maximum volume of inventory, the upstream process must stop production to avoid the waste of overproduction. *Kanban* is just one system that is used to avoid overproduction. Most other systems used to minimize inventory are based on limiting the physical storage space that the parts may occupy. This is simple and creates a very good visual management tool.

Lead-Time Reductions   Lead-time reductions are the essence of waste reduction in Lean. They give the process both the maximum flexibility and maximum responsiveness to changes; especially changes in demand either in quantity or model mix. Read about lead-time reductions in Chap. 5, with a specific case study in Chap. 15, which shows how you can break through the obstacles to flow and significantly reduce lead time.

Leveling   Leveling is spoken of in two terms. First, leveling is the concept to maintain a consistent nonvariant rate of production over time. It is also a waste reduction technique called model-mix leveling, that calls for the simultaneous production of multiple products, or models of a product, from a given production line. To do otherwise is to create a batch in the system. We have already stated that Lean is a batch destruction technique. In a perfect world, we should level production to the individual production unit level. In practice, this often is not practical and sometimes not desirable. Consequently, we will frequently level based on the packaging requirements. That is, if we package 60 units in one carton, we will run 60 of that model and then switch production to another model. For example, if a certain manufacturer produces 50 models of a given product, all in equal volume, and he has the ability to run all 50 models, one piece at a time, it would be easy to implement perfect model-mix leveling. However, let's say he packages 60 units to a box and the cycle time is 30 seconds, so it takes 30 minutes to fill a box. If this operation is run with perfect leveling, then at the packaging station there are 50 boxes being simultaneously filled, and every 25 hours a large batch of finished goods needs to be transported. If, however, the process is leveled so that one box is run at a time—this is called a "pitch"—then there is only one box at a time being filled. Quite frankly, this system of producing one pitch at a time makes the downstream handling more "level" and also makes the *kanban* system much easier to use. Considering the current conditions, leveling to a pitch is normally the optimum for any Lean system.

Kanban   *Kanban* is the revolutionary practice of using cards, for example, to smooth flow and create pull in a Lean system. It is also a continuous improvement tool. The cards represent and account for all the inventory in the system. By controlling the number of *kanban* cards, we control the inventory. *Kanban* is a technique used to control inventory, minimize overproduction and facilitate flow. The *kanban* cards are used to trigger replenishment. This will make the system more responsive to customer demand and shorten lead times because the signal comes directly from the customer and triggers replenishment. For a *kanban* system to be effective, all *kanban* rules must be rigorously followed. The Six Rules of *Kanban*, from *Toyota Production System, Beyond Large-Scale Production (Productivity Press, 1988)*, are:

- Later process picks up the number of items indicated by the *kanban* at the earlier process.
- Earlier process produces items in a quantity and sequence indicated by the *kanban*.
- No items are made or transported without a *kanban*.
- Always attach a *kanban* to the goods.
- Defective products are not sent on to the subsequent process. The result is 100 percent defect-free goods.
- Reducing the number of *kanban* increases their sensitivity.

Cells Cells are work areas that are arranged so the processing steps are immediately adjacent to one another. This lets parts be processed in near-continuous flow either in very small batches or in a one-piece flow. This, in turn, allows minimization of the wastes of transportation and inventory—in this case, WIP (work in process). The most common shape is the "Inside U" cell. This cell minimizes walking distance when standing operators are used. Cells have some natural advantages over the classic assembly line. First, the ability to use people for more than one activity in a cell allows the control of demand variations by staffing differently. For example, if a six-person cell were to cut production by 50 percent, it is commonplace to then staff the cell with only three people and have each person work two stations. This, of course, requires worker cross-training, but that is a staple of Lean Manufacturing. Second, cells are much more flexible. For example, in place of a 20-person assembly line, if we use four- to five-person cells we have a much greater model-mix capability without creating large batches and without having large time losses due to changeovers. But the coolest aspect of cells is that, although it is a very well kept secret, cells can be a natural variation reduction device. Cells are a very interesting topic, see Chap. 13, Cellular Manufacturing for more details on cells.

SMED/OTS SMED/OTS stands for Single Minute Exchange of Dies and One Touch Setups. SMED technology is a science developed by Shigeo Shingo and is designed to reduce changeover times. The problem is simple. Any machine that has long changeover times must have an excess capacity to account for the downtime of the changeover. Furthermore, to supply the rest of the downstream process during the changeover, a large batch must be stored up. Any effort to reduce the changeover times also reduces these two forms of waste: excess capitalization and overproduction. ("Single minute" means a single digit number of minutes that is less than 10.) In actuality, the objective is to reduce the changeover time as much as possible. In some refined cases, the changeover is handled by having multiple fixtures on the same basic machine, and by simply throwing a switch the changeover is made. This is called One Touch Setups (OTS), or sometimes One Touch Exchange of Dies (OTED). In his writing, Ohno refers to three basic elements of JIT. They are pull systems, operating at *takt* time with continuous flow. Those may be the big three but JIT is seldom practical without some application of SMED technology. It is a major batch destruction technique. The basic procedure of SMED is simple, it is a three-stage process:

1. Separate internal from external setup
2. Convert internal setup to external setup
3. Streamline all aspects of the setup operation

When a SMED application is first undertaken, we have found the best tool is the simple Gantt chart, showing all the steps in the changeover. Gather the knowledgeable people on the changeover, and then list all the changeover steps. Categorize them as internal setup, external setup, or internal but can be external; also list the conditions to make it external setup.

This is the basic starting point. From here you delete any unnecessary steps and simplify any steps you can. Next, you convert as much internal setup into external setup so it can be done with the machine running. With only internal work left, the technique is generally to create as many parallel paths as possible. At this point, you can get involved with intermediate and holding jigs, automatic adjustments, and a huge volume of imaginative approaches to shorten the changeover time.

SMED and *poka-yokes* are two of the Lean techniques that are truly for the imaginative. This combination is a powerful set of tools to use as we reduce lead times and more fully utilize our processing equipment,

Much has been made of making a video of a changeover. I support this and have found it to be useful, but generally it is best to do it after you have applied SMED techniques at least once. The reason is this: When you apply SMED, the entire process will change, so it is not very worthwhile to view the old process. You will get some minor improvement ideas, but the majority of the ideas come from the development of the Gantt chart referred to earlier. However, there is one large benefit to be gained from making a video: Watching the old technique is usually humbling if not downright funny, and feeling a little humility as well as a good laugh are both good for the soul. At any rate, doing a video is easier than it used to be, so I do not discourage it completely.

The application of SMED technology is a key batch destruction technique and should not be underestimated in terms of its potential. It is one of the major efforts that must be undertaken if Lean is your objective. For further study, I suggest you go directly to the author of the tactic, Shigeo Shingo. He has written two major books. One is *A Revolution in Manufacturing: The SMED System,* (Productivity Press, 1985), his landmark book on the topic. In his other book, *A Study of the Toyota Production System,* (Productivity Press, 1989), he expanded his coverage on parts of his SMED system. He has refined his three stages into eight techniques. It is good reading for the Lean professional.

Cycle, Buffer, and Safety Stocks   Cycle, Buffer, and Safety Stocks is the three-fold approach to inventory management used in Lean Manufacturing. Each of the three types of stock is calculated and marked separately. The common way to separate the stocks is to use color-coded *kanban*. For example, white cards are used for cycle, yellow for buffer, and orange for safety stocks. Red is normally reserved for emergency runs. Consequently, when a colored *kanban* shows up at the *heijunka* board, the production people are aware that something is abnormal and they usually have a specific protocol to follow. This description of inventory management is focused on finished goods inventories, but the concepts also apply to WIP inventories. Inventory volumes need to be reviewed periodically to assess possible waste reduction opportunities. Remember, any inventory beyond what is required to protect the supply to the customer, which is the next operation, is unnecessary waste.

Cycle stock is to account for the inventory built up between customer pickups.

Buffer stock, on the other hand, is the inventory kept on hand to cover the variations associated with external causes, including demand changes and such items as transportation variations. Any time this inventory is withdrawn, a note must be made in the warehouse operating log, at a minimum; it is desirable to institute corrective actions, as well. The warehouse log is a form of transparency.

Safety stock is that inventory kept on hand to cover the variations internal to the plant, including line stoppages, raw material stock outs, and anything else internal that hampers the ability to deliver the customer's demands. Any time this inventory is withdrawn, in addition to a note in the warehouse log, a corrective action report is initiated. (See Chap. 3 for more on inventory.)

Chapter Summary

The heart of Lean is its philosophy of long-term growth generating value for the customer, society, and the economy with the objectives of reducing costs, improving delivery times, and improving quality, all through the total elimination of waste. The key foundational strategies that support this philosophy are the investment in people and the stability of the processes that then yield a system that will produce a high-quality product. On this foundation of high quality is built the strategy of quantity control. The quantity control strategy is supported by two substrategies: *jidoka* and JIT. All of the strategies and substrategies are supported by a broad range of tactics and skills. Together we refer to the strategies, substrategies, tactics, and skills as the Tools of Lean. In Chap. 20, we have included a House of Lean that shows how all the tools of Lean work together to execute the objectives, which is simply "better, faster, cheaper," through the total elimination of waste.

CHAPTER 5

The Significance of Lead Time

Originally, I was not even sure if I should include a special chapter on lead time. Not that lead time isn't important—it's an extremely important Lean concept. But lead time is discussed in several chapters such as Chaps. 3, 8, 14 and 15, and though it is an extremely important topic, I didn't want to overdo it.

However, lead time, when it comes to useful information in Lean manufacturing is like onions in an omelet or cheese on lasagna—you just can't have too much of it. And so we have this chapter.

To cover lead time, we will:

- Look at its history
- Explain the benefits of lead-time reductions
- Review one case study
- Explain the seven techniques used to reduce lead time
- Explain why lead time is the "key" measure of Leanness

Some History of Lead Time

Until the JIT (Just In Time) movement got some traction in the U.S. in the late '80s, there wasn't much talk about lead time at all. One of the early books on JIT was *Zero Inventories* (APICS, 1983) by Robert Hall, and although he put together a good review of the Toyota Production System, he made virtually no mention of lead time. There was, however, an interesting treatment of lead time in *The Goal* (North River Press, 1984) by Goldratt and Fox. In addition, Richard Schonberger, in his book *World Class Manufacturing* (The Free Press, 1986), has some excellent information on lead time and lead-time reductions. These books were from the mid-1980s. But after Ohno and Shingo published their books in the U.S., and *The Machine that Changed The World* (Rawson Assoc. 1990) by Womack, Jones, and Roos was published, the topic of lead times became more prevalent amongst Lean professionals. This was about

> **"I** often say that when you can measure what you are speaking about, and express it in numbers, you know something about it; but when you cannot express it in numbers, your knowledge is of a meager and unsatisfactory kind...**"**
>
> Lord Kelvin

1990. However, many people did not know, beyond the conceptual level, what lead time really was, and few knew how to calculate it. Lead time made for good talk, but it wasn't really understood.

Then, in 1998, Mike Rother and John Shook published *Learning to See* (The Lean Enterprise Institute, 1998). I consider it a landmark book, not only because of the concepts they explored, but also because they showed the world how to do value stream mapping (VSM). In the process of teaching VSM, they also taught readers how to reduce two Lean metrics to numbers. These two metrics are the Percent Value Added Work and Lead Time.

Benefits of Lead-Time Reductions

As a Business Advantage

In his book, *The Toyota Production System, Beyond Large Scale Production* (Productivity Press, 1988), when asked what Toyota was doing, Ohno was quoted as saying "All we are doing is looking at the time line, from the moment the customer gives us an order to the point when we collect the cash … And we are reducing that time line by removing the non-value-added wastes." For Ohno, lead-time reduction is a key method in improving cash flow in the company. It clearly has this business advantage in addition to its advantages in the manufacturing system.

As a Manufacturing Advantage

Two key characteristics that few businesses measure in any form but that all businesses want are flexibility and responsiveness. They go hand-in-hand. With one, you also get the other. I know of no company that measures manufacturing flexibility or responsiveness and posts it with the other business metrics, but it is critically important nonetheless. Ask any planner what they would like to have more of, and second only to accurate forecasts is the ability to quickly change plans and still meet delivery dates. There is a good example of how reducing lead times helped the Bravo Line in Chap. 15. Read that now if you would like. It is made clear how lead time and lead-time improvements were turned into business advantages with huge gains. Two lead times are of critical importance and we will elaborate further on them. They are:

- First piece lead time, which is the time it takes for the first piece to finish and be ready for packaging. The primary benefit of this metric being short is that, typically, the last quality inspection is done just prior to packaging and so this is the response time it takes to confirm that either quality is good, or we need to change the process.

- Shipment lead time is the time it takes to complete the entire shipment. This, of course, is the key metric used in planning.

Table 5-1 shows the data from the Bravo Line. Take a look at the flexibility and responsiveness factors achieved by shortened lead times.

How might the advantage of flexibility work for you in your business? Well, let's say the customer calls up and wants to change the model mix. In the Original Case, we have to tell the customer that we have a batch in the production cell now, it will take about 6.2 days to clear the line, and then right behind that we can run their request

Time Impacts	1st Piece Lead Time, Cell 1	Shipment Lead Time
Original case	232 min (3.9 h)	149 h (6.2 days)
After lean improvements	6.5 min	28.4 h (1.2 days)

TABLE 5-1   Bravo Line, Lead-Time Improvements

which will run another 6.2 days, letting us ship it in 12.4 days. Now, any planner who values his life will add some fat to that because, if you recall, this line did not always run to schedule. So the planner will promise something like 15 days and probably will not sleep well until the shipment leaves.

On the other hand, with the Leaned out process, he tells them it will take 2.4 days, and since it runs on schedule more often than not, he not only tells them we will ship in three days, but he confidently delivers that message. But life is not always that kind. Problems can arise in the best of systems. In the short lead-time situation, if the current production is delayed, thus holding up the request from the new customer, it is known in one day, making some countermeasures possible. In the long lead-time case, it may take a week for the problem to surface. This is yet another type of flexibility inherent in a short lead-time production system; the ability to respond to abnormalities more quickly.

Please return to Chap. 2 and the section entitled, What is Lean? Here you will get a good dose of just what we mean when we say it is "emotionally much Leaner." That planner can proceed with confidence and, quite frankly, he will sleep better. Those examples abound in a Lean facility.

Responsiveness and flexibility are the life blood of a typical job shop. For them, these advantages can be achieved through the reduction of lead times. We have worked with a number of job shops and taught them the benefits of lead-time reduction by using Lean techniques even though the application of Lean techniques are not as straightforward in that environment.

Excalibur Machine Shop, Lead-Time Reductions

A look at the Excalibur Machine Shop will give us some insight as to the applicability of these principles to a job shop using batch type operations.

The Background

We were hired to train Excalibur in Lean principles. Although they knew little about the TPS (Toyota Production System) or Lean principles, they thought it might help them with some of their manufacturing problems. They described their problems as:

- Labor efficiency was only 56 percent compared to their goal of 80 percent minimum. This was a comparison of bid hours for a job compared to actual hours worked.

- They frequently had quality issues. No job went through without rework; most jobs had two or three episodes of rework.

- Even if they started jobs with plenty of time, they always seemed to need overtime to complete jobs. Even after working overtime, 35 percent of the jobs had unscheduled expedited freight charges that cut deeply into profit margins.

We took a plant tour and it was obvious that Lean techniques were something they desperately needed. Around their problems we designed a four-day training curriculum in Lean. During the class, we also did a group project. The task was to apply the Lean principles to one of their manufacturing applications. In the class they got so excited about what we had accomplished on paper, they wanted to apply it to the floor. They extended our contract and we went to the floor.

The Product

This plant was a metal fabricator. The product they had chosen to Lean out was a junction box used in the telecommunication industry. The external dimensions of the box were 36 inches high by 24 inches wide and 8 inches deep. It was made from 12 gauge precoated steel. In addition to the 80 rivets and 40 screws per assembly, there were 22 items on the bill of materials

The Process

The CNC Punch Press—Turret

The process consisted of using a Computer Numerical Control (CNC) punch press to stamp out a topside assembly consisting of the top and seven other smaller parts. The large sheets were manually loaded onto the punch press; each sheet made five assemblies. While it was cycling, the operator would separate the prior sheet, remove the protective coating, segregate the parts, and load them into containers to be transported to the deburring operation. After 100 assemblies were completed, the operator would transport the production to deburring, change the setup, and produce the bottom-side assembly. The bottom-side assembly consisted of the bottom and four other smaller parts. These parts were handled the same as the topside assembly, with protective coating removal, segregation, and placement into containers for transportation to deburring. After 100 assemblies had been produced, these too were transported to deburring and the operator would make a changeover to his next product. The machine cycle time was ten minutes, or two minutes per assembly, for both the top and bottom-side assemblies. The changeover time for this product was 36 minutes.

Deburring

The work in process (WIP) from the Turret was then transported to deburring. With automatic deburring machines, the cycle time per assembly was 36 seconds and no changeovers were required. The deburring operator was very lightly loaded. He would deburr a batch of 100 topside assemblies and transport them to the Press Break for bending. He would wait for the bottom-side assembly to arrive and then deburr and transport these to the Press Break.

The Press Break

At the Press Break, the operator would bend the pieces to the assembly and although there were more pieces and more bends to the topside assembly, both the topside batch and the bottom-side batch took 1.2 minutes. Even though there were several change-overs per batch, the changeover time was trivial since the tools were already loaded and the operator only needed to change the program in the computer's database.

The Assembly Cell

The completed WIP was placed in a holding area, awaiting time on the assembly line, which had eight work stations. When all the raw materials were available, the product would be scheduled on the assembly line.

The Analysis

The Lead Time Chart and Minimizing Lot Sizes

To begin the project, we had completed a lead time chart in the classroom, as shown in Fig. 5-1.

In this case, a large lot was 100 units and the batch lead time was 1220 minutes, or over 20 hours of production time. This batch of 100 was also about two months demand for this particular cabinet. We wanted to reduce the batch size and, of course, we would like to go to one piece flow, but one piece flow with one piece transfer lots was not possible at this time. Their use of large multipurpose machines made this impractical. But we still wanted to reduce the batch size. We selected a batch size of 20 because there were 20 finished units to a pallet and they would never produce less than a pallet, they said.

This would increase the number of changeovers for the punch press from two to ten for the 100 unit batch. Recall that the CNC punch press was used for cutting out both the topside assembly and the bottom-side assembly units. The CNC punch press was operated by one machinist who handled loading and unloading, as well as did the changeovers single-handedly.

Balancing the Assembly Cell

Next, we reviewed the operation of the assembly cell and completed a balancing study. There were eight operators in the cell and although there were only 26 minutes of work per unit, there was a bottleneck at station 4 lasting six minutes. Even though we only had 26 minutes of work per unit, with 8 operators and a bottleneck of 6 minutes, we needed 48 minutes of paid work per unit; a full 80 percent excess labor cost. Nearly all operations were manual, so by rearranging some work we were able to balance the work and reduce the bottleneck to 4.5 minutes. With this new constraint time, we only needed 26/4.5 or 5.8 people. We decided that although it would be possible to use only six operators in the assembly cell, we would start with seven; still a reduction of one person. We modified the work stations and work instructions and were ready in the assembly cell.

Reducing the Changeover Time at the CNC Punch Press

Next, from our prior work in the classroom, we thought there were opportunities in the changeover times for the CNC punch press. If we could not reduce the changeover time, we would add almost 300 minutes to the lead time. Again, in the classroom we did a SMED (single minute exchange of dies—that is, quick changeover) analysis. Recall that the changeover time was 36 minutes. In this 36 minutes, we found over 11 minutes of external work; this left 25 minutes of internal work to analyze. It was just 25 minutes of changing tools for the most part, so we acquired two additional operators and put

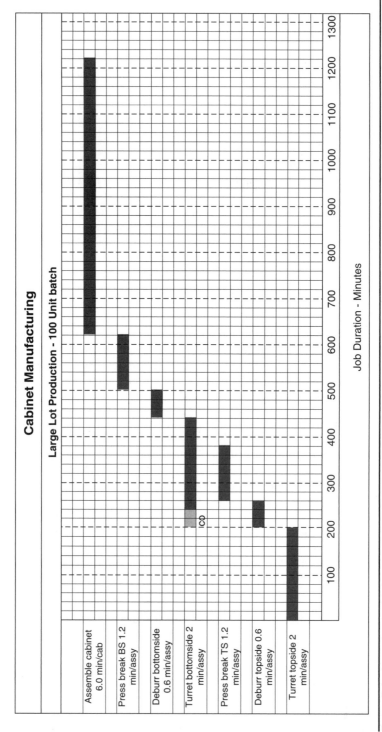

Figure 5-1 Large lot production.

them to work. One operator was the deburring operator who was lightly loaded, while the other one was the person freed up from the assembly cell. We just divided the 25 minutes equally among the three people and gave them the necessary training to change out the tools on the punch press. This allowed us to perform the tool change-outs in three parallel paths and so we reduced the changeover time to just over 8 minutes, on paper. However, in our planning we used 10 minutes as the time for a changeover. Actually, when we did the first changeovers, they took about 10 minutes. With three people all gathered around this punch press, access was an issue. Although we were confident we could make further reductions, we made no more changes at this time. Notes were taken on possible time reduction opportunities and saved for future use.

Making the First Run

Starting Production

We were now ready and started the production. It went well and we were able to produce small lots (per Fig. 5-2, notice the change in scale from Figure 5.1)... almost.

Quality problems surfaced and we could "see" the type of quality problems they were experiencing. During this initial run, we also learned why these quality problems had such a large effect on their labor efficiency.

Encountering Quality Problems

When we started up the assembly cell, which was about 2.5 hours into the run, we found a problem on the topside assembly. One small hole had been made, on the CNC punch press, with the wrong recess dimensions. Luckily, we found this after only 40 units had been produced. We were able to rework the 40 units offline without disrupting the assembly cell. On the next changeover of the CNC punch press, we installed the correct tool. In addition, we found that one bend had been missed on the bottom-side assembly at the Press Break. These were returned to the Press Break and finished. The overall loss of time at the assembly line was about one hour.

The Process Smoothes Out

Other than that, the process proceeded smoothly. About three weeks later, they needed to make another run of 100 and it went flawlessly. In both cases, they were extremely pleased.

The Results

Labor Efficiency Skyrockets

In the first run, using smaller batches, their labor efficiency was calculated to be 118 percent, and in the subsequent run three weeks later, labor efficiency rose to 146 percent. Go figure. (I did a paper *kaizen* study and found that with no capital, this labor efficiency metric could be improved another 40 percent to over 180 percent.) As so often is the case, we redesign the system, improve it to create a new present case condition, and then again find further opportunities to improve. The cycle seems endless—and it is!!

Changeovers Are Increased, with No Out-of-Pocket Costs

It is interesting to note that there was some resistance to increasing the number of setups on the CNC punch press since they stated that this tied up labor and did not increase

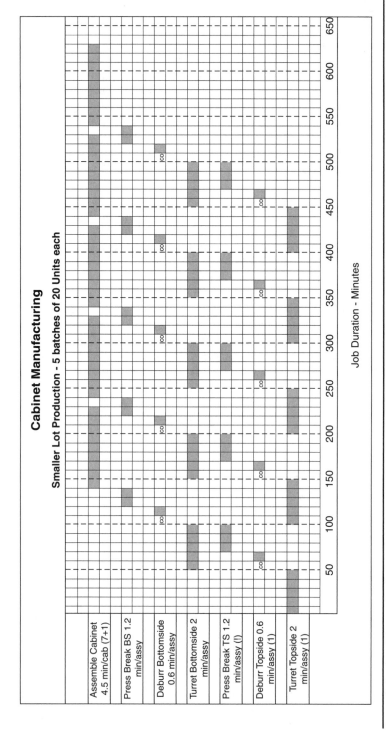

Figure 5-2 Small lot production.

production. They considered it waste. However, a quick review of some related operating practices showed the following.

First, in an attempt to find quality problems early, they had instituted a practice of doing 100 percent inspection of the first production piece from the punch press. This was done using an automatic optical inspection tool. In addition, it was their policy to stop producing after the first sheet until the quality tech gave them first-piece approval. This inspection process step was to take 12 minutes. However, the inspection was often a holdup and took far longer than 12 minutes. I reviewed some records and found that over the last two months, the inspection delay averaged 46 minutes (above the 12-minute inspection time allocation). Everyone was aware of this, and no one liked it, but the bottom line was there was no action or even a proposal to reduce it. They simply accepted this time delay but would baulk at the time delay caused by another setup.

Second and even more pointedly, the CNC punch press was only scheduled to run 73 percent of the time, maintenance included, and it was staffed 100 percent of the time. So they were paying the same for it whether it was running or not. In this case, at this time, changeovers were free on this machine, yet they baulked at doing them.

So it is easy to "see" that they have some serious wastes in the system that are larger than the cost of a changeover, yet these go "unseen." This is not unusual and we find this in many businesses. Often, it is due to the paradigms these businesses live with, and these paradigms typically go unchallenged. The paradigm is not the problem—we will always have paradigms. The problem is the unwillingness to challenge the status quo. Herein lies the advantage of having a consultant (or *sensei*) to assist you. He/she will "see" these opportunities much quicker than the people within the business will see them. Consequently, your consultant will bring forth opportunities that you might be blind to. Your consultant will help you in your "Learning to See." This is invaluable help and the following paragraphs will describe how that advantage was turned into solving business problems and, in a phrase, making some money.

Lead Times Are Dramatically Reduced

Table 5-2 shows the key Lean metrics of first piece lead time and batch lead time.

First piece lead time, a key measure of responsiveness, is cut by 75 percent, and total time to deliver the completed lot is cut in half. All of this was achieved by simply:

- Reducing the batch size
- Applying SMED
- Balancing the assembly cell
- Implementing a type of *jidoka*

We were able to do it faster, better, and cheaper. And the most encouraging lesson from this story is, we are just getting started!!

	First Piece Lead Time	Total Lead Time
Large lot, 100 units	620 minutes	1220 minutes
Smaller lot, 20 units	140 minutes	630 minutes

TABLE 5-2 Lead Time Changes, Bravo Line

How Did the Shortened Lead Times Positively Affect Those Concerns Stated Earlier?

First, for labor efficiency, they had rather liberal estimates for the needed work; and then both planned and performed poorly compared to the estimates. Labor efficiency on this lot was 118 percent compared to the 56 percent they had been achieving. Labor efficiency more than doubled—and at no cost! For them, that meant we could do the work with less than half the manpower. I would call that a huge gain!

Regarding their quality problems, it is clear they had huge problems with rework. The shorter lead time model allowed them to correct the problems. For example, with the quality problem on the CNC punch press, we were able to catch the problem after the production of only 40 units; 150 minutes into the run. We were able to promptly feedback the quality problem and fix the problem with the changeover procedure and this also led to reworking a much smaller volume. In the large lot production situation, a review of the schedule shows that the CNC punch press would finish the entire lot after 460 minutes and probably be on another job when the problem was found in assembly after 620 minutes. In this case, there is no possibility to correct the problem and our only choice is to manually rework all 100 units. Whatever the quality problem, whether it is scrap or rework, the long lead-time model only exacerbates the magnitude of the problems. Again, the shortened lead time allowed us to correct the problems because they could be found in a timely fashion and reduced the labor to rework the problems. That was a huge gain for this company since you will recall that it is common to perform rework two or three times, on each job!

As for the scheduling concerns, this by now should be obvious. It is very difficult to plan your way out of problems that you have managed your way into. A review of past production for this model showed that in no case were they able to start and complete this product in the same calendar week! Even with their large lot approach, this job should be done in less than three shifts. It is easy to see that with the large lot production model, quality problems and the delays associated with correcting them were bound to be large and very punishing in terms of lead time. Yet in the reduced lot size we were able to produce the lot in less than 11 hours. It was done the same day! We reduced the theoretical time to produce from 22 to 11 hours, and we reduced the actual time to produce from over seven days to less than one day. I would also call that a huge gain!

How Well Did We Implement the Lean Concepts in This Batch Operation?

Let me give you a word of both caution and encouragement. We were not able to get the essence of Lean incorporated into this production process since we could not really set up pull systems, operate at *takt*, and flow one piece at a time—and we may never be able to, considering their business.

> **P**oint of Clarity   Do not let the things you can not do prevent you from doing the things you can!

However, in the end we were Leaner than when we started because we were able to produce the same product using less labor, less space, and do so with a much shorter lead time.

Regarding the Applicability of Lean at Excalibur

So although this is not a textbook example of the application of the principles of the TPS, Excalibur Manufacturing clearly was able to reduce the waste in their process and produce a superior product in a more flexible and more responsive manufacturing system. They were able to apply the Lean principles and produce their products with less waste. I call that a huge success! What do you think?

How Lead Time Works for a Job Shop

There is one last point to make about lead time in a job shop environment, and it is absolutely critical to understand. Unlike the typical automobile supplier, job shops often do not have a promise of three years worth of work on the horizon; usually they live from job to job. Typically, their jobs are competitively bid, and very often quoted delivery time is a crucial decision making criteria, second only to cost in the final bid analysis. If you are not able to deliver on time, even with the low bid, you will lose the job. So lead time is crucial to these entrepreneurs. Although short lead times will not guarantee success, long lead times will almost certainly guarantee failure. Short lead time is equal to future business to these dynamic businesses.

Lead Time as a Basic Tool in Variation Reduction

Reducing the lead time is not only a result of reducing variation, it is a variation reduction technique of extreme power. Reducing the lead time directly allows inventory reductions in the process. This reduces exposure to environmental factors and possible damage and deterioration.

Also, there is the variation introduced to the process as a direct result of the planning process. I find the planning process to be a huge source of variation in its best form. Anyone who has tried to manage a production floor using MRPII (Manufacturing Resource Planning Two) or any of the other planning models will attest to this. It is not practical. The shop floor is operating at a speed that the typical planning processes are not able to achieve—even under the very best of circumstances. Usually, the planning cycle has a weekly update, but the floor is changing hourly or faster. It makes no difference if the planning process is updated daily; it still isn't fast enough to manage the production floor.

In addition, the farther out you need to project your plan—that is, the greater the planned lead time is—the more uncertain the plan is. This uncertainty is simply variation by another name. To make sure all contingencies are covered, planners pad their estimates and add just a little, "just to make sure" or "just in case" (JIC). Now we have a JIC planning process trying to guide a JIT (Just in time) production process. When the "padded" plan is given to the tier 1 suppliers, to cover their contingencies they "further pad" their estimates, creating another layer of JIC. It proceeds this way down the entire planning chain. This adding over time is amplified, and the further out the schedule goes, the more the amplification. This happens along the supply chain and is a major source of variation. Or simply put, the shorter the lead time, the more accurate is the forecast.

Planning programs such as MRPII are still needed to do long-term planning and raw materials handling as well. They also function well as an interface to accounting, for example. But make no mistake about it, to those who do not understand a JIT system, MRPII makes sense. To those who understand JIT, MRPII is a large waste generating, tool—when it is used to trigger production on the floor.

> **P**oint of Clarity   Lead time IS the basic measure of being Lean!

Techniques to Reduce Lead Times

Seven basic techniques can be employed to reduce lead time and improve flow.

Reducing Production Time

Reducing production time is a combination of:

- Eliminating unnecessary processing steps
- Reducing production defects
- Changing the current conditions so those processing steps which are currently necessary, but not valued added, can be eliminated

Reducing Piece Wait Time

Piece wait time is reduced by balancing, so the flow is synchronized.

Reducing Lot Wait Time

Lot wait time is the time that a piece, within a lot, is waiting to be processed. To reduce lot wait times, shrink lot sizes and level the model mix. The goal of minimum lot sizes is one-piece. When lot wait time is reduced, first piece wait time is also reduced. This time is often overlooked but is incredibly important. There will always be quality and production issues and these issues must be uncovered quickly so they can be solved. First piece lead time is often the key to being responsive to quality problems.

Reducing Process Delays

Process delay is the time an entire lot is waiting to be processed. Often it is called queue time. To eliminate this, we must level production quantities and processing capacity and synchronize the flow in the entire plant. The most common causes of these delays are mismatched capacities and batch production. This can also be caused by lack of synchronization and by transportation delays.

Managing the Process to Absorb Deviations and Solve Problems

Many sources of deviation increase production lead times, such as machinery break-downs and stoppages for quality problems, to name just a few. All these deviations cause inventories to rise, and inventories are the nemesis in Lean manufacturing—we want to reach zero inventory if we can. Where we have variation in the system, do not add inventory. Instead, attack the variation. One of the key tools to manage the process is the concept of transparency. If the condition of the process is transparent, then Rapid Response PDCA (Plan, Do, Check, Act) can be performed.

Reducing Transportation Delays

One piece flow, synchronization, and product leveling all place emphasis on transportation, which (if you recall) is a waste. To reduce this waste, several strategies can be employed. *Kanban* is the first thing most people think of, but *kanban* has inventory and creates a second delay, the delay of information transfer, so it is a double waste in itself. Thus, try to avoid *kanban* systems by instead using close coupled operations, such as those used in a cell, or use conveyors.

Reducing Changeover Times

Whenever a machine has multiple uses, we must changeover between production runs. To maintain production before and after the machine, we install inventory buffers that, while they allow continuous production, slow down the overall flow. Consequently, if we reduce the time for the changeover, we can reduce the inventory both before and after the machine. This lead time reduction and productivity technique, known as SMED (single minute exchange of dies) will often have dramatic effects on inventory reduction and consequently improve lead time and flow significantly.

Why Lead Time Is the Basic Measure of Being Lean

Short lead times and lead-time reduction is such a basic tool in Lean that you will find it to be a strong measure of Leanness. In addition, if a company has short lead times, several other inferences can be drawn about that company. Almost without exception, you will find that they have:

- Good inventory management
- Good quality
- Good delivery performance
- Good machine availability
- Good problem solving
- Low levels of variation
- Stable processes

Chapter Summary

Lead time as a metric was not discussed much at all until the book *Learning to See* (Lean Enterprise Institute, 1998) heightened people's awareness of this powerful concept. The business benefits of improved responsiveness and flexibility due to reduced lead times are now well understood. The story of the Excalibur Machine Shop cited here showed in detail how reducing lead time enhanced their production as well as improved their ability to find and solve problems. In addition, this is a good example of how Lean tools can be applied in the job shop environment. The seven techniques to reduce lead time were explained, highlighting not only that lead time is the key metric to evaluating whether a facility is Lean, but that those firms with short lead times have other natural competitive benefits as well.

CHAPTER 6

How to Do Lean—Cultural Change Fundamentals

This chapter and Chap. 11 center on cultural issues. We have made a concerted effort to steer away from many of the deeper cultural issues, but any time you make a change, you will need to consider how the culture will need to change. Consequently, we will discuss only those few cultural issues you need to understand and properly address so you can make your initiative a success. In this chapter, we focus on the three fundamental issues in any cultural change—something you might call Cultural Change 101. These three cultural issues are leadership, motivation, and problem solving, and they provide a foundation for the efforts described in Chap. 7. In addition, we also discuss interdependence, the need to focus on foundational issues, and the *jidoka* concept as key cultural concepts that are deserving of special thought before the implementation plan is completed.

Three Fundamental Issues of Cultural Change

Any time any major culture-changing initiative is undertaken, such as the implementation of a Lean initiative, three fundamental issues must be addressed. Stated in question format, these are:

- Do we have the leadership to make this a success?

- Do we have the motivation to make this a success?

- Do we have the necessary problem solvers in place to make this a success?

> "...**M**anagement must awaken to the challenge, must learn their responsibilities, and take on leadership for change...."
>
> **W. Edwards Deming**

Leadership

We will address each question, with the issue of leadership being the first and the most important to explore. (See Chap. 14 for a good example of leadership in action.)

Just what is a leader? Simple: It is someone who has a following. His followers are willing to do what he says, just because HE says so. They may follow him because of his character or they may follow because of his position, or maybe it is his personality, or his competence, or just maybe a critical combination of these four aspects of who he is. But the critical factor is that they are willing to follow him, they are willing to act, and many times they are willing to undergo severe hardship and sacrifice for this leader. If he/she says to do something, they do it. They need nothing more.

How does this leader get his following to do these things? There is a great deal in the literature on leadership, and some of it is outstanding—Robert Greenleaf's being some of the best. Some of the literature puts the topic of leadership completely out of context, yet others touch on all the major points but do not emphasize the very critical few necessary characteristics. I do not intend to fully explore leadership. That is not necessary right now, but what we will discuss is the critical few characteristics that all leaders must have. They are also the skills that the leader of this Lean initiative must have. Depending upon the type of cultural change undertaken, or maybe where or when it happens, the leaders of those efforts may need other skills as well, but these are the critical few skills that are absolutely necessary for ANY leader. If he/she is lacking in any of these characteristics, the effort—be it a Lean initiative or the overtaking of a nation—will almost surely fail.

> **P**oint of Clarity One can lead through:
>
> - Competence
> - Character
> - Position
> - Personality
>
> Our Lean leader must lead primarily through competence.

Think of a leader, any leader. It could be a political great like Churchill, Martin Luther King, or even John F. Kennedy. It could be a sports leader like Vince Lombardi or maybe a spiritual leader like Gandhi or Christ. All these people were great leaders. Their individual fields of endeavor were different, yet they shared some common abilities that led to their success. That said, what are the common characteristics shared by these leaders? By all leaders?

First, all leaders have a plan. Martin Luther King's plan was immortalized by his "I have a dream" speech. Even today, some 40 years later, many people quote his words, "…one day live in a nation where they will not be judged by the color of their skin, but by the content of their character." What about Gandhi's plan of passive resistance? Maybe the grandest of all was the plan of Christ. Even having died over 2000 years ago, his following still grows. They all had a plan.

> **P**oint of Clarity It is not possible to lead without a plan.

Your Lean initiative leader must also have a plan. Lacking this person or his plan, my advice is strong and clear. *Don't even start*. If you attempt to undertake this initiative without strong leadership and a strong plan, you will fail. Furthermore, you will have raised the expectations of your employees, and their expectations will be crushed. This only makes the next effort more difficult. Certainly, do not rush into this effort. Find a good leader and develop a solid plan before you start.

The second necessary characteristic these men all shared was the ability to articulate their plan so people could understand and buy into it. They have the ability to get people engaged. It is not coincidental that all these examples were also great orators. I'm not so sure about Gandhi, but the others could mesmerize a group with their speeches. About Vince Lombardi, his great middle linebacker Ray Nietzsche once said,

"When coach Lombardi said to sit down, I didn't even look for a chair." Their speeches were an extremely strong tool that allowed them to attract, motivate, and activate a following. Often, the content of their speeches was very motivating in its own right—but make no mistake: the choice of words and the type of delivery made a huge difference. The leader of this Lean initiative must have this skill also. They must be able to make the Lean initiative come alive to those involved. Not only must they encourage the hands and feet of their followers to move in the right direction, they must also motivate the people and engage their hearts as well.

Some examples exist of successful initiatives where the person who put together the plan was not the person to articulate it. This type of brains and mouthpiece combination will not work in our Lean initiative, however. Both skills must be resident in the same person since they will have to be on the floor on a daily basis—observing, dealing with problems, and interacting with all the people in the facility. The leader must be able to handle questions and problems quickly—in short, they must be able to lead from competence.

> **P**oint of Clarity   The leader must be able to translate their concepts into behavioral traits that will support the execution of those concepts.

The third, and final, requisite characteristic of a leader is the willingness to act on the plan, at the exclusion of all else. He must exhibit the skill to not lose sight of the goal and stay the course in spite of roadblocks, obstacles, and resistance. He must be willing to act, at all costs, to reach the necessary objectives. This ability to act is a huge need. Many great plans have failed not because the plan was not needed or the plan was not good enough, but because the leadership did not have the courage and character to make the difficult decisions. In the end, this compromised the entire effort, and the plan, simultaneously. We have seen this all too often.

To act properly, they must first be able to recognize exactly what the situation is, and they must be acutely aware of happenings in the facility. Second, they themselves must be excellent problem solvers. They must be able to sort through the options and properly apply the values needed. Finally, they must have the wisdom, courage, and character to act when action is required, and conversely, should use those same skills to hold back when thoughtful inaction is the appropriate behavior.

> **P**oint of Clarity   It is not possible to lead unless one is willing to act, especially in the face of adversity. This sounds almost trivial, but it is the key failing of leadership.

So the leader of your Lean initiative—be it you or someone else—must have these three requisite skills:

- The ability to develop a plan
- The ability to articulate this plan and engage others
- The ability to act on the plan

Carefully choose your leader, and if you do not have someone with these skills, find a person who does. There is no substitute for leadership. Any compromise on this topic will guarantee failure.

Our leader and all the key persons in the Lean initiative will be well served if they have an abundance of some of the blessings we humans are endowed with. These include the blessings of:

- **Awareness** The ability to fully see and appreciate the reality of all that is happening around us.

- **Imagination** The ability to see things that may not, but could be there; and the ability to conjure up options that translate into opportunities.

- **Conscience-driven values and principles** These are the things, ideas, and ideals that we consider important. They become the basis for prioritizing our options.

- **Choice** The independent ability to apply our priorities and consciously decide what to both do and not do.

Motivation to Change

> **"I**t is not the strongest of the species that survive, nor the most intelligent, but the one most responsive to change.**"**
>
> **Charles Darwin**

It has been our experience that almost all successful Lean initiatives are driven by one of two motivating factors. The first factor is evident when the company is looking survival square in the face, and is on the verge of going broke. Under these circumstances, it is easy to get people's attention. However, the most common factor occurs when your customer says you must implement Lean—that is, they say, "If you wish to continue doing business with us, you must implement a Lean Manufacturing System." In the end, both are about the same issue: survival.

On some occasions, we run across companies that want to implement a Lean initiative because some visionary has decreed it so. This choice is often an informed choice that is part of an overall business strategy. Normally, the direction comes from the home office, which is often a long way from the affected manufacturing facilities. In my experience, these efforts generally proceed far slower and with much less success than those that are survival motivated. Seldom do I see a strong buy-in at the plant level, and the degree of success is inversely proportional to the distance the plant is from the visionary.

So in the end, if your concern is not that of your immediate survival, the issue is likely one of *long-term* survival. For you see, you will learn that the competition is improving, and if *you* do not improve, you will not survive. Our Lean initiative is the most aggressive form of improvement we can create—so, in fact, it is a survival issue. It will be important to carry this message to the entire facility so they will have the proper motivation to make this initiative a success. Do not be surprised if you find resistance to this issue, since what we are talking about is cultural change—and with cultural change you *always* get resistance.

This is true of my experience with virtually any type of a cultural change initiative. All cultures seek stability. This means all cultures naturally resist change.

This resistance to change is also seen in the human body—something called *homeostasis*. It is the body's desire to seek a position of equilibrium. If we wish to change our body, we must force it to go beyond its limits. We must make it uncomfortable. Take, for example, someone who wants to become a great athlete—a soccer player, let's say.

In addition to the many skills a soccer player must develop, he must have tremendous cardiovascular fitness—in a phrase, he must be able to run all day. To acquire this ability, he must push his body beyond its normal limits time and again; he must undergo discomfort and even pain; he must exhibit the mental strength to fight through the pain and discomfort and be willing to do more next time, knowing full well it will be painful again. He must resist the desire to take a day off or go partying because, even though he would like to do those things, he knows it will not be in the best interest of his goal to become a great soccer player. With his eyes clearly focused on his goal, he can maintain his regimen in spite of what both his mind and his body tell him. To measure fitness levels, he can chart his two-mile run time. When he sees this improving, he is further motivated by his mini-success, and then when he sees how this better fitness translates into better play on the field, he feels justified in putting his body through the discomfort and pain he had to endure. Frequently, his feelings at this point go beyond justification: He is proud of his efforts and often extracts a serious and mature sense of accomplishment.

In spite of this, he will encounter other distractions and challenges. Others will tell him he is working too hard; he needs a day off or he is taking the project too seriously. Frequently, they tell him this not because it is in his best interest, but because it furthers some issue of theirs. Nonetheless, it sounds good and his inner voice says, "Well, I have been working hard. Maybe I do deserve a day off." Well, maybe he does and maybe he doesn't, but these distractions test his resolve and he must decide which is the best course of action. Those who want to be great will summon up their courage, fight through the challenges, make the sacrifices, and will not be deterred from their goals. Often, this makes them less popular or puts other social pressures on them, but those who are focused keep their eye on the goal and proceed.

> **P**oint of Clarity It takes great courage and character to say no to the things we would like to do so we can focus on the things we must do in order to attain those goals worth achieving.

Well, the changes we must undergo in our Lean initiative are much like the changes an athlete must undertake. Like the athlete we must:

- Have a clear goal in sight.
- Recognize that we need to change to reach the goal.
- Recognize that the changes will be uncomfortable and even painful at times.
- Recognize there will be forces within ourselves, and outside ourselves, that are driven by different motivations. These forces will resist the necessary changes and often coax us in the wrong direction.

In spite of all this, we must focus on the goals and persist!

Here, in particular, leadership is crucial. The leader must know all this. Furthermore, they must prepare the people for the changes, and the constancy of the change that is to come. The leadership must realize and make all those involved understand that the entire manufacturing world is improving and that if we stay where we are, we are really regressing. If we wish to survive, we must improve,

> **"W**e are made to persist. That's how we find out who we are. **"**
>
> Tobias Wolff

and to improve we must change. Change becomes a given, and in fact if we want to become more competitive, we must improve faster than our competition. Hence, we must change more and probably more rapidly than our competition. So, to survive we must change, and to prosper we must increase our rate of change so it exceeds our competition. That is not an easy sell for any leader.

> **A Lean Paradox**   In Leanspeak, we talk of the removal of variation in many forms. One such technique to reduce variation is the standardization of work. We continually look to standardize things so they are all done the same; we do not want even one operator to vary from the other in work methods. We seek standardization. On the other hand, we tell everyone that the system must change, and we must foster creativity to improve. So we must remain the same, yet we must continually change. It is a major paradox worth thinking about.

> **"I**n times of change, learners inherit the Earth, while the learned find themselves beautifully equipped to deal with a world that no longer exists. **"**
> Eric Hoffer

In many non-Lean plants, there is minimal impact from this paradox since the people doing the changing are typically engineers and managers, while the workers generally follow the instructions and seek to not vary from them. However, Ohno in his writings promotes that the problem solving—in fact, the changing—should be done by all: the worker, supervisor, engineer, and manager alike. There is a clear effort on his part to increase worker involvement in the change cycle. It is seen by quality circle efforts, employee involvement programs. and other sincere efforts to tap the skills of the workers to leverage the ability of the plant. In those locations, which more fully utilize the workforce, this paradox needs to be understood and utilized.

So our leader has a daunting task in front of him. He feels he must create within the organization a willingness to change. Quite frankly, we believe this is impossible; human nature is such that people do not really wish to change. Rather, the leader must develop a culture that is mature enough to recognize that change is needed. Furthermore, he must then equip the culture with the necessary skills so it *can* change. Causing this change will always be a challenge for the leadership and the management. The desired state they must reach is similar to the mental state of the athlete. The leadership must develop a mindset, a culture, that says, "We are willing to undergo the temporary discomfort of the change because we know that this discomfort will go away, and only by going through this discomfort can we reach our goals."

> **P**oint of Clarity   Great leadership will move us into and through the needed pain so we might change and improve. Sometimes the leadership technique used is one of nurturance; other times it is simply force.

Talented Problem Solvers

The third requirement to begin any initiative is the presence of talented problem solvers. First, let's make sure we are on the same wavelength here. Just what is a problem? For this, I rely on the problem-solving methodology popularized by Charles Kepner and Benjamin Tregoe (KT Methodology) in their book, *The New Rational Manager* (Princeton

Research Press, 1981). They define a problem as, "the difference between what is and what should be." Furthermore, they break down what most of us call problems into three types of concerns. These three concerns are problems, decisions, and potential problems. It is great reading and I recommend it to all. And by all, I mean *all*, not just those interested in Lean.

Back to problems for a moment… Once a leader develops his plan, he has just created a whole series of problems.

Just how has he done that? As soon as he creates goals, he now has created a new "should be." For example, if OEE is 60 percent and the goal is to achieve 85 percent, the OEE "should be" 85 percent— *et voilà!*—the manager has created a problem for someone else.

> **P**oint of Clarity   A key role of a good leader is to create problems where no problems previously existed.

In a Lean plant, problems can be broken down into three categories. We will have problems when we have:

1. No standard

2. A standard that is not met

3. A standard that is not ideal

Problems of type 1 and 3 fall generally within the province of management to solve or resolve. The most typical problem is type 2 and these problems need to be solved by everyone. Such problems include the typical customer complaint, the production demand that is not met, the quality standard that is not achieved, and the delivery date not met. In addition, type 2 problems include internal problems such as OEE (Overall Equipment Effectiveness) not achieved or cycle time degradation. Early in the Lean initiative, it must be made clear that everyone is responsible to solve problems. It is the challenge of management to engage everyone in problem-solving activities. As the initiative develops, it can be determined how this is approached. For example, some have used small group activities such as *quality circles* with great success. At a minimum, all employees should be taught the "5 Whys."

It is not this large group of problem solvers that is most crucial to the initial phases of a Lean initiative. Most crucial is a small cadre of very talented problem solvers. Even in a facility of 500 people, only three or four are usually required. Many problems, especially in the early days of implementation, are easily solved by a wide range of personnel, including group leaders, production supervisors, technicians, and of course engineers. However, some problems will crop up that require more technical skills than the typical group leader, production supervisor, or technician will have. In addition, some of these problems require significant dedicated time to do the data gathering and analysis. Many production workers, even if they have the skill to solve these problems do not have the block of time so they can do the necessary data acquisition, reduction, and analysis. These three or four talented problem solvers should be versed in plant operation, as well as a wide range of problem-solving techniques.

There is yet another issue with problem solving that is not well understood by many—that is, the process of standardization is just another name for the process of problem solving. So those who are good at problem solving are also good at standardization. The opposite is also true: Those who are weak at problem solving will be weak at standardization. In this book, the emphasis on standardization and the emphasis on reduction of variation of all forms is repeated over and over. It is a hallmark of Lean manufacturing to have standardized processes, and there is no substitute for it. So our problem solvers are now doing double duty. First they are fixing problems—and now you learn that the same skills are required to execute the techniques of standardization. This concept is developed more fully in Appendix A at the end of this chapter. Right now, though, let's discuss the skills needed by this cadre of problem solvers.

The first and most important technique is logical problem solving. I know of no technique superior to the Kepner-Tregoe (KT) Methodology. These skills can be acquired by attending a training session taught by Kepner-Tregoe—this is really the very best training investment you can make. If for some illogical reason this is not in the cards, pick up their book, *The New Rational Manager* (Princeton Research Press, 1981) and teach yourself. KT also has a less intensive program named "Analytical Trouble Shooting," which is also excellent. Several years ago you could have taken a class at the Ford Training Center entitled, "TOPS: An Acronym for Team-Oriented Problem Solving." In the class, you learned the KT Methodology as part of completing Ford's 8D. However, when the Automotive Industry Action Group (AIAG) was formed and methods in the auto industry became standardized, the TOPS program disappeared. This was a large oversight and should be corrected. In my experience, the KT Methodology has been proven to have the broadest applications and the highest success rate of problem solution. It does have one weakness, though: a scarcity of statistical techniques to assist in the quantification of variation and in statistical decision making.

This is where the Six Sigma statistical problem-solving techniques are so powerful. Actually, even though we say Six Sigma is a problem-solving methodology, it lacks many of the powerful logical tools inherent in the KT Methodology. Instead, it relies on the simple DMAIC methodology (define, measure, analyze, improve, and control). The Six Sigma tools are extremely powerful when making statistical decisions and in understanding, mathematically, the risks involved. These Six Sigma tools include statistical and decision-making skills such as multivariate analysis, hypothesis testing for both variation and averages, correlation and regression plus SPC (Statistical Process Control), MSA (Measurement System Analysis), DOE (Designs of Experiments), and Response Surface Analysis. Even if you do not have any Six Sigma Blackbelts, someone should be well-versed in DOE. It is not very time consuming to acquire the basics of DOE, and the applications abound in most manufacturing plants. Those trained in Six Sigma skills are much like the "Deming Statisticians" which are often talked about. For a good description of that group, refer to *Out of the Crisis* (MIT, CAES, 1982), specifically Chap. 16, by W. Edwards Deming.

A much overlooked set of skills are those required for group facilitation. These skills are not only helpful with groups, they are helpful any time a problem solver must interact with another person on any problem, such as getting information from the line workers. So these skills are very powerful in making a problem solver more efficient and effective. There are various places to acquire this training, and there is even a touch of it in Six Sigma training. However, I recommend you send those people who need these skills to the training as is provided by Oreil Incorporated, (formerly Joiner Associates). Alternatively, buy *The Team Handbook* (Joiner Associates, 1988) by Peter Scholtes and teach yourself.

Finally, especially if you are in the discrete parts industry, especially electronics, I have found the so-called Shainin tools to be valuable. They have been published in a book by Keki Bhote entitled *World Class Quality* (AMA, 1988). If you want to pick up this book, do an out-of-print search and try to find the first edition if possible. It's superior to later editions and usually less expensive, as well.

Which trainings to do first? The following three are a must!

- Kepner-Tregoe problem solving (clearly the #1 choice)
- Statistical problem solving (such as Six Sigma or at least DOE) is a must
- Group facilitation training

It would be good if one or more of your group of problem solvers was an accomplished Six Sigma Blackbelt, and if one or more had strong group facilitation skills and were skilled and experienced enough to do a facilitated "spin-around" on a broad variety of problems (described in the accompanying sidebar).

Brainstorming Rules, the Facilitated Spin-Around

The facilitated spin-around is a very powerful way to gather information and resolve several types of problems. For many data-driven, on-the-line production problems, this is not a good technique, however. First, it is normally not needed, second, it simply takes too long. It is good for issues with "soft data," such as opinions and for decisions that are emotionally charged, such as "should we require all employees to wear a uniform?" It is a technique that will reach very good decisions, but more importantly, because of the process itself, the members will have a strong commitment to implementing the group decisions. The facilitated spin-around is often required to reach consensus. It is critical that your organization be able to use these techniques and have facilitators capable of leading these groups. The technique goes like this:

1. The facilitator introduces:
   a. Himself
   b. The topic very briefly
   c. The objectives of the group
   d. The agenda
   e. The planned timeframe for this meeting
2. If meeting ground rules need to be discussed (such as cell phone use, bathroom breaks, and so on), this is done at the beginning.
3. The facilitator reminds them of the brainstorming rules, which are:
   a. One item per person per turn.
   b. Each person in turn (hence "spin-around" the room).
   c. The documenting and posting of each item is done on flipcharts. (I recommend you use flipcharts. They are much more personal than using a computer projected onto a wall.)

d. When an item is stated, no value judgments by others are allowed (at this stage). No agreement and no disagreement is permitted. All items are taken at face value.

e. Discussion can proceed on any item, but only "to the point of understanding." Once it is understood, discussion ceases.

f. No other talking or work of any nature is allowed; no side discussions are permitted. If there is a question, it must be directed to the group. This requires a great deal of both attention and patience, not to mention respect for the people and the process.

g. No piggy-backing or modifying of any item is permitted without the author's agreement.

h. Pass if you do not have an idea on this turn.

i. Spin-around the room until all items are exhausted and documented.

4. The session then proceeds when the facilitator documents the first item and the spin-around continues.

The preceding procedure requires good facilitation, which is a topic in itself, but is unfortunately outside the scope of this book. I recommend you refer to *The Team Handbook by* Peter Scholtes, (Joiner Associates, 1988) for further information.

> **P**oint of Clarity We will have a JIT plant, and thus we will need JIT problem solving.

In addition, this small cadre must be available whenever the plant is running. Generally, this entire group is not readily available, but it needs to be managed in such a way that at least one of these problem solvers is readily available.

Problem solvers who have requisite skills to solve problems are rare indeed. They must be able to:

1. Readily grasp a situation.

2. Turn it into a meaningful problem statement.

3. Know how to gather and sort through the data.

4. Analyze the situation, including doing a statistical review.

5. Utilize root cause problem analysis.

6. Create a list of possible solutions.

7. Sort through the options, comparing the options to the needs of the business, and weighing the risks of each solution.

8. Decide which one is the best solution.

9. Use project management skills to turn this solution into action plans.

10. Show the leadership to implement those plans, turning them into improved performance for the facility.

It takes an awesome complement of skills to be an accomplished problem solver. Most managers believe the scarcity of good problem solvers is solely a result of not having the necessary skills—by that, they mean the technical skills; and by implication they believe that all of the requisite skills can be taught. These managers then believe that all that is required is to train these people in the necessary technical skills.

Our experience does not support this logic. While it is true that few people have the technical skill inventory listed earlier, there are some who have the requisite technical skills yet are not effective problem solvers because they lack other requisite traits. So what are these other traits that are required?

First, we must define some methodology to solve problems. Various ones exist, but they all take the form of:

1. Some observation/evaluation occurs and it triggers the thought that "we have a problem." Someone then defines the problem.

2. Observations about the problem area must be made.

3. Evaluations must find the cause of the problem.

4. Solutions must be imagined, created, and compared to values.

5. Decisions are made.

6. The decisions are turned into action plans.

Second, let me say a word about individual personality development. C.G. Jung developed a theory of personality that is widely accepted today. In short, he said there were two major aspects of how we incorporate and handle information (see Fig. 6-1, Personality Types diagram).

1. The first aspect was the "Observing Scale." This is the way in which individuals accept information. It is done by "sensing it," by touch, feel, smell, and so on. We call these people "sensates." Or this information can come in via the process of "intuition." This is characterized by such things as "my gut feeling" or "I just sensed that…" Although to most people intuition is a lesser form of accepting information, its value cannot be underestimated, especially in problem solving.

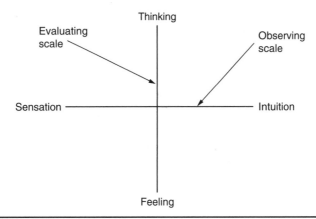

Personality Types

FIGURE 6-1   Jung's scales for personality types.

The twin sister to intuition is imagination, which you will find is incredibly important in problem solving, as is intuition itself.

2. The second aspect in Jung's personality model is the "Evaluating Scale." This evaluation gives meaning to the observations obtained. It has two extremes. One is "thinking" and the other is called "feeling"—although empathy may be a better English word. Pure thinkers make evaluations based on logic and cold hard data. While those with a strong "feeling" function will make decisions based on what we call human values, such as community, human-worth, and quality-of-life to name a few. As you might expect, most businesses are filled with "Thinking-Sensate" personality types. If we plotted them on the scales in Fig. 6-1, the thinking-sensates would be in the upper left-hand quadrant.

The dynamics of personality development are such that at a very early age people start to use a certain type of observation technique and evaluation format. Then, when they find one that works for them, they work to refine it so life is better for them. This process of personality development is almost entirely unconscious, but it is still easily characterized, even at a very early age. It is a dynamic of personality development, that as one end of the observing pole tends to work, for you, then the other end of the scale becomes subordinated. Hence, people tend to become sensates at the expense of the skill of intuition, and one type of information gathering dominates your personality. Sensing becomes the conscious way to accept information, and intuition is driven into the unconscious. It will remain there until something occurs to cause it to surface. Most people will develop a dominant way to accept information and to evaluate. This then becomes their personality type. For example, one could be a "thinking-sensate" or a "feeling-sensate," to name just two. Most people do not develop—at least until they are older—more than two of these dimensions.

Now here is the rub. Once a person develops a personality style, they use it and refine it. In fact, they can become very rigid and structured. In some aspects of life, this rigidity of personality will serve them well, but in time the very nature of life will expose us to different challenges, and this rigidity will harm their ability to resolve some of life's issues. For example, early in life a young man can be a very rigid "thinking-sensate." Then this fellow gets married and finds that this cold, hard, rational thinking does not work so well with his wife and kids and needs to adjust.

> "**A**t 20 years of age the will reigns, at 30 the wit, at 40 the judgment..."
>
> Benjamin Franklin

This dynamic is one reason why there are few really good problem solvers in our business who are also young. More on that later.

So how does this fit with problem solving? Unlike many aspects of work, while solving problems a person must have both strong sensing skills to make objective observations, but they also must have intuitive skills (maybe a better term would be "imagination skills") to foresee what could *possibly* happen, even if it is not currently happening. This is a crucial aspect of the observing process that must occur in problem solving. If a person lacks this intuitive skill (imagining), he can only envision what is occurring at that moment in time, and so if the problem just happens to manifest itself in the present context, he will see it—otherwise, he will be hampered by some level of blindness. At the level of quality demanded by most companies, it is very rare that we actually see the problem when it occurs, hence those who do not have this intuitive skill are hampered as problem solvers. This intuitive skill is not only needed at the observing stage but is certainly needed in the "creating possible solutions" aspect of problem resolution.

In short, what happens is this. When needed, most businesses hire individuals with a strong "thinking-sensate" personality type. Since they are strong sensates, they are usually deficient in intuition—this is natural, predictable, and largely unavoidable. Unfortunately, when it comes time to solve a serious problem and intuitive skills are needed, the workforce is, alas, weak in this area. So the company, with the best of intentions tries to teach problem solving. Much of the methodology can be taught, but the intuitive skills can't really be acquired in the classroom. So in the end, real problem-solving skill development proceeds slowly. Since most companies think problem solving can be taught, they are disappointed and usually shrug their shoulders. In fact, the truth is that some problem-solving skills can be taught in the classroom, but for those "thinking-sensates" the intuitive skills of observation and solution conception are largely taught by life. Sometimes that means just enduring more of life's experiences to broaden your personality—or in other words, it's a part of getting older, and wiser.

Hence, my point is that few people have the total complement of skills to be really good problem solvers. Some of the skills can be taught in the classroom, but some skills come to work with the person's personality and are largely unteachable.

This polarization of skills within an individual has given rise to the concept of group problem solving. Given a group as small as five or six individuals, frequently we will have all four of the personality traits present in one or more of the people. What then happens is that those who are strong "sensates" are very active in the problem definition stage, but may be less involved in the highly intuitive stage of imagining possible solutions. On the other hand, a person with a strong intuitive tendency could assist in the step of creating possible solutions yet be less involved in the problem definition stage. With a small group, it's very possible to get all four of the poles adequately represented. Now, if the group also has the technical knowledge, it is likely a good problem solving effort will be achieved.

I have taught groups and done controlled experiments with groups and find them to be very effective in finding superior solutions to many problems. But as with so much of life, if you get something good, it often comes with some baggage. And the group, when used for problem solving, does have some baggage—a situation that is threefold, at least.

- First, to be successful the group must be well facilitated; hence, yet another skill is required, that of facilitation, a scarcity in its own right.

- Second, although the process is thorough, it is not fast. Groups need to form and develop. Then there is the whole issue of meetings. Solutions tend to be weeks or even months away. This is anything but JIT problem solving.

- Third, often the issues in Lean implementation have to do with very technical issues, so the group may not have the requisite technical knowledge. Where intimate technical knowledge is an issue, one well-informed person working alone is always more efficient and usually far more effective.

We have a great deal of information on group problem solving, group management, and group dynamics, and what we find is that these skills are not the most important ones to work on during the early stages of a Lean initiative. Typical problems are usually more Lean-specific, and the answers are needed quickly. These characteristics of a problem generally mean they are better solved by an individual.

On balance, it is much better to develop a small group of problem solvers and turn them loose to solve the problems individually. If you have three or four persons, it is easy to teach them technical skills, as well as statistical skills. Keep your eyes open for the select few who have a developed enough personality to be excellent problem solvers, and you will likely have a strong solid group. Save the group problem-solving efforts for a later stage of the initiative.

An Example Application of the Three Fundamental Issues of Cultural Change

The three fundamental issues of leadership, motivation, and problem solvers are described well in Chap. 14. This story illustrates in graphic detail how these three issues were so wonderfully managed. As a result of the proper handling of these three issues, the plant succeeded, grew, and prospered beyond all expectations. On the other hand, Chap. 17 showcases an example of what happens when the three fundamental issues of cultural change are not managed well. I suggest you skip ahead and read these descriptive stories before continuing with this chapter.

Some Cultural Aspects of a Lean Implementation Worthy of Further Thought

A Lean implementation initiative has several unique aspects. None of the items listed next are unique in concept, but all are unique in the intensity in which each subject must be approached when compared to a typical project. These three aspects are:

- The interdependence of activities
- The emphasis on foundational issues and basics
- The implementation of *jidoka*

Some people will see this list and say, "Well we're aware of that" and just move on. If that is your attitude, I can almost guarantee your initiative will fail. On the other hand, if you are one of those enlightened managers that exhibit humility, curiosity, and insight and you spend time, observing, understanding, and acting on these three items, you will obtain long-term gains you had not anticipated. These three issues, or more importantly, the depth to which these three issues are addressed, is often driven by some deep cultural issues, including:

- Respect for the workforce
- The natural maturation process in going from dependence to independence, and then ultimately to interdependence
- The need to avoid the "convenience of compartmentalization" and "simplistic thinking"
- Fundamental business and managerial humility
- The need to properly balance the long- and the short-term needs of the business and the culture
- An awareness of human needs, system needs, and business needs ... to name just a few

The Level of Interdependence

The level of interdependence of activities is a truly amazing phenomenon, and especially so in a Lean implementation. Most people would like to think the systems of the world work independently and are simple and linear in nature. It makes them easier to understand. Seldom is this true. More often than not the systems of the world interact with interdependence rather than independence. For example, as you begin to reduce the variation in production rate, implementing such Lean techniques as load leveling using *heijunka* boards, many other aspects of production will sympathetically change. For example, the workload on individuals will stabilize and they will become more comfortable. As they become more comfortable, they make fewer mistakes and as they make fewer mistakes, scrap is reduced, buffers reduce in size, and the workload variation and the production rate variation are further reduced.

The changes resulting from interdependent causes are often large and hard to foresee by the novice and are sometimes unique to your circumstances. They are an area where the best advice I can give you is twofold.

- First, look for them, so when they appear you will be prepared.

- Second, on these topics listen very carefully and defer to your *sensei* since many of these interactions are not only counterintuitive, they are paradoxical.

For example, we have already discussed the paradox of change in Lean. In addition there is the paradox of *jidoka*; which is "*We* shut down the system so the system can run continuously."

The Emphasis on Foundational Issues and Basics

Support by top management on the foundational issues must be applied to a degree seen by no other initiative. There are few things a manager does that are so culture changing as the implementation of Lean, and when the culture needs to change, management must lead the way. I can say with certainty that, next to inadequate leadership, the most common reason facilities do not reach their goals is a marked weakness in addressing the foundational issues.

- If you find a weakness in one of the foundational issues, immediately fix it. Don't ponder, don't budget, don't meet to discuss it, and don't organize—just fix it.

- If you have concerns that one of the foundational issues may be weak, immediately improve it. Again, don't ponder, don't budget, don't meet to discuss it, and don't organize—just fix it.

- If you see a problem appear that should not be present, and if this problem is related to a foundational issue, attack it with a 24/7 approach until it is fully understood, and then aggressively implement the corrective actions necessary.

Spare no effort on this topic.

All deficient foundational issues are crises, requiring immediate attention, and need to be managed as such. They are emergencies and do not lend themselves to the time-consuming decision-making of optional activities. They are not optional, they are crises. In effect, the patient is bleeding uncontrollably, and the bleeding needs to be stopped. In such cases, we don't seek out advice from others, we don't request permission

> **P**oint of Clarity Toyota started with *jidoka* before they started with either SPC (Statistical Process Control) or DOE (Designs of Experiments). More accurately, they started with *jidoka* before either SPC or DOE *had been invented!* So it's not surprising they are among the world leaders in *jidoka* implementation and that the rest of us simply need to catch up.

to proceed, we don't think about the cost of the bandaging—*we just do it*. This is the attitude that must be taken with foundational issues. I call it the Nike attitude: Just Do It!

The Implementation of *Jidoka* Is *Always* a Fundamental Weakness

I have yet to see a Lean initiative where *jidoka* was the leading strategy employed. It always lags behind other strategies in its implementation, and lags behind other strategies such as JIT (Just In Time) in both application and depth of application. There are a variety of reasons for *jidoka* lagging behind rather than leading the effort: none of them good. The best I can come up with is that it is hard to do, that it's hard to see progress, and that it takes a lot of time and effort to do it well. Or to put it in more flowery terms, it is not quite as sexy as JIT, TPM (Total Productive Maintenance), or *kanban*—or in more practical terms, it is not seen as important by management.

An Example of the 5 Why is Technique…
Jidoka is a weakness in Lean implementations.

1. **Why** is *jidoka* a weakness in Lean implementations?
   • Because, quality problem solving, an integral part of *jidoka*, is hard to do.

2. **Why** is quality problem solving hard to do? Because:
   • We avoid these problems rather than embrace them,
   • We do not have a sound continuous improvement philosophy,
   • We are not skilled at problem solving
   • We do not have time to work on problem solving
   • Our data are not good enough for solving quality problems

3. **Why** are our data not good enough to solve quality problems?
   • Because the data are total reject data only, not stratified by type of defects.

4. **Why** are the data total reject data only and not stratified by type of defects?
   • Because we only need total reject data to answer the questions we are asked by management.

5. **Why** are the total reject data adequate to answer the questions from our management?
   • Because the questions we get from our management focus on production and yield and not quality.

6. **Why** does our management ask questions about yield and production and not quality?
   - Because management is more interested in yield and production than they are in quality (solution statement).

A couple of points about the 5 Whys

- First, the 5 Whys technique is seldom a straight line-linear process. For example, you will note that the second "Why?" has five possible answers and we only addressed one of them. We have two options when we have multiple causes.
  - We could answer why to each of these, which would give us a very branched problem solution. Quite frankly, all five branches might converge on the same conclusion.
  - However, the normal advice is to quantify the issues and follow the branch with the largest impact.
- Second, as is the case here, it is arguable if you have reached a root cause, which is the objective of problem solving. For example, why is management more interested in production than quality? Maybe it's because of their bonus structure, or just maybe it's the best thing for the company today. Nevertheless, with problem solving of human systems, often the root cause is not really found, and sometimes it isn't necessary. What is necessary is to find an "actionable cause **that reasonable people agree** should be changed?"

Let's check this possible solution statement with the "therefore" technique. If "management is more interested in yield and production than they are in quality" is a good answer to our problem, then the "therefore" technique will logically connect the "Whys" in reverse order, starting with the solution statement.

An example of the "Therefore Technique," which is a check of the 5 Whys.
   Management is more interested in yield and production than quality (the solution statement).

1. **Therefore**, the questions we get from management focus on production and yield, not quality
2. **Therefore,** we only need total reject data to answer these questions.
3. **Therefore,** the data are total reject data only and not stratified by defects.
4. **Therefore**, the data are not good enough for solving quality problems.
5. **Therefore**, quality problem solving, an integral part of *jidoka*, is hard to do.
6. **Therefore**, *jidoka* is a weakness in Lean implementations.

And it works, which tends to confirm the validity of our solution statement.

Jidoka is hard to do, but the reason it is hard to do is caused by two other issues.

- First, management does not emphasize it.

- Second, as shown in the preceding example, a number of foundational issues must be in place to make *jidoka* work. These include a quality test and inspection system, a quality data system, and root cause problem solving by all, to name just a few. And why aren't these foundational issues taken care of? Just go back to reason one—that is, it was not emphasized by management in the past.

What Happens When *Jidoka* Is Not Done Well?

When *jidoka* is not done well, production problems appear. Most typically, defects advance in the system, causing quality production problems. Almost always, the source of the problem is the failure to properly implement some foundational issue. For examples, refer to the 5 Whys exercise done earlier. However, now you have two problems: the production problem and the poorly executed foundational issue. Consequently, we must go back and repair the foundational issue before we can improve on the *jidoka* system.

This is like finding that the roof on your house is not level and you search for the problem and find the concrete foundation is not level. To correct the roof issue, you must first correct the foundation issue. However, as soon as you modify the foundation you find it affects other things like the walls and floor. The same is true of your Lean system. Once you go back and clean up your operational definitions, for example, you need to clean up your work instructions, your training instructions, your training matrix, your visual displays on the floor, and the list goes on. All of this is rework, wasteful, and is only necessary because the foundational issues were not handled well. All of this is waste of some of the most precious commodity we have: the thinking doing problem solvers in the business.

Recalling *Jidoka*

Do not forget about the purpose of *jidoka*: It is there to prevent defects from advancing in the production system, and it is a continuous improvement tool. Since both of these tactics are crucial to Lean implementation, it is clear we must have some kind of *jidoka*, even if we can not capture the most mature forms of *jidoka*, which include line shutdowns by operating personnel followed by "Rapid Response PDCA (Plan, Do Check, Act)" problem solving.

We must keep in mind the definition of *jidoka*:

- It is a 100 percent inspection technique that will prevent defects from advancing in the production process.
- It is done by machines not men.
- It uses techniques such as poka-yoke (error proofing),
- It will prevent defects from advancing in the system by:
  - Isolating bad materials
  - Implementing line shutdowns

- It is a continuous improvement tool because, as soon as a defect is found, immediate problem solving is initiated, which is designed to find and remove the root cause of the problem.

- In the design case, the line does not return to normal operation until it has totally eliminated the defect-propagating situation.

When Line Shutdowns Are Not Practical

First, and without delay, we need to work on all the foundational issues mentioned earlier. They need to be analyzed, prioritized, and aggressively brought to maturity. If this is not done, there is no hope for real progress.

Second, while working on those foundational issues, we need to do some other things.

1. Change the attitude toward defective materials. Make sure the concept of defective material advancing in the system is unacceptable and that it must be reduced. Make sure "continuous improvement" is understood by all.

    A. Teach it, preach it, and do it.

    B. Measure it and post it.

    C. Make sure that all know we cannot send scrap to the customer; the customer is the next step.

2. Improve quality responsiveness by reducing first-part lead time (see Chap. 5).

    A. Implement all seven techniques to reduce lead time.

    B. The most powerful initial activity will be to reduce lot sizes to a minimum. Very often this is easily done. If this cannot be done everywhere, do it where you can.

    C. The next most powerful activity is inventory reductions of all kinds.

    D. Usually, the third most powerful tool is the reduction of changeover times.

3. Emphasize JIT problem solving: our objective is Rapid-Response PDCA. This, very likely, will require a huge cultural change, but you should do some of the following.

    A. For example, if your engineers are the problem solvers, move them to the production line.

    B. Post a flipchart at the line and list the problems to resolve. Keep the flipchart up to date.

    C. Get the managers to spend time on the floor, evaluating, observing, and just being available.

    D. Have a top manager be at the daily production meeting where the problem solving is reviewed.

    E. Be brief, but make problem solving a part of the meeting, which should only be a few minutes if it is run well, even with a problem-solving review added.

    F. The point is to be proactive in problem solving: "Do something."

> **"Do** not let what you can not do, prevent you from doing what you can do..."
>
> **John Wooden,**
> former basketball coach at UCLA (Wooden won ten National Championships as a coach, and one more as a player at Purdue University)

4. For all process cells and lines that are highly people-dependent for quality results, begin the process of line shutdowns for quality issues. If you can't shut down for all quality problems, shut down for some problems, and prioritize your activities. Start this practice, and then expand it. There is no substitute for it.

5. Processes that are highly machine-dependent, such as high-speed pick-and-place machines, ovens, and other continuous process equipment are a little more difficult to manage in this way.

   A. If you can't shut them down, use *andons* to signal abnormalities and make sure the andons are responded to.

   B. Make those responsible for the machinery also responsible for the defects the machinery creates.

6. Aggressively categorize all defects as to the most likely source. Assign the appropriate groups to take on the issue of problem solving for defect reduction. Support them; hold them accountable.

7. Become more introspective in your problem solutions.

   A. Cease completely the philosophy of using inspection as the means to achieve quality of product and process.

      i. Eliminate visual or human inspection as an acceptable quality control technique; emphasize *poka-yokes*.

      ii. Institute *poka-yokes* widely; emphasize process improvements.

   B. Eliminate improved procedures—done by humans—as an acceptable quality improvement technique. Change the process design so it can't be done wrong; "*poka-yoke*" the process.

The point is that you can do a great deal. So start now! Show the commitment, and show the initiative.

Jidoka is a crucial element and is absolutely necessary to make Lean work.

Chapter Summary

Any time any major culture-changing initiative is undertaken, we must answer some questions. These are "Do we have the leadership, motivation, and necessary problem solvers in place to make this a success?" "Do the leaders have the three requisite traits of being able to put together a plan, sell the plan, and then act on the plan?" These leaders must then show the strong character and courage needed to make tough decisions and act in the face of adversity. You will need this from your organizational leader and your *sensei* as well. We must develop the motivation to change. This will be a key task of the leaders. They must be able to convince the workforce that the changes are not only necessary but desirable, and although there may be some short-term discomfort, the long-term gains more than justify it.

Finally, we will need a JIT problem-solving mentality among the workforce. We will need to engage the entire workforce and also have a small cadre of skilled problem solvers who are well versed in Kepner-Tregoe logical skills, statistical problem solving, and facilitation skills. This small group will be very instrumental in our success, so they must be chosen wisely.

In addition to the three questions mentioned earlier about the culture, we will need to pay particular attention to three other cultural characteristics. First, the degree of interdependence that all the changes will have. As we modify one aspect of our culture—one aspect of our business—other parts will change sympathetically. These are pronounced in a Lean initiative and may be changed for the good or the detriment of the initiative. Second, we can not stray from our commitment to emphasize the foundational issues. Finally, the key issue that is likely to get less attention than needed is *jidoka*. It will be a key to our future vitality and we will need to implement it aggressively, starting now.

Appendix A—Problem Solving and Standardization: How Are They Similar?

Concerns and Problems

Many things bother us. We have many concerns but not all concerns are problems. A problem is defined by the Kepner-Tregoe Methodology as a "concern" that meets three criteria:

- There must be a difference between what "should be" and what "is."

- It must be a present issue.

- It must have a root cause that is unknown, and the root cause must be eliminated.

If our concern meets all three criteria, it is a problem and needs problem-solving logic to resolve it. Problem solving is the finding and removal of the root cause of the problem.

Let's say warrantee returns, by our standards, should be less than 50 ppm but are currently 220 ppm. This is characteristic number 1. It is a present issue so it meets the second criteria, and if we do not know the cause, it meets the third criteria. Now we have a problem to solve. That seems clear enough.

Let's look at something a bit more mundane. For example, each time you take your ten-year old car in for servicing, the mechanic finds $200 worth of necessary extra work. This should not be, you are thinking. So, though it should not be, it *is*, and it is happening right now. But can I find and eliminate the root cause of the problem that is manifest by the extra costs? Probably not. So in this case, without some possible corrective action on a root cause, we do not have a problem. But what we do have is a concern in the form of a decision to make—that is, do you buy a new car or continue to pay for the repairs?

On the other hand, what if—instead of having high maintenance costs right now— you are worried about them happening in the future? In this case, even though it is a concern, it is not a present concern, so you do not have a problem. However, what you do have is a threat or a possible future problem.

Each of these concerns: threats, decisions, and problems require a different logical approach for their resolution. To study the various methodologies here is beyond the scope of this text, but with a quick trip to the bookstore you can buy *The New Rational Manager* (Princeton Research Press, 1981) by Kepner and Tregoe and you can learn all of these powerful techniques. I highly recommend it.

Concerns?

- Threats
- Decisions
- Problems

Each requires its own methodology to resolve.

So that briefly describes one key part of problem solving, but what does this have to do with standardization?

Standardization

Standardization is an attempt to get all the parties who perform some activity to perform it using the same skills and actions. It is an attempt to eliminate the variation that exists in any activity.

Just how does this interact with problem solving? Maybe a real live example can help clarify this question.

For example, at the Jayaroot plant in Juarez, Mexico, SPC (Statistical Process Control) was implemented to better manage the process and to monitor the quality of the finished products. This plant made electric lighting ballasts. The final process step, prior to packaging was to perform a 30-minute infant-mortality burn-in. Over 1000 ballasts were produced per hour and the normal defective rate was about 1.1 percent. The data was recorded hourly on a P (Percent Defective) control chart, see Fig. 6-2, and the process was statistically stable.

This process was statistically stable and had normal variation about the mean of 1.1 percent defects. Then at 1PM, it went out of control. Note that the 1PM data is below the lower control limit (LCL). I noticed this and asked what had been done about the change? Francisco the supervisor said, "We did nothing because, although the process went out of control, the process improved; our defects have decreased to practically nothing."

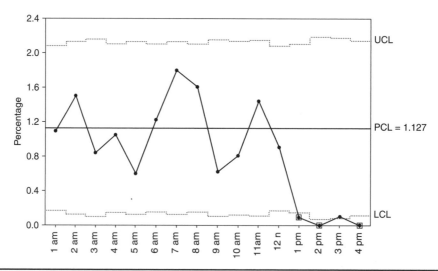

Figure 6-2   P chart for defectives.

The science of SPC is designed so that when a process goes out of control (that is, a plotted point on the control chart lies above the upper control limit or below the lower control limit), we know with 99.73 percent certainty that the process has changed. It is then a requirement that the cause of this change must be found and understood. These causes have a unique name in the science of SPC. They are called special causes of variation. Then, if the special cause created an undesirable change (in this example, that would be for the defects to increase and hence the plotted point would be above the upper control limit), you then have a problem to solve. However, if the special cause creates a desirable change, as in this case, the proper action is to standardize. See Fig. 6-3.

In this case, there are several possible solutions to this problem. However, the logical one, consistent with the Company's objectives at that time was to go to a single source of circuit boards and purchase only boards from supplier C. A little more on that later.

Why did Francisco treat the change incorrectly? When a process goes out of control, it means the process has changed. Well, entropy being what it is, the change is usually bad, in the form of deterioration. In the preceding several months, since they had started with SPC, Francisco had spent so much time chasing root causes of problems in which the change had occurred in the "undesirable" direction that he understandably had overlooked this "avalanche of diamonds" that was given to him. In this case, the change signaled that some desirable change had occurred.

So Francisco and I investigated the process to find the root cause of the change ... and what we found was amazing.

This line had three different suppliers of circuit boards, which we shall call suppliers A, B, and C. The process change that was seen on the control chart coincided to the minute (taking into account the lead time) when a new batch of circuit boards was introduced to the line. These boards were manufactured by supplier C. Upon further investigation we found that all three suppliers met all quality standards but the hole placements on supplier C's boards had a Cpk (Process Capability Index) of 2.2, while the Cpks for supplier A and B were 1.37 and 1.43, respectively. We found that, although all three suppliers met the minimum standard of Cpk >1.33, supplier C was a superior quality supplier. Since two suppliers barely met the spec and this created over 1 percent scrap, it is only logical to conclude that the specification was too wide and should be reduced; however, that was not our problem—at least not at that specific moment.

We had found the root cause of this "desirable effect": The change to supplier C had caused the reduction of defects. This reduction of defects was not a problem since it was

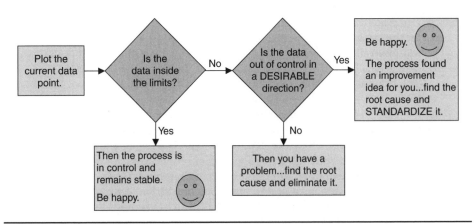

Figure 6-3 The logic of SPC.

not an undesirable effect. Rather, it was just a change that was both unexpected and serendipitous. So our plan was to standardize on supplier C.

How Is Standardization Similar to Problem Solving?

In both problem solving and standardization, it is required to find and eliminate the root causes of variation. Consequently, the skills required for good problem solving overlap almost 100 percent with the skills of standardization.

A Little More for Your Amusement and Amazement

I told you earlier that there would be a little more … and so now I'll tell you "the rest of the story," as it's said, which goes like this.

Francisco, having been buoyed by the knowledge that we could reduce defects a full 1 percent simply by single-sourcing our circuit boards to supplier C, was energized. You see, the company was attempting to implement a Deming type management system, and Point No. 4, of Dr. Deming's 14 Obligations of Management has to do with supplier development and going to single sources of supply. Francisco was energized beyond belief, and the first thing he did was calculate the possible gains. They exceeded $1,000,000/year for the quality losses alone. Next, he investigated this supplier a little. He found that supplier C was relatively new to Jayaroot Co. but was also the low-cost supplier for this board. Wow! Even more potential gains. Higher than a kite and filled with new-found energy, Francisco contacted the purchasing department and found out the following:

- Their objective was to go to a single source of supply.
- Although supplier C was the low-cost supplier, they had already been eliminated. In the future, boards would be supplied by A and B only. We were told that the decision had been made at central purchasing.

Well, not too discouraged, Francisco contacted the manager of the central purchasing department. He confirmed what Francisco had already been told. Upon questioning this manager, Francisco found the decision had been made, and no amount of data or logic would deter purchasing. Supplier C was out; A and B were in. Francisco argued so long and hard he almost lost his job over it. But the low-cost superior-quality supplier was eliminated and we needed to find another way to eliminate these defects.

I never actually found out why this decision was made this way, but the general manager for the plant just shook his head and said it did not surprise him. He went on to explain that they had similar problems in the past, nothing quite so clear as this, but in the past there was little they were able to do—just as in this particular case. Subsequently, I learned that this facility was strictly a cost center and all raw materials sourcing, costs, and metrics were managed at the home office in Chicago.

This example, more than most examples, points out the futility and destructive nature of centralized control taken to an extreme. We had transformed the manufacturing facility to operate with pull systems using cells operating at *takt*, and it was very Lean by any measure. From a manufacturing standpoint, it was a large success. However, it was not surprising that two years later, I was on the front steps saying goodbye to many fine workers as this facility was closed. You see, although it was a manufacturing success, it was a business failure.

The notice in the paper said they were moving the facilities to China to take advantage of the low-cost labor market there. This product sold for about $17, and had less then $0.12 direct labor in the cost to produce. The home office burden was 11 percent.

CHAPTER 7

How to Do Lean—The Four Strategies to Becoming Lean

I n this chapter, we will cover in simple format the four strategies to make a value stream lean. In support of the strategies, we will detail the diagnostic and analytical tools used to reduce the seven wastes. We will cover the *takt* calculation, the time study, the balancing analysis, the spaghetti diagram, and value stream mapping. I have attached five appendices to this chapter, one to cover each technique.

Overview of the Lean Implementation Strategies

The Overall Lean Approach

The overall approach to make a value stream lean consists of four key strategies. They are:

1. Synchronize supply to the customer, externally.
2. Synchronize production, internally.
3. Create flow.
4. Establish pull-demand systems.

The Diagnostic and Analytical Tools

To apply the four strategies, there are five basic diagnostic tools you will need to use in evaluating a value stream. They are:

1. The *takt* calculation (see Appendix A)
2. The basic time study (see Appendix B)
3. Balancing analysis (see Appendix C)
4. A spaghetti diagram (see Appendix D)
5. The present state value stream map and the future state value stream map (see Appendix E)

Eliminating the Wastes

The goal is to apply the four strategies using the five diagnostic tools to eliminate the seven wastes, which are (remember TWO DIME):

- Transportation
- Waiting
- Overproduction
- Defects
- Inventory
- Movement
- Excess processing

Implementing Lean Strategies on the Production Line

Strategy 1, Synchronize Supply to Customer, Externally

Conceptual Discussion

To synchronize externally is to supply the product to our customer at their needed demand rate, normalized to our production schedule. We want to supply all of the customer needs but we do not want to overproduce and create excess inventory. These tools allow this balance to be achieved.

In order to properly synchronize to the customer we need to meet the contractual volume demand and, in addition, we will need to handle the normal variations in both supply and demand. In a mature make-to-stock production system, with good raw materials supply, reliable production equipment, stable cycle times, and high quality yields, our supply variation should be low. However, we will still have supply variations, therefore we will need a safety stock inventory to compensate for these variations. In addition, there will be demand variations to contend with if we wish to be synchronized to the customer. This variation will require buffer stock inventory.

Tools Used

- **The Takt Calculation** will allow us to understand at what rate the customer will normally wish to have product supplied. This is the basic starting point for all production rate calculations. This is often referred to as rate-leveling. We want to avoid the ups and downs of normal production and rather stabilize the rate.

- **Cycle, Buffer, and Safety Stocks** are inventories, but they are the definition of necessary inventories. Cycle stock is necessary to assure normal pickup deliveries are in place, buffer stocks will handle the demand variations, and safety stocks will take care of internal supply variations. In this way, we will assure we meet demand, but have the minimum inventory on hand. These inventories are designed to handle normal variations in both supply and demand and therefore allow the production process to stay at *takt* and remain as stable as possible. (Chap. 3 has an example set of inventory calculations.)

- **Leveling of Model Mixes or Products** is used when more than one product is made on a given production line. The goal of leveling is to avoid making a batch of model A and then a batch of model B, and instead make both products, simultaneously, one at a time at the demand rate of the customer. We are trying to synchronize externally to the demand rate of the customer. To level production in amount and by model mix, we will frequently use a *heijunka* box. If leveling is not achieved, cycle, safety and buffer stocks must be much larger and, of course, we want the inventory to be a minimum.

Wastes Reduced

Overproduction   Overproduction is the waste targeted here. However, when overproduction is reduced, all other wastes are reduced as well; especially the waste of inventory. In addition, this strategy to synchronize externally is the key to on-time delivery. It will allow smooth production line operations so the line can produce at a constant rate, using the safety and buffer stocks to take up the supply and demand variations. In addition, it will allow the supply to be both more flexible and more responsive.

Summary of Synchronize Externally

To establish the production rate at *takt* is absolutely crucial. This, coupled with the establishment of Lean inventories will allow you to maintain supply to your customer and run the process at a level and stable rate. This is always the first, and the most important, step. It is often difficult to redesign the work stations for leveling of the model mix since it often requires changeovers, and so on. If that is the case for you, then do the model mix leveling later, but make sure you are producing at *takt* and also have inventories set up to protect your supply to the customer. A word of caution is in order. If you found that your production rate was unstable while doing the systemwide evaluations, carry extra inventory. Seek advice from your *sensei* as to how much. This inventory then becomes the protection that will allow you to both assure supply to the customer as well as reduce the variations on your line from your planning systems.

Frequently, you will want to install a *heijunka* board and a production *kanban* system right away. You may have *kanban* training scheduled for later. However, implement *kanban* here anyway, if at all possible. Have your *sensei* make the calculations and do a short training for just those involved in this system. This type of JIT training will often be necessary and it is not unusual to improvise from time to time like this.

Strategy 2: Synchronize Production, Internally

Conceptual Discussion

To synchronize production internally is to divide the necessary work in processing steps such that each processing step takes the same time. The ideal is that all processing steps perform at a cycle time equal to *takt*. The following Lean tools are used:

Tools Used

- **Balancing** is done by completing the basic time study and then designing the work at each work station to be the same. Normally, some accommodation is made for OEE (Overall Equipment Effectiveness) to account for production losses caused by availability issues, quality dropout, and cycle-time losses. The end result of balancing should be work stations that are synchronized.

- **Standard work** is the technique used to review the performance, including the cycle time of a production process, production cell, or a production work station. It is a key tool in evaluating and assisting the production process to achieve synchronized production.

Wastes Reduced

Waiting is the key waste removed, and while inventory is often reduced, the goal is one-piece flow.

Summary of Synchronize Internally

The key tool used to synchronize internally is the basic time study coupled with the balancing study and chart. The balancing chart will show at a glance three major aspects of the process.

- The relative cycle times of each step: the balance
- The waste of waiting due to the imbalance
- The process bottleneck

From this balancing chart, the work begins and is comprised of two major steps. First, the production cycle time must be calculated. It is normally the *takt* time multiplied by the actual OEE, if that is known. So if *takt* is seven minutes and OEE is 0.80 or 80 percent, the cycle time of the process will need to be 5.6 minutes. With this known, it is now time for the second task, which is to design all work stations to have a 5.6 minute cycle time. With these two steps we have balanced the process steps (synchronized internally) to a production cycle time that will achieve *takt* (synchronized externally).

To synchronize the process internally, always create a huge to-do list.

In the application of this strategy, it is best to go to the end of the value stream and work backwards, as much as practical. For example, if there are three work cells in series that comprise the value stream, work on the cell closest to the storehouse first. This order, of starting at the customer and working backwards in the flow, should also be employed with the following two strategies of creating flow and installing pull-demand systems.

Strategy 3: Create Flow

Conceptual Discussion

The concept of flow is such that we do not want the production units to stop, except for value-added work. The flow concept has both overall measures and local measures. The local measure would be cycle time. That is the increment of time between consecutive production units. If work is done, one piece at a time, it is also the processing time at the work station. The overall measure of flow is production lead time. It is the overall time it takes for a unit to complete the entire production process. In every case, if we can reduce cycle time or if we can reduce lead time, we will make process improvements. Obstacles to flow include:

- Inventory
- Batches and batch processes
- Distance
- Any defect-creating process
- Variation
- Process steps with mismatched cycle times
- Changeovers
- Non-valued-added work steps

Creating flow is "the basic condition" and this strategy has some strategies of its own, including:

Point of Clarity   These eight obstacles to flow will be the focus of the majority of your lean improvement activities from this day forward.

- Rate balancing of all steps in the value stream from the customer all the way through the raw materials supply
- Removal of inventory
- Reduction of distances between stations
- Elimination of defects, which we call *jidoka*
- Elimination of non-value-added work

Tools Used

- **Minimum lot sizes** with the ideal being one-piece flow.
- **Cells** and other techniques to close-couple process to achieve short transportation distances and one-piece flow.
- **SMED** (Single Minute Exchange of Dies, quick changeovers) to reduce changeover times and the needed inventory to sustain production.
- *Jidoka* (see the section at the end of this list).
- **Problem solving by all,** for the elimination of defects and to achieve process improvements. The goal should be Rapid Response PDCA (Plan-Do-Check-Act).
- **CIP (Continuous Improvement Philosophy) and** *Kaizen* to organize the problem-solving activities.
- **5 Whys** is the key problem-solving tool used.
- **Reduction of variation** is a key tool used in inventory reduction.
- **OEE** is a key metric to use in prioritizing if **quality yield, availability,** or **cycle time performances** must be addressed to achieve and increase flow rates.
- **Availability** improvements through the use of **TPM** (Total Productive Maintenance).

Jidoka   *Jidoka* is not only the most important strategy to implement, it is also one of the most difficult. However, it is the one I find most often slighted. The danger is that early in the implementation, other effects will be much more pronounced in terms of achieving goals. There is always significantly more quantity control reduction to be made in reducing the wastes of inventory, batches, and transportation than the waste caused by defects. Hence, these aspects get more attention, at the expense of the *jidoka* concept. In addition, most companies, although not very effectively, have been working on defect reduction for some time. However, I have yet to see even one company that had a *jidoka* concept in place—at this point in the implementation—that even remotely resembled the TPS model. However, later, this weak *jidoka* system will undermine literally all other activities. I cannot stress enough that *jidoka* must be given top priority, even though the current gains may not seem to warrant it. Return to Chap. 6 and the section titled, "The Implementation of *Jidoka* Is *Always* a Fundamental Weakness" for further information.

Wastes Reduced

- **Transportation** is reduced by the reduction of distances traveled by using cells, for example.

- **Waiting** is reduced because the production lines, while flowing, have little downtime.

- **Overproduction** is dramatically reduced on a local basis since local inventories and batches do not need premature replenishment.

- **Defect** reduction is the objective of jidoka.

- **Inventory** reduction is achieved by several of the techniques, including problem solving, SMED, and minimum lot sizes, to name just a few.

- **Movement** is reduced when transportation is reduced and distances are reduced. In addition, since availability is up, there are fewer instances of reassignment and wholesale personnel movement.

- **Excess processing** is reduced as all non-value-added activities are reduced.

Summary of Creating Flow

This strategy is simply removing all inventory possible, moving process steps as close together as possible, and eliminating non-value-added work—plus, the most important aspect of implementing *jidoka*.

Strategy 4: Establish Pull-Demand Systems

Conceptual Discussion

Pull systems have two characteristics. First, they have a fixed inventory, so the cycle stock, plus the buffer and safety stocks need to be determined. Second, they are activated when product is removed and this signals the upstream process to produce—no signal, no production. All *kanban* systems provide this function. However, for some simple systems such as pull systems within a close coupled cell, for example, the most effective pull signal often is the "*kanban* space."

With a *kanban* space, when the customer removes the upstream production, the customer has "opened" the *kanban* space—this is the pull signal. Afterward, the upstream process produces more product, but not before. It is the perfect "take one make one" system. Operationally in pure pull systems, it means you do not send anything anywhere. If it leaves, someone came to pick it up. However, it is not possible to have a pure pull system in all cases. Wherever there is inventory, this inventory will delay the pull signal from the customer. This is the basis used in *kanban* design.

In *kanban*, the *kanban* card, for example, is removed from the product as product is consumed. The card is then placed in a *kanban* post and the *kanban* card is then transported back to the *heijunka* board to signal replenishment. Second, since we can not always use pure pull signals, the pull signals need to be time responsive to the needs of the customer. The time it takes—from the receipt of the signal by the customer (product is removed and the *kanban* card is placed in *kanban* post) until the replacement product arrives at the storehouse—is called the replenishment time. This is effectively "pull signal delay." We strive to minimize pull signal delays. Simply stated, we work to minimize replenishment time.

The perfect pull system is "take one, make one." For more information on replenishment time, refer to Chap. 3, the section "Finished Goods Inventory Calculations" and Fig. 3-1. The opposite of a pull system is a push system. In a push system, product is made at an upstream station and then "pushed" to the downstream station independent of the need of the downstream station. Push system allows local machine optimization at the expense of overall system optimization. They overproduce and create not only excess finished goods inventory but excess WIP also.

Tools Used

- *Kanban* is the second most important tool used in creating a pull system. The most important tool is training. It is critical that all employees understand the concept of pull production. For example, in a close coupled manufacturing cell, there are no *kanban* cards, but pull is practiced fully.

- JIT (Just In Time), support of all types, especially JIT staff planning support.

Wastes Reduced

Overproduction and inventory (in the form of WIP) is reduced.

Summary of Pull-Demand Systems

In most pull-demand systems, we will establish a signal to produce and then work to reduce the time in the replenishment cycle.

Almost surely, if you have central planning or even a local MRP program designed to do the scheduling, this will need to be altered. This will be a huge cultural change. To bypass this planning step is normally not very hard technically, but it must be done carefully since you will uncover all kinds of non-Lean activities going on. You will find that the planning program does not work as designed and requires a great deal of human interaction to make it work. This human interaction is often just variation by another name.

The first thought most engineers have is to implement *kanban*, or rather the first thought should be training. It will be a huge cultural change to get people out of the practice of delivering. It may be some time before you can set up good delivery loops with materials handlers to take care of WIP. Consequently, often in the early stages of implementation, operators are still moving goods from one station to the next. This needs to be changed from delivering to picking up.

For raw materials delivery, and finished goods pickup, many facilities already have a skeleton *kanban* system in place. If that is the case, teach these people the proper use of *kanban* and implement *kanban* properly—employing all six rules.

Early on, get the planning people involved so they can begin the integration of your planning system with the new Lean tools. They have much work to do.

Chapter Summary

The process used to make a value stream Lean consists of implementing four strategies, which are:

- Synchronize supply to the customer, externally
- Synchronize production, internally
- Create flow
- Establish pull-demand systems

To apply the strategy, the basic diagnostic tools include:

- The *takt* calculation
- The basic time study and balancing analysis
- A spaghetti diagram
- Present state value stream map
- Future state value stream map

The goal is to apply the strategy, using the diagnostic tools to eliminate the seven wastes, which are:

- Transportation
- Waiting
- Overproduction
- Defects
- Inventory
- Movement
- Excess processing

Appendix A—The *Takt* Calculation

Background

Takt is the rate at which a customer would pick up a product if he picked product up uniformly during the day, while you produced it. It is the true one-piece flow, pull-demand concept. The key waste that the *takt* concept strives to avoid is the waste of overproduction, the greatest of all wastes.

The equation for *takt*, or *takt* time, is the *available work time* divided by the *customer demand* for that work time interval.

An Example Calculation

For example, let's say we produce two ten-hour shifts and each shift includes a 30-minute lunch and two ten-minute breaks. So the available work time is 20 hours – (2 × 50 minutes) = 18.33 hours. Our normal work schedule is five days per week and we have nine holidays, so our work year is 365 – (2 × 52) – 9 = 252 days per year. The customer has a contractual agreement to purchase 500,000 units per year.

To calculate *takt*, let's use the work week, since that is a common planning interval (If you use the day or the month, the answer will be the same—try it).

- Available time is 18.33 h/day × 5 d/wk = 91.67 hours, which equals 330,000 seconds/wk

- Customer demand is 500,000 units/yr ÷ 52 wks/yr = 9615 units per week

- *Takt* = 330,000 seconds/wk ÷ 9615 units/wk = 34.3 second per unit

In simple terms, we need to produce one good unit every 34.3 seconds to stay in step with our customer's demand. This is the synchronization time, to synchronize supply externally. If it is met, the first strategy has been executed successfully.

How to Handle Model Mix Leveling

It is common on many cells to produce several models of the same basic production unit. These models, taken as a whole, are often referred to as a family of products because they use many of the same parts and many of the same processing steps. In that case, the *takt* equation remains unchanged. It is still available work time divided by customer demand but must be calculated for each model. The complication is not in the *takt* equation; rather, it is setting up the cell so the units can be produced simultaneously.

Frequently, the concept of model mix leveling is avoided. The typical logic used to avoid doing the leveling goes like this, "Since my customer comes for his pickup on Friday, and as long as I have the entire shipment made by then, it makes no difference whether I make the models at a uniform rate during the week or in a batch, just as long as I have all the models completed by pickup day."

However, let's look at a specific example. For example, you produce five models, A,B,C,D, and E and it takes exactly one day to produce the contractual volume of each model. Your work schedule is five days per week and his pickup is first thing Monday morning. So you make A on Monday, B on Tuesday, and so on, finishing with E on Friday, and all five models are ready for pick up the following Monday. This causes spike demands of raw materials as the various models—that is, batches go through the system, but under normal circumstances it does not sound too compelling to force model-mix leveling. But what if something abnormal happens? For example, you get a call on Wednesday and your customer says, "By the way, we want to change our pickup for next Monday. We will not need Model A, B, or C but still need the normal weekly volume but with a mix of 50 percent D and 50 percent E. Well, you now have produced only A, B, and C and the customer only wants D and E. In this case, if you work Saturday and

Sunday you still can't meet their demand changes. However, had the model mix been leveled, you could just switch to D and E, and since you already have some on hand, you could work on the weekend and have everything the customer wanted for pickup on Monday.

The fact is, if model mix leveling is properly employed, the business is just more flexible and more responsive. These are two huge business advantages that are hard to come by and serve you well in the competitive battle to survive and prosper—and leveling augments both.

How Does Cycle Time Relate to *Takt* Time?

Cycle time has many meanings, but generally people mean one of two things, one relating to the product, one to the process. Production cycle time is the time interval between two consecutive production units at the end of the production process. Process cycle time is the amount of time the unit is being worked on at any given production step. If the process cycle time in each processing step is the same, we say the process is balanced: it is synchronized internally. However, this cycle time must not only be synchronized, it must be synchronized to *takt* to stay in compliance with strategy number one: synchronize externally.

This has practical limitations since sometimes the line is not available to produce because of machine failures, stock outs, cycle-time problems, or defective parts. If the production process would be designed to operate at *takt*, then each problem mentioned earlier would result in a customer supply shortage and necessitate overtime or some other countermeasure. Since, it is practically impossible to avoid all these problems, we normally calculate the desired cycle time to be:

Cycle time = *takt* time × OEE or in this case if OEE = 0.88 (OEE is Overall Equipment Effectiveness, defined in Chap. 4), Cycle time = 34.3 seconds × 0.88 = 30.18 seconds

Hence we would design the production system to be synchronized to 30 seconds cycle time. We would design all stations to perform to 30 seconds, or stated another way: For a one person work station, we would balance the work so each work station has 30 seconds of work. Now when we have the normal problems of production, manifest by our OEE = 0.88, we will still end up producing the equivalent of one good unit each 34.3 seconds. When we calculate the cycle time thusly, we now are compliant to both strategy number one and two.

OEE is a manifestation of the reality that all problems can't be fixed right now, but we still have to supply the customer. A measure of a Lean system is the difference between *takt* time and cycle time. Among other things, this extra time is the waste of manpower that we must pay for because our system has losses. To improve systems, a first pass is always focused on OEE and the three losses of quality losses, cycle-time losses, and availability losses. When we are able to reduce the losses of OEE, we are now able to produce more using less:

- Less space

- Less manpower

- Less capital

- Less raw materials

OEE is a very descriptive and powerful metric.

How Do We Handle Variations in Supply and Demand?

Having this cycle time to *takt* time relationship will take care of our production problems in the long run, but what about the short-term problems we might have? Those short-term problems are usually of two types: underproduction and short-term increases in demand. Say, for example, that the customer calls on Monday and informs you that for their Wednesday pickup they need twice their normal volume. Normally, there is no way to handle this with cycle-time changes, so we hold buffer inventory for exactly this situation: demand variation. Take a second example in which your customer picks up daily and you have a problem with one of your parts and can't make production. How do you make the shipment? We will hold a volume of safety stock to cover this variation. Hopefully these do not happen often, but in both cases the answer is that we protect the supply to the customer with finished goods inventory.

Appendix B—The Basic Time Study

Background

The most fundamental tool to synchronize flow and analyze the work is the basic time study. The Zeta Cell Time Study, shown in Fig. 7-1, is included for your reference. The study can be done in many ways, but several cycles should be observed, and then a rational decision made on the times to use in the study. Notice that, although the average value is employed, several times where the average was overstated due to problems found during the study, a lesser number was used. In addition, those problems encountered were taken care of with *kaizen* activities.

The Time Study

The form used has a place for the:

- Process step number
- Flow chart identification
- The work element
- Eight time studies
- Summary statistics
- Final time selected

It is important that the times measured are for the work only and do not include the wait times. If you wish to catalogue wait time, simply modify the form to have a work column and a wait column for each cycle.

It is important to evaluate several cycles. In this case, we did eight. Ten is very common, and five is too few.

The Time Study Analysis

In this case, we put the data into a tool I call a Time Stack of Work Elements, see Fig. 7-2, which shows the cumulative cell time on the left, with the work steps defined and notation of which operator performs the work. Here, there was one operator per work station.

These data are then used in the balancing study. See Appendix C.

Process to Monitor Rayco 43-27 Date 3/9/2005, 2 shift

Station: Zeta Cell Done by: J. O. Bengineer

Step No	FC Id.	Work Element	Cycle 1	Cycle 2	Cycle 3	Cycle 5	Cycle 6	Cycle 7	Cycle 8	High	Low	Range	Average	Final
1	10	Cut bracket	3	4	3	2	5	11	3	11	2	9	4.4	3
2	20	Assy bushing (3)	11	10	13	12	13	19	12	19	10	9	12.9	12
3	30	Install o-ring and clip	9	6	6	8	7	8	7	9	6	3	7.3	7
4	40	Place in jig, glue	7	8	9	11	10	10	9	11	7	4	9.1	9
5	50	Press in magnets (2)	4	5	6	5	4	7	17	17	4	13	6.9	6
6	60	Insert o-rings, cap, grease	14	12	12	13	19	13	14	19	12	7	13.9	13
7	70	Install support	7	8	7	8	8	9	7	9	7	2	7.7	8
8	80	Install o-ring and clip (2)	6	7	8	9	23	7	8	23	6	17	9.7	8
9	90	Apply epoxy, 3 locations	12	13	15	14	14	14	13	15	12	3	13.6	14
10	100	Install control capacitor	7	8	9	9	8	8	7	9	7	2	8.0	8
11	110	Apply epoxy, topside	7	6	5	9	6	5	5	9	5	4	6.1	6
12	120	Install retainer ring	9	8	9	8	9	7	8	9	7	2	8.3	8
13	130	Install covercap	6	7	7	8	6	7	7	8	6	2	6.9	7
14	140	Unload/load machine (2)	2	3	3	2	12	3	3	12	2	10	4.0	3
15	150	Apply final sealant (1)	22	14	15	28	14	15	16	28	14	14	17.7	15
16	160	Final test, wrap leads	16	19	17	18	22	17	18	22	16	6	18.1	18
17	200	Package	12	10	28	12	13	11	12	28	10	18	14.0	12
18														
19														
20		Total	154	148	172	176	193	171	166	193	148	45	168.6	157

Notes	
1	Gun required unplugging hence long times, place on PM program
2	Long cycle time was due to dropped parts, attention to details
3	Long cycle times were due to dropped parts, operator needs surgical gloves
4	Hard to do study, so much inventory and lots of movement plus lots of wait times
5	Numerous units dropped on the floor
6	Transportation times not taken
7	
8	

Figure 7-1 Zeta cell time study.

Time Stack of Work Elements, Zeta Cell					
Time in seconds	**Activity**	**Original Plan**	**Time in seconds**	**Activity**	**Original Plan**
80, 79, 78, 77, 76, 75, 74, 73, 72, 71, 70, 69, 68, 67	Apply epoxy 14 secs	Operator 5	157, 156, 155, 154, 153, 152, 151, 150, 149, 148, 147	Package 12 secs	Operator 10
66, 65, 64, 63, 62, 61, 60, 59	Install O-ring and clip 8 secs	Operator 4	146, 145, 144, 143, 142, 141, 140, 139, 138, 137, 136, 135, 134, 133, 132, 131	Final test, wrap leads 18 secs	Operator 9
58, 57, 56, 55, 54, 53, 52, 51	Install support 8 secs	Operator 4			
50, 49, 48, 47, 46, 45, 44, 43, 42, 41, 40, 39, 38	Insert O-ring, grease insert, place in cap. 13 secs	Operator 3	130, 129, 128, 127, 126, 125, 124, 123, 122, 121, 120, 119, 118, 117	Apply final sealant 15 secs	Operator 8
37, 36, 35, 34, 33	Press in magnets 6 secs	Operator 2	116, 115, 114, 113, 112, 111, 110	Unload-load machine 3 secs	Operator 7
32, 31, 30, 29, 28, 27, 26, 25, 24	Place in jig, glue 6 places 9 secs	Operator 2	109, 108, 107, 106, 105, 104, 103	Install cover cap 7 secs	Operator 7
23, 22, 21, 20, 19, 18, 17, 16	Install O-ring 7 secs	Operator 1	102, 101, 100, 99, 98, 97, 96, 95	Install capacitor retainer 8 secs	Operator 6
15, 14, 13, 12, 11, 10, 9, 8, 7, 6, 5, 4	Assembly bushing to bracket, install cap. 12 secs	Operator 1	94, 93, 92, 91, 90, 89	Apply epoxy 6 secs	Operator 6
3, 2, 1	Cut Bracket 3 secs	Operator 1	88, 87, 86, 85, 84, 83, 82, 81	Install control capacitor 8 secs	Operator 6

FIGURE 7-2 Time stack of work elements.

Appendix C—The Balancing Study

Background

A balancing study is completed to see how well the actual work elements will fit into the desired cycle time. It is easy enough to calculate the desired cycle time, but often the work elements do not allow a perfect distribution of work. In addition, at the Zeta cell the design was an outside U cell, with stationary operators so the operators could not be moved as a technique to balance the work. Instead, the work needed to be moved to the operators. Nonetheless, we came up with a nice balance for a first pass (see Fig. 7-3 and Fig. 7-5).

The Present Case Balancing Graph

On the balancing chart, you can quickly see three things: how much time is wasted, the degree of balancing achieved, and the bottleneck.

- First, the vertical distance from the *takt* line to the station cycle time represents the waiting time, which is time wasted for that work station.
- Second, by comparing the heights of the bars, you can see at a glance if the process is unbalanced and what results rebalancing will yield.
- Third, the highest bar is the bottleneck, which is now obvious.

The balancing chart for the original ten-operator layout looked like that shown in Fig. 7-3.

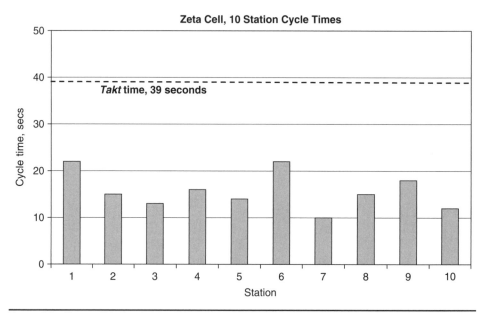

FIGURE 7-3   Zeta cell balancing graph.

The Present Case Analysis

Let's analyze the chart for the three simple reviews available from this type of a graph, specifically.

This line balance chart shows graphically the time wasted, which is the distance from the top of the bar graph for any station to the *takt* line—that is, if the production could be made at *takt* rate. You will find in Chap. 16 that this cell is performing so poorly it takes over two shifts, rather than the design of one shift, to produce the volume. So even though this graph shows clearly there is a huge time waste, it also grossly understates the waste. The actual wasted work time is much worse than this chart shows, and this chart is very bad! (The actual paid time was over 78 seconds per unit × 10 operators or 780 seconds, yielding 623 seconds per unit of waste; over 80 percent wasted labor time.)

The balance is poor since the tallest bar is 22 seconds and the shortest is 10. This is not good balance; the cycle times should be very similar. The bottleneck, the longest cycle time, occurs twice. Both stations 1 and 6 are 22 seconds.

The Redesign: Synchronize to *Takt*

Next, we rebalanced to takt by redistributing the work at the work stations. Recall that there were 157 seconds of work and we need a *takt* of 39 seconds, which yields: 157 seconds of work ÷ 39 seconds/station = 4.02 theoretical stations. With one person per station, we can't have 0.02 persons, so we will check out five stations, which gives us 157 ÷ 5 = 31 seconds per station. And if OEE is 90 percent, we could design for a cycle time of 39 × 0.90 = 35 seconds. This will work as a starting point. So our balance will be based on five operators and a design cycle time of 35 seconds.

With that in mind, we redistribute the work as shown in Figure 7-4.

			Zeta Cell, Original Design			New Balanced Proposal		
			10 Operators			5 Operators		
Step no	FC Id.	Operator No.	Work element	Final time	Time per operator	New design, operator No.	Time per operator	
1	10	1	Cut bracket	3		1		
2	20	1	Assy bushing (3)	12	22	1	31	
3	30	1	Install o-ring and clip	7		1		
4	40	2	Place in jig, glue	9	15	1		
5	50	2	Press in magnets (2)	6		2		
6	60	3	Insert o-rings, cap, grease	13	13	2	35	
7	70	4	Install support	8	16	2		
8	80	4	Install o-ring and clip (2)	8		2		
9	90	5	Apply epoxy, 3 locations	14	14	3		
10	100	6	Install control capacitor	8		3	28	
11	110	6	Apply epoxy, topside	6	22	3		
12	120	6	Install retainer ring	8		4		
13	130	7	Install cover cap	7	10	4	33	
14	140	7	Unload/load machine (2)	3		4		
15	150	8	Apply final sealant (1)	15	15	4		
16	160	9	Final test, wrap leads	18	18	5	30	
17	200	10	Package	12	12	5		
			Total	157	157		157	

FIGURE 7-4 Zeta cell rebalancing calculations.

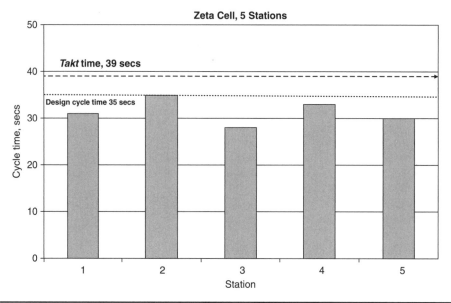

Figure 7-5   Zeta cell balancing graph, redesign.

Redesigned Case, Balancing Graph

Figure 7-5 shows the Zeta cell balancing graph, redesign.

The Redesigned Case: Analysis

To analyze this line balance chart, we find:

- The waste is much smaller than the ten persons. Useful work is 157 seconds per unit, paid work is 39×5, or 190 seconds per unit, so waste is 33 seconds per unit.

- The balance is reasonable. Actually, it's quite good considering the restraints we had to deal with. The longest cycle time is 35 seconds, right at max design, and the shortest is 28 seconds. Clearly an improvement in balance, and quite frankly pretty good.

- The process bottleneck is now station 2 at 35 seconds.

The Results

The entire story of the Zeta Cell is told in Chap. 16. Read it there and see how employing the Four Strategies created the process gains shown in Table 7-1.

Metric	Original Case	Leaned Process	% Improvement
First piece lead time	4.5 hrs	9 min	97% reduction
Batch lead time	20 hrs	8.5 hrs	58% reduction
Space utilization	425 sq. ft.	160 sq. ft.	62% reduction
Operators per cell	10	5	50% reduction
Labor costs/unit	15 min/unit	3.19 min/unit	79% reduction

Table 7-1   Process Gains Summary

Appendix D—The Spaghetti Diagram

Background

The spaghetti diagram, see Fig. 7-6 and 7-7, is a simple yet powerful tool to visualize movement and transportation. When the transportation paths are seen, it is often easy to spot opportunities to reduce these wastes. A spaghetti diagram is normally hand-drawn on a simple floor layout. The example here is of the movement of a motor as it progresses through the processing steps.

Present Case: Spaghetti Diagram

From this simple diagram (Fig. 7-6), you can see that the motor makes an excessively long transportation from the Polish step to the CNC Machining process and then back to Assembly. Also note that finished goods staging is a long ways from the finished goods storehouse. See the changes shown in Fig. 7-7.

Future Case: Spaghetti Diagram

Note how the floor space has opened up by moving the CNC Machine and placing FG Staging near the aisle. In this case, this floor plan should be reviewed and a new layout considered. There is excess space and the work areas are not well laid out. On the list of

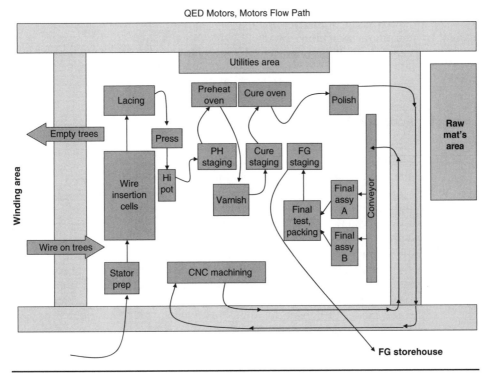

FIGURE 7-6 Spaghetti diagram, QED Motors, before.

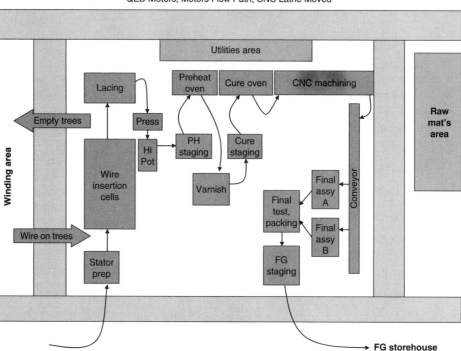

QED Motors, Motors Flow Path, CNC Lathe Moved

FIGURE 7-7 Spaghetti diagram, QED Motors, after.

future *kaizen* activities for this project is an item to revise the layout and for no capital the engineer believes he can reduce the floor space another 46 percent.

The story of QED Motors is covered in detail in Chap. 16.

Appendix E—Value Stream Mapping

Background

Value stream mapping (VSM) is a technique that was originally developed by Toyota and then popularized by the book, *Learning to See (The Lean Enterprise Institute, 1998)*, by Rother and Shook. VSM is used to find waste in the value stream of a product. Once identified, you can work to eliminate it. The purpose of VSM is process improvement at the system level.

Value steam maps show the process in a normal flow format. However, in addition to the information normally found on a process flow diagram, value stream maps show the information flow necessary to plan and meet the customer's normal demands. Other process information includes cycle times, inventories held, changeover times, staffing and modes of transportation, to name just a few. The typical VSM is called a "stock to dock" or "door to door" value stream map since it normally covers the information and process flow for the value stream at your facility. VSMs can be broader and cover any part or the entire value stream. However, functional responsibility often precludes the ability to take actions on these larger value stream maps. If actions are not the result of the value stream map, then the map has lost most of its effectiveness.

The key benefit to value stream mapping is that it focuses on the entire value stream to find system wastes and tries to avoid the pitfall of optimizing some local situation at the expense of the overall optimization of the entire value stream. The strength of value stream mapping may also be its weakness. It is not uncommon to find large wastes in cells, for example, which are not detailed on VSMs. If this is the case, large wastes can go unnoticed. This is a problem to those who only use VSMs in their battle to find and eliminate waste. Value steam mapping is only one tool in the battle for waste reduction, and to truly attack wastes, many tools are required.

VSMs detail information and the two key metrics highlighted are value added work and production lead time.

Value Stream Mapping (VSM) Common Sense

First, VSM is a tool to assist you in the battle to reduce waste. Don't get too hung up on the rules. The general ideas will guide you nicely, but if there are some relevant data you think would be helpful but you haven't seen it on some primer on value stream mapping, don't hesitate to use it. The objective is waste reduction—don't lose sight of that. Make the VSM do what you want it to do. See Figs. 7-8 to 7-10.

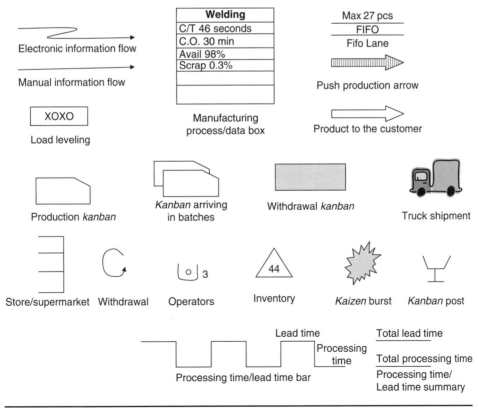

FIGURE 7-8 Typical value stream mapping icons.

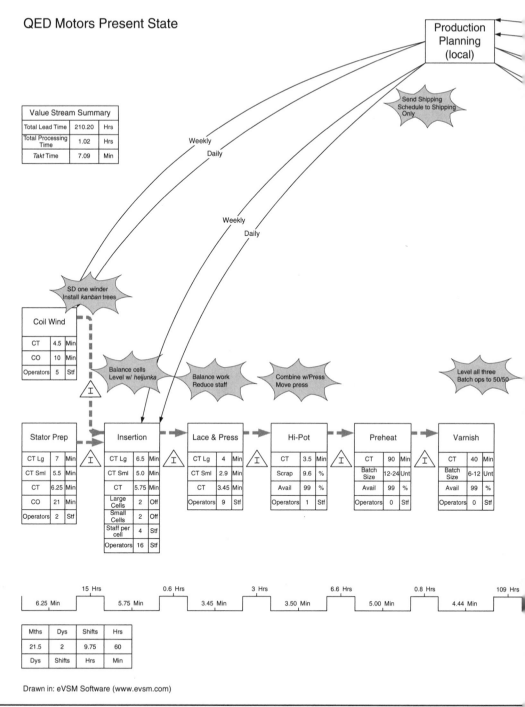

FIGURE 7-9 Present state value stream map, QED motors.

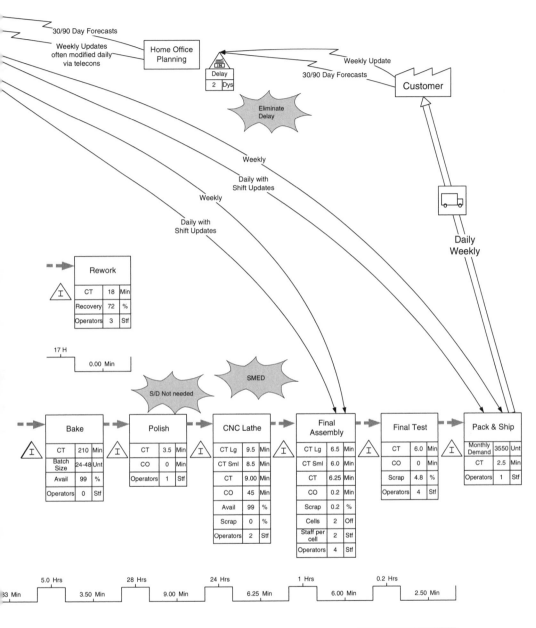

30/90 Day Forecasts

Weekly Updates
often modified daily
via telecons

Home Office Planning

Delay	
2	Dys

Eliminate Delay

Weekly Update

30/90 Day Forecasts

Customer

Weekly

Daily with
Shift Updates

Weekly

Daily with
Shift Updates

Daily
Weekly

Rework

CT	18	Min
Recovery	72	%
Operators	3	Stf

17 H

0.00 Min

S/D Not needed

SMED

Bake

CT	210	Min
Batch Size	24-48	Unt
Avail	99	%
Operators	0	Stf

Polish

CT	3.5	Min
CO	0	Min
Operators	1	Stf

CNC Lathe

CT Lg	9.5	Min
CT Sml	8.5	Min
CT	9.00	Min
CO	45	Min
Avail	99	%
Scrap	0	%
Operators	2	Stf

Final Assembly

CT Lg	6.5	Min
CT Sml	6.0	Min
CT	6.25	Min
CO	0.2	Min
Scrap	0.2	%
Cells	2	Off
Staff per cell	2	Stf
Operators	4	Stf

Final Test

CT	6.0	Min
CO	0	Min
Scrap	4.8	%
Operators	4	Stf

Pack & Ship

Monthly Demand	3550	Unt
CT	2.5	Min
Operators	1	Stf

5.0 Hrs   28 Hrs   24 Hrs   1 Hrs   0.2 Hrs

33 Min   3.50 Min   9.00 Min   6.25 Min   6.00 Min   2.50 Min

Summary		
Total Lead Time	210.20	Hrs
Total Processing Time	1.02	Hrs

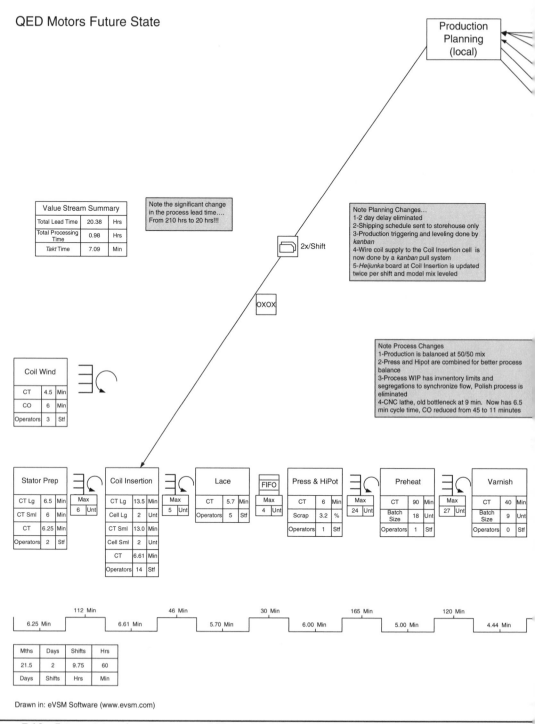

QED Motors Future State

Value Stream Summary		
Total Lead Time	20.38	Hrs
Total Processing Time	0.98	Hrs
Takt Time	7.09	Min

Note the significant change in the process lead time.... From 210 hrs to 20 hrs!!!

Note Planning Changes...
1-2 day delay eliminated
2-Shipping schedule sent to storehouse only
3-Production triggering and leveling done by kanban
4-Wire coil supply to the Coil Insertion cell is now done by a kanban pull system
5-Heijunka board at Coil Insertion is updated twice per shift and model mix leveled

Note Process Changes
1-Production is balanced at 50/50 mix
2-Press and Hipot are combined for better process balance
3-Process WIP has invnentory limits and segregations to synchronize flow, Polish process is eliminated
4-CNC lathe, old bottleneck at 9 min. Now has 6.5 min cycle time, CO reduced from 45 to 11 minutes

Coil Wind		
CT	4.5	Min
CO	6	Min
Operators	3	Stf

Stator Prep		
CT Lg	6.5	Min
CT Sml	6	Min
CT	6.25	Min
Operators	2	Stf

Coil Insertion		
CT Lg	13.5	Min
Cell Lg	2	Unt
CT Sml	13.0	Min
Cell Sml	2	Unt
CT	6.61	Min
Operators	14	Stf

Lace		
CT	5.7	Min
Operators	5	Stf

Press & HiPot		
CT	6	Min
Scrap	3.2	%
Operators	1	Stf

Preheat		
CT	90	Min
Batch Size	18	Unt
Operators	1	Stf

Varnish		
CT	40	Min
Batch Size	9	Unt
Operators	0	Stf

Max 6 Unt — Max 5 Unt — Max 4 Unt — FIFO — Max 24 Unt — Max 27 Unt

2x/Shift   OXOX

Production Planning (local)

112 Min — 46 Min — 30 Min — 165 Min — 120 Min

6.25 Min — 6.61 Min — 5.70 Min — 6.00 Min — 5.00 Min — 4.44 Min

Mths	Days	Shifts	Hrs
21.5	2	9.75	60
Days	Shifts	Hrs	Min

Drawn in: eVSM Software (www.evsm.com)

FIGURE 7-10 Future state value stream map, QED motors.

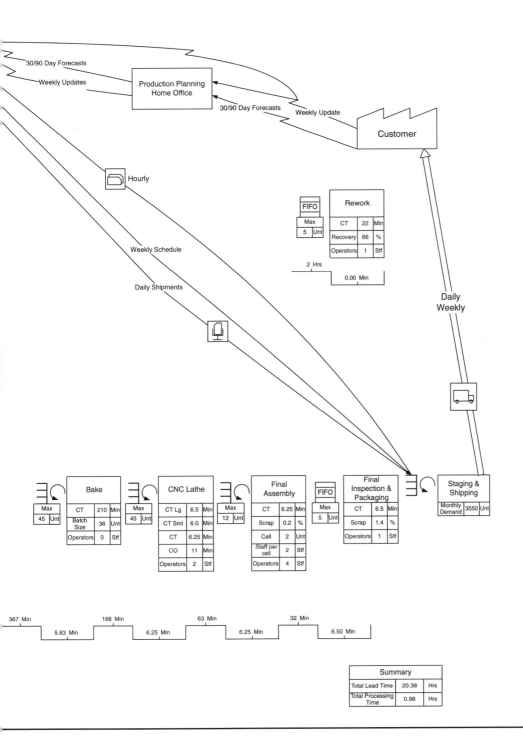

30/90 Day Forecasts

Weekly Updates

Production Planning
Home Office

30/90 Day Forecasts   Weekly Update

Customer

Hourly

FIFO
Max
5 | Unt

Rework
CT	22	Min
Recovery	66	%
Operators	1	Stf

2 Hrs

0.00 Min

Weekly Schedule

Daily Shipments

Daily
Weekly

Bake
Max
45 | Unt
CT	210	Min
Batch Size	36	Unt
Operators	0	Stf

CNC Lathe
Max
45 | Unt
CT Lg	6.5	Min
CT Sml	6.0	Min
CT	6.25	Min
CO	11	Min
Operators	2	Stf

Max
12 | Unt

Final Assembly
CT	6.25	Min
Scrap	0.2	%
Cell	2	Unt
Staff per cell	2	Stf
Operators	4	Stf

FIFO
Max
5 | Unt

Final Inspection & Packaging
CT	6.5	Min
Scrap	1.4	%
Operators	1	Stf

Staging & Shipping
| Monthly Demand | 3550 | Unt |

367 Min   168 Min   63 Min   32 Min

5.83 Min   6.25 Min   6.25 Min   6.50 Min

Summary
| Total Lead Time | 20.38 | Hrs |
| Total Processing Time | 0.98 | Hrs |

Second, we often deal with three types of value stream maps, Present State value steam maps (PSVSM), Future State value stream maps (FSVSM), and Ideal State value stream maps (ISVSM). Each has its own merits and problems.

The PSVSM is just what it says. However, a word of caution: finish them. Very often I see early efforts of PSVSM get derailed by starting to make changes, even before the mapping is done.

Next are the Ideal state value stream maps. Quite frankly they are the best and worst of value stream mapping. They are the best because they ask you to think about the ultimate in waste removal, which will achieve the best percentage of value-added work and the absolute shortest lead times. They are the worst because they are the least defined—and if you are not careful, a waste of time. For example, what technology do you use for the ideal state? Do you have unlimited capital? Can you move the plant? These are serious boundary questions that must be answered. Well, my answer to that is to simply state the assumptions on the ISVSM, and go from there. I have seen a number of seemingly impossible ideas reach fruition because someone spent some time making an ISVSM. The trick is to stretch the boundaries of your paradigms, but still have some possibility of achieving, at least in part, the ideal state.

Finally, the real driver to most improvement activities is the FSVSM. They should not be "ideal." Rather, they should be "achievable," and generally in a reasonably short, say three- to six-month time frame. My experience is that I have yet to see a FSVSM completed as it was drawn. Halfway through the *kaizen* activities, new ideas emerge, a new PSVSM is made—along with a new FSVSM—and the cycle starts all over again.

Third, the VSM is designed to be a tool to highlight activities. In Leanspeak, we call them *kaizen* activities, for waste reduction. Once highlighted, the purpose of a VSM is to communicate the opportunities so they may be prioritized and acted upon. Hence, the prioritization and action must follow the VSM. Otherwise, it is waste just like any other waste.

Fourth, who should make values stream maps? Well, since the purpose is action, those involved in the action decisions need to make the maps—or at a minimum be on the team that makes the map. There is no rationale for having a VSM specialist make all the maps—that is counterproductive. The benefits of value stream mapping do not come solely from the creation of a map. Instead, they come from the interaction of the people making the maps, with the process and making the observations on the floor, which are necessary to gather the information for the value stream maps. In short, there needs to be a management presence in the value stream map construction process.

Fifth, how do I start to make my VSM? Well, walk the value stream, starting closest to the customer and working backwards up the value stream.

Sixth, should these VSMs be handmade or computer generated? I make mine by hand. Most serious practitioners make them by hand. There are a few advantages of having them on software, such as transportability, quick update on calculations, and easier use in presentations. However, what I have seen is that the people then tend to get stuck in the office, which totally undermines the benefit of the VSM. On balance, I believe handmade—in pencil—is a clear choice.

Two value stream maps are included. The PSVSM and FSVSM from QED Motors. Read the story of QED Motors in Chap. 16 for all the details and see if you can make the map from the process information … everything is there.

CHAPTER 8

How to Implement Lean—
The Prescription for the
Lean Project

In this chapter, we will describe how the various evaluations will create a comprehensive list of action items that can then be turned into a project for completion. This is the implementation plan.

> **"F**ailing to plan ... is just planning to fail. **"**
>
> Unknown

An Overview on How to Implement Lean

The implementation of our Lean initiative is outlined here for ease of understanding. It will be managed and executed like any large project. The action items for this project will be compiled from two sources. The first set of data is gathered from the "systemwide evaluations," while the second portion will be from the "specific value stream evaluation." Both sets of evaluations will be compiled, prioritized, and lead to action items: *kaizen* activities. Together, in an eight-step process, all the evaluations and action items will be included in our plan and documented on some type of project documentation, whatever you are most comfortable with. We normally use a standard Gantt chart. *This is THE PRESCRIPTION on how to implement the Lean project.*

Steps 1–3: Systemwide Evaluations and Action Items

1. Assess the three fundamental issues to cultural change.
2. Complete a systemwide evaluation of the present manufacturing system outlined in Chap. 19.
   a. The Five Tests of Management Commitment to Lean Manufacturing.
   b. The Ten Most Common Reasons Lean Initiatives Fail.
   c. The Five Precursors to Implementing a Lean Initiative.
   d. Process Maturity.
3. Perform an educational evaluation of the workforce.

Steps 4–8: Specific Value Stream Evaluations and Action Items:

4. Document the current condition of the value stream.

   a. Prepare a present state value stream map.

5. Redesign to reduce waste. (This is simply a summary of Chapter 7.)

   a. Prepare a future state value steam map that will:

      i. Synchronize supply to customer, externally.

      ii. Synchronize production, internally.

      iii. Create flow.

      iv. Establish pull-demand systems.

   b. Create a spaghetti diagram.

   c. Show all *kaizen* activities on the Gantt chart.

6. Evaluate and determine the goals for this line:

   a. Determine critical process indicators.

   b. Set specific goals for this line/product. (Goal #1 is to protect the customer.)

   c. Document all *kaizen* activities found in this analysis on the Gantt chart.

7. Implement the *kaizen* activities.

   a. First implement finished goods inventory controls to protect the supply to the customer.

   b. Implement your *jidoka* concept.

   c. Implement all other *kaizen* activities on the Gantt chart.

8. Following the changes, evaluate the new present state, stress the system, and then return to step 4.

The Implementation Plan Will Be Documented on a Gantt Chart

We have spent little time in this book on project management. Since this book is written to those in manufacturing, like engineers, performing projects is something we assume you are already skilled in. If you are not, a number of excellent references are available. I recommend *Project Management DeMYSTIFIED* (McGraw-Hill, 2004) by Sid Kemp. At a minimum, your Gantt chart—or project management tool, whatever it is—should have certain data. For example:

- Show all *kaizen* activities.
- Show key milestones.
- Show completion dates and responsibilities.

A Key Question to the Implementation

Value Stream by Value Stream or Facility-wide?

To install a Lean manufacturing system for all value streams in a large complex plant simultaneously is very difficult. Although this "clean sweep" approach is surprisingly popular, it usually is an inferior technique to one-value-stream-at-a-time implementation. The "clean sweep" has a couple of merits that I will discuss in a minute. However, there are at least four very large reasons to implement one value stream at a time.

- The initial learning curve is very steep, and no matter how well prepared you are, there are many unforeseen results. Sometimes they are positive results, other times negative, but it is best to experience these issues on a smaller scale and learn from them. This learning from the first line is invaluable in executing the conversion on the subsequent lines.

- There is the matter of resources ... Specifically trained eyes and ears to evaluate implementation, evaluate progress, and assess problems on the value streams being converted. No one has a plethora of such people and they are invaluable. These few people can move from line to line with the implementation and provide expert advice and training so the gains may be sustained.

- Each change has an effect on subsequent and past changes. It is not uncommon to Lean out Line A and then move on to Line B for its implementation and then find some aspect of Line B implementation that has an unexpected effect on Line A. Sometimes the effect is a positive one, other times it is detrimental, but either way this approach allows you to better understand these effects so they may be exploited more fully on all lines .

- I have seen a number of line-by-line efforts progress quite nicely, but all the global implementation efforts I have seen have struggled to meet their objectives. A common postmortem comment on a global implementation effort is "In hindsight, we should have done this one value stream at a time."

There may be a few cases where plantwide implementation is the best choice. I can think of two.

- In a few cases I have seen where the customer has decreed that they will only work with facilities that are Lean, top to bottom.

- The second case is rare but there are some operations where the manufacturing is trivial compared to the overall operation. Packaging is one such case I have encountered. In this instance, 19 different products were packaged and shipped to different countries. Hence, the packages were labeled in many different languages. The net effect was that there were over 400 individual part numbers. The manufacturing was not complicated. Rather the complications, and hence the variation, and hence the waste, was in the interaction with the storehouses. First, there were the issues of pickup, delivery, and movement of raw materials from the raw materials storehouse to the production lines. Second was the handling of the over 400 part numbers on the floor and in the finished goods warehouse. In this case, the handling of the storehouses was the key issue and the implementation had to be done globally.

In the absence of these two items, I can find no reasons more compelling than the four mentioned earlier: learning curves, resources, lessons learned, and historical success. Consequently, we have found it usually the best choice to implement one value stream at a time.

Step 1: Assess the Three Fundamental Issues to Cultural Change

Do We Have the Leadership to Make This a Success?

The first step is to select the leader for our initiative. If that person is probably you, that step is done. If it is not done, carefully select a leader, taking into account all the information on leadership detailed in Chap. 6.

Next, you will need to find a *sensei*. The leader and the *sensei* are far and away the two most important people in the effort. All too often, some high-level manager appoints someone to lead the effort and expects him to act as *sensei* also. In our experience, they do this because good *sensei* do not come cheaply. Quite frankly, a good *sensei* is the best bargain you will ever have in this Lean effort. His experience is invaluable. He will guide you away from failure and toward success. Many of the paths you would like to take in a Lean initiative are counterintuitive and most managers reject some important ideas because they simply do not know this. For example, the order in which you attack the transformation is of the utmost importance. Many managers want to dive in head-first and begin to reduce inventories. Almost always that is not only a bad idea, it is counterproductive. Your *sensei* will make sure the path you take is not only logical but is the best path for your particular situation. I have yet to see any Lean initiative that was a carbon copy of a previous one. They each have their uniqueness, which may be unnoticed by the novice.

In addition, the *sensei* is a set of eyes with some "distance" from the inner workings of the plant. This distance will give him objectivity that is invaluable when any evaluations are made. Just as there is no substitute for good leadership, on this journey there is no substitute for a good *sensei*. Choose carefully and don't worry about the costs for these two people. If you have chosen well, this will be your smartest investment, without question.

Do We Have the Motivation to Make This a Success?

Motivation is a complex cultural issue, so we will not delve into it deeply at this point. However, most managers can determine, at least at the gut level, if the organization is properly motivated for this initiative. Two things must be done, at a minimum.

First, one of the great motivators is "to be in the know"—that is why rumor mills are so popular. So, to "get them in the know," the Lean effort will need to be publicized and all that Lean is should be open and discussed. That puts a stress on a lot of cultures, which are very closed. Characteristics of closed cultures include lots of secrecy, even about the most mundane of topics, and managers who do not appear to be forthright. For example, in these closed cultures, many questions get avoided and there is an unstated but understood air of secrecy that forbids some questions being asked. Those elements will destroy a Lean initiative, so they must be avoided. So tell them what is going to happen, when it is going to happen, and what you expect the results to be. Be sure to get them "in the know," and above all else be open and honest.

Second, for the group to be motivated, the leaders must be motivated. This motivation must be obvious by their actions. If the leaders are not motivated—not just motivated, but literally exuberant about the future of this Lean initiative—either you have picked the wrong people or they simply do not know the power of a Lean initiative. Quite frankly, in the world of manufacturing, there is no cultural change initiative that can create so much hope as the proper implementation of Lean!

Do We Have the Necessary Problem Solvers in Place to Make This a Success?

Do we have enough and do we have the needed level of training? Make sure you have earmarked the problem solvers and they know their roles. It is common to have a small cadre dedicated to problem solving under some sort of Lean support staff. Also, generally assigned to each value stream are what most organizations refer to as process engineers. Train and use these two groups fully. If they are not fully trained, have a plan to get there. It is very common to have a lot of the training done by the *sensei* as you progress. I call this JIT training. Some offsite training is also common. Whatever the current status, this is a critical aspect because the initiative will only proceed as fast as you can solve the problems—and the problems will come rapidly as soon as you start making the big changes when the four strategies to become Lean are installed on the first value stream.

Step 2: Complete a Systemwide Evaluation of the Present State

The Role of the *Sensei*

Now it is time for the leader and the *sensei* to make an evaluation of the present state of the entire manufacturing system. They will make four evaluations of the manufacturing system. They are:

- The Five Tests of management commitment to a Lean Manufacturing
- The Five Precursors to Implementing a Lean Initiative
- The Ten Most Common Reasons Lean Initiatives Fail in part or totally
- Process maturity

In this step, the *sensei* is absolutely critical. This step cannot be done by a good manager who understands Lean techniques. This must be done by a seasoned veteran in Lean applications. If you, for some odd reason, have decided not to employ a *sensei* and are doing this with a seasoned manager with solid engineering skills, but who is not seasoned in Lean, go hire someone with experience to assist you. I have never seen a good plan come out of anything but a person seasoned in Lean.

In the absence of a *sensei* or a consultant, *quit this effort and move onto something else.* I cannot state this more clearly. The likelihood of partial failure is 100 percent, and the likelihood or total failure is significant. The bottom line is, management is not committed (they will fail questions 2, 4, and 5 in the commitment test).

Well, now that you have either a *sensei* or at least experienced help, let's proceed to the assessment. In Chap. 19, I have included four assessment tools. I will explain each. The four assessment tools are:

The First Management Commitment Test

This management commitment test is the first and most important of the commitment evaluations. The key questions are, "Do we have the level of commitment necessary to make this a success?" And, "If we do or do not, what are we going to do about that?"

Before we discuss the evaluation, let's explore exactly what commitment is. (In Chap. 14, we have the story of the Alpha Line, which is a good example of management commitment.) A lot has been written about commitment and many confuse it with involvement. Those walking around, making casual comments, and even helping out are involved. Commitment goes much deeper. The best description I have ever heard uses the metaphor of eating ham and eggs for breakfast. In this case, we can say the chicken was *involved*; the pig, however, was *committed*.

Now, back to the evaluation. This can be done two ways: quick and dirty or thoroughly. I suggest you do both, for different reasons and at different times. First, you and your *sensei* should do a quick-and-dirty evaluation:

- List the key players who are necessary to make this initiative work.
- Make the best evaluation you can using The Five Tests of Management Commitment to Lean Manufacturing.

Make sure you do your evaluations *based on the behavior you see* from the key players. Don't just go by what they say. At this point, this evaluation may be very rough but it will point out any obvious issues and will get you thinking about this issue. At some point, or at several points, individual or group commitments, or a lack thereof, will be significant issues, and it is best to flush them out early. So an early awareness of problems is a good first step.

From this evaluation, develop potential action items. At this point, this is just an opinion, so treat it very carefully. As you proceed in the early steps of the initiative, take careful note of any commitment issues. Be complete and document the who, what, when, where, why, how often, and how much, for each example. However, unless there are obvious and very serious problems, take no action on this list of items at this time. Save it for later.

The Second Management Commitment Test and Group Dynamics

Next, do a thorough evaluation of the management commitment—this can not be done initially. For this second evaluation to have significant meaning, there must have been some context developed so the key players can more fully understand what types and levels of commitment the Lean initiative will require. It is best to do this evaluation after the key players have been trained in your Lean initiative. Also, there must have been several major changes made in the basic operation so the key players can understand, firsthand and by experience, the type of changes required. Generally, this can be done as early as three months after kickoff, or as late as 18 months afterward.

This is a crucial time in the program. For the first few weeks of program implementation, particularly if the initiative was sold well by the Lean leadership, there will be a lot of positive energy and the majority of those involved will be very enthused about the initiative and its future prospects. This is how nearly all groups form and develop—lots of positive energy and seeming agreement early on. Life seems pretty great during this phase, which we will call the "cocktail party phase."

Unfortunately, some time after starting the program, some tensions develop due to the necessary and resultant changes. This tension is unavoidable. What happens is that people begin to jockey for position, reassess their positions, and question things like, "How can I do this and still meet my goals?" In short, individual goals conflict with group goals and differences surface. Even if they do not, the worry exists that they might. This creates some conflict that was not present on day one. What started as a conflict-free group with seemingly common goals and a large amount of agreement is now breaking down.

It always happens. It is not only inevitable, it is necessary. This is the second phase of group development: let's call it chaos (a term coined by M. Scott Peck). This chaos has developed because the realities of life have been imposed on the group, and differences that they initially did not envision are popping up all over the place. These differences are now creating problems. Furthermore, having just left the cocktail party, so to speak, they don't really know how to handle the problems.

This is the earliest time we would want to do the second evaluation. Without this chaos and some understanding of the chaos, there is no real context to have a deep and meaningful evaluation. That chaos is often felt as early as three months after kickoff, but more typically in the 6- to 18-month period afterward. It is at this time that it is appropriate to do the second evaluation of management commitment. This will be a very sensitive evaluation and it will require good planning by both the Lean leader and the *sensei*. All too often, because it is sensitive and brings up problems most groups work hard to avoid, it is frequently avoided. Doing this evaluation will be a test of the courage, character, and resolve of the Lean leadership and the facility's management.

Just so you are aware, two more phases of group development exist. These are called forming and performing. In forming, the group must confront the differences, and then by using honest open communication and problem solving, they can resolve the differences. This is easier said than done, however. After accomplishing this stage, the group can then move into performing. In the performing phase, the group will be able to effectively and efficiently execute its mission. (There is some very good information available on group dynamics. I recommend *The Different* Drum (Simon and Schuster, 1987) by M. Scott Peck and *The Team Handbook* by Peter R. Scholtes (Joiner Associates, 1988) . Both are excellent.)

Since the second commitment evaluation may be more than a year off, I have attached it as Appendix A at the end of this chapter.

The Five Precursors to Lean

Background
The Five Precursors to Implementing a Lean Initiative have an interesting history. In the late '80s and early '90s, when the first U.S. firms were implementing JIT (Just In Time) systems, a large number were encountering problems. Many firms were able to reduce inventory volumes significantly but often other problems developed. Frequently, the production rate would drop—this was the worst and also the most common of the problems. Other less serious issues cropped up as well. Once these unexpected problems surfaced, we would then be asked to assist these firms as they tried to work out of the JIT mess they had so carefully managed themselves into. After a few experiences, we found that the reasons these groups had failed could be classified into a few categories. In addition, about this time Ohno's book and several others became available, which more fully explained the TPS (Toyota Production System) and the JIT portion that so many firms were trying to copy.

Upon reading Ohno's book, and with further study of the TPS, it became obvious that the TPS was vastly superior to the production system in most North American firms, even if these North American firms had already implemented the JIT system. We quickly realized that the majority of the problems were related to this fact—that is, the TPS is a superior manufacturing system, with or without JIT. We also concluded that when Ohno began his quantity (note: that is QUANTITY) control and seriously undertook his JIT effort (in about 1961), his production system was superior, at that time, to the manufacturing systems we are working with now, nearly 30 years later. This was not only sobering to most, it was also depressing.

We understood the problem and began preaching that the failing of a JIT effort was not due to the JIT effort itself, but the fact that the facility did not have the needed foundational elements in place to start a JIT effort. These needed foundational elements we named "The Five Precursors to Implementing a Lean Initiative."

Today we find, without exception that all companies must work on one or more, or often all five, of these precursors to have a fighting chance of implementing quantity control measures. However, these issues need not be totally corrected before quantity control can be initiated—often they can be done simultaneously. Let's say, for example, we have a process that has very high variation in hourly production. Although the stability needs to be addressed and it needs to be one of the first things done, we have seen instances where the solution to the rate variability problem was solved by a *kanban* system. So voilà! We get better stability and quantity control at the same time. This is something that is frequently not seen by the novice, but your *sensei* can give guidance as to:

- Which precursors need to be addressed
- What order these precursors must be addressed
- What amount must be addressed

Each of the five precursors should be evaluated and the results of this evaluation will often become a significant part of the implementation plan. A description of the five precursors follows. Refer to the Matrix in Chap. 19 for clarifications.

Stability and Quality

High levels of stability and quality in both the product and the processes are the most basic of standards. Ohno says that flow is the basic condition. It is the foundation of the Toyota Production System (TPS). He says this only because he takes it for granted that process stability is a given. (Return to Chap. 3, and reread the section "It Is Not a Complete Manufacturing System.") Absolutely nothing is more basic to quantity control than process stability. So it is necessary to review all aspects of the product and process stability and make a list of items to include in the goals of your Lean initiative. First, evaluate and make sure the quality and the production rate of your product and processes are statistically stable. This can easily be done on a simple control chart—most often an Xbar-R or an XmR chart. Check the stability of the production rate: day by day, shift by shift, and hour by hour. If there are instabilities, put them on the list of items to address. For both the product and the process, check each quality characteristic for both stability and levels; evaluate both. Each can be checked using control charts. List all product and process quality characteristics that are not statistically stable. If they are variables data, list all those that have Cpk below your threshold value, usually most organizations start with a minimum of 1.33. If they are attribute data, work to improve the process to eliminate the need to do the evaluation. If this is not practical in the short term, work to correlate the attribute characteristic to variables data, and then strive to reach the threshold Cpk for this as well. If there are processes that do not meet Level 2 criteria, this constitutes a crisis and should be addressed immediately. The minimum goal for nine months is to have all product and process quality levels to Level 3 on the matrix; Level 4 can be a goal for 18 months.

Machine and Line Availability

Excellent machine and line availability is frequently a very large problem that has gone unattended for years. For those companies without a formal TPM (Total Productive Maintenance) program, it is common for this factor to reduce OEE by 25 percent. In practical terms, this means we must run the line 25 percent more than cycle time would predict, with all the attendant costs of running the line. This problem alone will frequently make a product unprofitable. Usually low levels of line availability are due to two major factors: materials issues (usually stock outs or late deliveries) and machine downtime. Frequently, materials issues can be alleviated by a Lean initiative and the quantity control aspects of the TPS. However, machine downtime is different and usually must be addressed by a concerted TPM effort. If a TPM initiative is not already in place, almost all firms need to develop a new database to keep track of machine uptime and train the personnel in the use and manipulation of this database. There is some commercially available software for TPM and machinery uptime, although I find most can develop a good Excel spreadsheet and make it quite serviceable. Again, if you find in the evaluation that Level 2 criteria cannot be met, this is a crisis requiring immediate action. Regardless, there is usually a large list of materials and machinery issues that are created by this evaluation. The goal to reach Level 3 should be set for six months, with Level 4 scheduled to be attained in one year.

Problem Solving Talent

In several sections we have discussed problem solving and the need for problem solvers. The need for these skills cannot be underestimated; it literally is the vitality of the Initiative. Furthermore, to complete this five-part evaluation, the skills of MSA (Measurement System Analysis) and SPC (Statistical Process Control) are required. Hence, anything less than Level 2 is an absolute crisis and must be corrected. If these skills are not available onsite, hire trainers right now. A six-month goal is to reach Level 3, while Level 4 should be met by 18 months.

Continuous Improvement Philosophy

Mature continuous improvement philosophy is something that is often talked about but infrequently reduced to a process that can be taught to and understood by all. A copy of one we created is shown in Fig. 4.1. If the evaluation is less than Level 2, we again have a crisis and it needs to be taken care of immediately. Reaching Level 3 is a reasonable six-month goal.

Standardizing

Strong proven techniques to standardize is the foundational of all foundational issues. This skill is so important it received its own chapter in this book: "Sustaining the Gains." Anything less than Level 2 is an immediate crisis and must be addressed. Reaching Level 3 is a reasonable goal for 18 months.

As part of this evaluation of the Five Precursors to a Lean Initiative, a list of issues must be compiled. This wish list will usually make up the majority of the effort for the first pass of the Lean Initiative. After the first six months, more and more quantity control efforts can be implemented, but in our experience most firms have a number of foundational issues, which must first be addressed.

The Ten Reasons Lean Initiatives Fail

These ten reasons listed in the evaluation in Chap. 19 were developed from actual data I have accumulated over the years, and I include them here for your reference and use. I suggest you go through this list with your *sensei* and address each for your situation. Take notes and develop action items.

Process Maturity

The Process Maturity document is included so you might analyze the issues with your processes with more specificity. It is a refinement of the first of the five precursors: high levels of stability and quality in both the product and the processes (it contains more specific auditing criteria). The vitality of your initiative will be highly dependent upon your ability to improve your processes. This document is designed for just that purpose.

Document the System Evaluation in the Plan

These four evaluations have given you a rather large group of action items that will generally fall into one of three categories.

- System action items
- Educational and training action items
- Line-specific action items

What Is Next?

In about two minutes we will do a more thorough educational evaluation and a series of value stream evaluations. This will add considerably to your list of action items. These action items now form the nucleus of your implementation plan. As you recall, it is simply a project—something we engineers are good at accomplishing. To execute this project, the first and most important tool is, guess what? The plan—the first of the requisite tools of a leader. Your Lean implementation plan will normally be expressed in the form of a Gantt chart. I normally use MS Project. It works nicely, but I am sure others out there are equally good.

You can now start your Gantt chart. Fill in the System, Educational, and Line by Line Action items you have from the System evaluation.

Step 3: Perform an Educational Evaluation

Formally Introducing the Issue

A formal introduction is often a key to getting started well. Remember, the second requisite skill of leadership—the ability to articulate the plan so all can understand it. It is worthwhile to tell the entire facility, "We are going to make a change and that change is to implement the concept of Lean manufacturing."

Many facilities make this a monster effort, with special invitations, a formalized meeting attended by all the top management, coupled with meals and motivational speakers galore. I find this degree of effort is not needed. In the end, the most important aspect of selling the issue of changing to Lean is dependent upon the continued actions of the Lean leaders and top management. If they talk Lean and do not walk the talk, no amount of up-front selling will work. On the other hand, if the Lean leaders and top management do really walk the talk, then no large selling effort is required. Either way, a mega-effort at selling is generally a waste, hence I do not recommend it.

Rather, I recommend you have two types of Lean introductory training sessions, as described in the following paragraphs.

First and foremost, an introductory session to management will need to be prepared. This group should be the key decisionmakers, usually what is referred to as "top management." This first training is the classic "Who, What, Where, When, Why" training with special emphasis on the: "What," which is the House of Lean that was created with the help of your *sensei*; and the "Why," which is the motivating force behind the effort. This training is often more than a one-day event. It could be the basis for a two- or three-day retreat for the management team. This group needs to especially understand the House of Lean. The training should include exercises on variation reduction (the dice game in Chap. 19 works nicely), *takt* calculations, OEE calculations and line balancing (for these, my Lean Kit is a great tool and is available at my web site: www.qc-ep.com). This training session should be hands-on and instructional, and it must take into account both the present state of the facility and the desired state of the facility. Change, including the types and amounts of changes needed, should be openly and honestly discussed. At this point, it is unlikely that all four of the systemwide evaluations of the present state have been done, but still your Lean implementation manager and your *sensei* will be able to quantify this concept for the purposes of this training.

The second type of training is informational in nature and should be given to all employees on all shifts. It should first be given to those reporting to top management, next supervisors, and then to the general plant population. This is typically a PowerPoint presentation describing the House of Lean and the implementation schedule. Including ample time for a Q&A session, this usually takes about four hours, and groups can be as large as 40 and still be effective.

Specific Skills Training

The systemwide evaluations almost always create a very large list of needed training to teach the strategies, tactics, and Lean skills. The composite list will largely follow directly from the Five Precursors to a Lean Initiative, which have already been added to your Gantt chart. In addition, as you do the assessment of each value stream, training topics will almost surely be found. Once combined, these trainings will almost always include problem-solving training, training on statistical tools, and facilitation training for all in leadership positions. In addition, Lean-specific trainings are available in skills such as line balancing, SMED methodology, *takt*, and *kanban* calculations, to name just a few. It will be necessary to inventory the needed skills and teach them as they are needed to those using the tools.

Just another word on education and training: It should be focused and JIT. For example, during the implementation, if you choose to change the plant one value stream at a time, train just those people involved. Often, it is not that simple and some people may need to be trained prior to the implementation of their product; it is never perfect. The point here is to avoid the global mass training of individuals that makes good use of the training resources, yet provides the training either too early or too late. Efficiency of the training organization is not of paramount importance when compared to training effectiveness. If there is long time between the training and the implementation of that training, a large fraction of the learned material is forgotten. Consequently, it will not be effective and it is then waste, the very item we are trying to eliminate, not create.

At this point, you can also add these introductory training items to your Gantt chart.

Step 4: Document the Current Condition

Preparing a Present State Value Stream Map (PSVSM)

This document will be used to gather current information of the present state conditions for the entire value stream. This will be a door-to-door PSVSM—that is, we will start at the shipping dock and document the value stream up to the raw materials supply.

Step 5: Redesign to Reduce Wastes

Prepare a Future State Value Steam Map (FSVSM) That Will:

- Synchronize supply to customer, externally.
- Synchronize production, internally.
- Create flow (including the *jidoka* concept).
- Establish pull-demand systems.

This will analyze current conditions and redesign the process flow to eliminate waste. Refer to Chap. 7 for the specific technical details of this step.

Creating a Spaghetti Diagram

This diagram will show the movement of the assembly as it is constructed, and show the movement of both the people and the product. Work to reduce the movement and transportation wastes and free up floor space. Do this on a plot plan, made to scale.

Document all *kaizen* activities determined in Step 5 on your Gantt chart.

Step 6: Evaluate and Determine the Goals for This Line

- Determine critical process indicators (for more details on goals, see Chap. 9).
- Set specific goals for this line/product (goal #1 is to protect the customer).
- Document all *kaizen* activities found in this analysis on the Gantt chart.

Step 7: Implement the *Kaizen* Activities

- Implement finished goods inventory controls to protect the supply to the customer.
- Implement your *jidoka* concept as defined in Step 5.
- Prioritize and implement all other *kaizen* activities on the Gantt chart.

Step 8: Evaluate the Newly Formed Present State, Stress the System, Then Return to Step 1

Some Clarification on Step 8

As part of the project prescription, you will:

- Evaluate the newly formed present state.
- Stress the system.
- Return to Step 1.

To make a system Lean is a never-ending process. Each change brings about a new present state that then gets evaluated for improvement activities, which creates more changes and the cycle starts all over again. On many occasions, the system will stress itself through the unexpected appearance of quality or availability problems, for example. Sometimes demand changes will put a stress on the system. All of these are opportunities to improve the robustness of the system. Although it sounds a bit crazy at first, it is wise to stress the system yourself to see what other process opportunities may be present. A typical "stressor" for the system would be to remove a few *kanban* cards and see what the system response will be. The primary tool you will use to protect you from system failure will be system transparency. Remember when we said we need to create a culture that embraces change? This may be the clearest manifestation that we have changed the culture, when we start stressing the system to make it better. Recall the metaphor of the athlete in Chap. 6. How did he get better? Isn't this the same concept?

Lean Goals

Lean goals (Step 6, item 2) are always an interesting topic. The name Lean came about because, in the end, the process can run using less manpower, take less time, consume less space, and use less equipment and material investment. So often when evaluating the success of a Lean initiative, these terms are used and calculations of space utilization, and even distance traveled, are used. In the long run, these are not very meaningful measures since they typically are not a good subset of the plant goals, nor do they readily translate into key business parameters such as profits or return on investment (ROI).

Most plants already have good measures of manpower utilization. For the other Lean measures, the ones that typically get woven into the general plant goals are inventory management measured as inventory turns, and the lead time, measured as manufacturing lead time. If OEE (Overall Equipment Effectiveness) is not already a line goal, this is clearly to be added. In total, there should be five to seven goals that are the metrics to measure how the line will supply the product, with better quality, with shorter lead times, and do so less expensively.

What about Goals for the Lean Initiative Itself?

We do not favor any specific goals here beyond the goals of the project that are included in the schedule. This is for two very sound reasons:

- At some point these goals will clash with the plant goals. It is best to simply weave them into the plant goals. If the plant goals do not reflect the need to be Lean, *change the plant goals.*

- We want to make as much effort as we can to weave the Lean initiative into the normal workings of the plant and not make it a "New Thing We Do." Rather, it should not be a new thing, but a *new way of doing the things we need to do.* We want to begin immediately weaving Lean activities into the culture, which will start the needed cultural change to sustain the gains. There is no better point to start than right here.

What to Do with the Plan?

Management Review

The plan needs management review, discussion, and acceptance. This should be done in a formal meeting. This formal review is done for four reasons.

- It will show, in one document, what is going to happen and when.

- It will give top management, the movers and shakers, an opportunity to see the entire effort. They can see and comment on those things in their areas of responsibility and also those changes outside their areas, but these changes still might affect them. In short, they will have an opportunity to bring up questions.

- Any plan includes the topics of objectives, timing, and resources. This meeting will allow a check on not only those three topics, but their interrelationships as well.

- How they respond to the plan will be a reality check on the commitment of the top management. This is most important.

It is necessary to make sure, at this meeting, that everyone understands that the next step is implementation. You want to leave the meeting with the understanding that the top management understands and will support the plan, because in five minutes you will implement it.

Publish and Follow-up

Immediately following the meeting, publish the plan and put it into action.
    Let the Fun Begin!

Chapter Summary

The book *How to Implement Lean Manufacturing* is summarized in this chapter. First, we make evaluations using the following tools:

- The three fundamental issues to cultural change, outlined in Chap. 6.

- The fourfold evaluation of the present manufacturing system, including the commitment evaluations, five precursors to Lean, ten reasons Lean initiatives fail, and process maturity (found in Chap. 19).

- The educational evaluation of the workforce.

- Specific value stream evaluations, as detailed in Chap. 7.

These evaluations, and the countermeasures, then create a huge list of *kaizen* activities that can be included in a Gantt chart or an appropriate project planning and tracking tool. We can then evaluate and determine completion dates for *kaizen* activities in the project, set specific goals for the value stream, prioritize the activities, and implement the *kaizen* activities. Almost always, our first two goals will be to implement finished goods inventory to protect the supply to the customer and to implement our *jidoka* concept. Finally, we will present this to management for review, and once reviewed and accepted, we will begin to immediately eliminate the waste.

Appendix A—The Second Commitment Evaluation of Management Commitment

The Process of the Second Commitment Evaluation

The second commitment evaluation should be done in a facilitated session, which your *sensei* should be capable of facilitating. It is not unusual to take a full day for this evaluation and problem-solving session.

As preparation for this session, make sure each manager has a copy of The Five Tests of Management Commitment to a Lean Initiative. In addition, ask them to do a personal evaluation, and also an evaluation of the management group as a whole. When you ask them to do these evaluations, tell them the results from the tests are personal and will not be shared.

What you want to do is uncover the issues, frame the issues into problems, and begin the problem solving to eliminate the issues. Actually, most of the issues will be known, but most will be difficult for people to discuss and bring out in the open. We find that once the problems are on the table, 70 percent of them solve themselves. However, you will also find that some problems are more difficult to solve.

The agenda could be something like:

1. Group icebreaker on teamwork.

2. Break into small groups of five to seven and ask each group to answer the question, "What is commitment?" You and the *sensei* should move from group to group to make sure they are on track. Give them about 15 minutes for this exercise. This is for the small group's purpose only.

3. Follow this by bringing everyone together in a group discussion facilitated by the *sensei* about "How do we measure commitment?" This must be done in a "spin-around" brainstorming session using strict brainstorming rules. Document this discussion and all subsequent steps on flipcharts that you will post on the walls of the meeting room. Beyond posting the flipcharts, nothing more needs to be done with this session. Very likely, you will want a break here.

4. Again facilitated by the *sensei*, do a brainstorm of, "What are our problems with commitment?", or tackle a similar question. Document all the comments, concerns, and issues on flipcharts and post them. Make no effort at this point to discuss, validate, or reduce the list in any form. Just let it sit there. I cannot

emphasize those last two sentences enough. This step, to be effective, must be totally nonjudgmental.

5. A group exercise on values may be appropriate at this point. Many exist. I like the "Lifeboat Decision." In this story, ten people have clambered onto a lifeboat, but it will only hold seven. Thus, three must be thrown overboard and they will die. If you do not throw three overboard, all ten will die. For instance, in the boat are a young child, a blind man, a priest, a prostitute, a mother and her baby, a grandmother, a convicted killer (who is strong and muscular) ... well, you get the picture. There are no right answers. It just forces the group to discuss values. It is a great exercise.

6. At this point, if there are many problems on the list you made in step 4, discuss each item in turn until the group is reacquainted with the list. The discussion must be "to a point of understanding"—we *still* do not want to pass judgment on the projects. We only want to understand the context of the problem.

7. If the list is large, we will need to select the most critical problems. It is likely that we have a number of very good projects, so almost any from the top of the list will be productive to solve. To find a few of the better projects, use nominal group technique or multivoting to select say the top four or five. Once this is done, we need to select just one of these projects to work on. Probably the *sensei* could say something like, "We will eventually solve all these problems, but for the first problem I would like to see the group solve the problem of..."

8. For the first problem, brainstorm. "What are the thoughts, concerns, issues, and so forth about this problem?" When the group has all the issues down on paper, do not reduce this list. Instead, proceed to the next step.

9. For this first problem, brainstorm again: "What are the possible solutions?" When the group has all the possible solutions down on paper, do nothing more with this list for the moment. Instead, proceed to the next step.

10. Conduct another brainstorming session on "What are the key criteria we will use in our decision-making process?" Discuss this and reach a consensus. This second step of reaching consensus on the criteria is often difficult. If you are not familiar with consensus, it is not disagreement, nor need it be 100 percent agreement. It is the concept whereby everyone involved can say, "I may agree with the group decision or I may not agree with it, but regardless, I recognize it is in the group's best interest and I will give it my 100 percent support and commitment." Like I said, it is not easy, but to reach consensus on this step is crucial.

11. Select the best of the possible solutions, and then develop action plans with responsibilities and due dates. Very often after the consensus on the criteria is reached, the process moves very rapidly.

12. Return to step 7, select the next problem, and move on.

It is important that the group completely resolve at least one problem. If they can and want to take on more than one problem, that is even better. However, here the major benefit comes not from the problem solution itself. Instead, it comes from the process of solving the problem. The Spin-Around technique requires:

- Good listening
- Understanding problems from different perspectives

- Nonjudgmental discussion
- Introspection
- Patience and respect shown by all, to all

Frequently, these are behavioral traits that are not found in abundance in the typical manufacturing plant environment. Consequently, and nearly always, the results of the process are more both important and more lasting than the actual problem that was solved. Following this, facilitated spin-around—the managers will be better equipped to work together to solve their problems. In addition, they now have a behavioral model they can take back to their individual groups to use in resolving their own internal issues.

It has been my experience that this evaluation is a very sensitive one—everyone thinks they are committed, but this is simply not the truth. I wish I could give you a prescription on how to do this comfortably, but I can't. The best advice I can give you is to do it—but do it carefully.

Simply because there is a possible downside is no reason to avoid it—yet avoid it is precisely what most people do. Unfortunately, when either fear or denial sets in and begins to rule the culture, the progress stops and the end is in sight. There is no substitute for simply fighting through these two problems of fear and denial, because they will appear again and again. Many of these issues test the courage and the character of the Lean initiative leadership. If they waver, the effort will suffer.

There is a wonderful quote from a movie where the protagonist, who is only 17, has thousands of dollars of video and sound equipment that he purchased with the profits from his marijuana sales. When he was asked by a friend if his father knows how he financed the purchases, he says, roughly, "My Dad thinks I can afford this on my minimum wage job," and then adds, "never underestimate the power of denial." This is true of denial, and the same maxim applies to fear as well.

Both fear and denial are two extremely powerful detractors that will rear their ugly heads time and again as you pursue this journey into Lean. The leadership has to be aware of these issues and must handle them in a professional and open fashion. This is necessary for the success of your Lean effort … well, for any effort you might embark upon.

CHAPTER 9

Planning and Goals

I have included this chapter on planning and goals for two reasons. First, for your Lean initiative you will need to develop both an implementation plan and very likely a plan beyond that. However, I have included far more information than is required to accomplish your Lean objectives. I do this because all too frequently I find that goal development and planning is a huge weakness in many businesses. All too often I find it is only a superficial perfunctory exercise—some do it better; yet very few do it really well.

Of all the tasks of management; I can think of none that take so little time to do well, yet are so powerful in terms of driving the plant to higher performance levels. In addition, I find that the completion of plant goals is not only beneficial for the performance of the plant, I find it is beneficial to the individual for several reasons. Plants with well thought-out goals guide employees better, so they proceed with more comfort and confidence in their day-to-day activities. Well thought-out goals also provide a future state—in a word, hope, for a plant. For these reasons, goal setting and goal deployment have become a passion of mine—and as you might expect, there is some background behind that statement.

Some Background

As a young engineering manager, I was assigned the job of managing a group of engineers whose task was to design and install numerous capital projects. In fact, at the refinery where I worked, there were several such groups. As luck would have it, one day a man who chose to be my mentor, out of the blue asked me about a project in his area: "What's your plan, man?" So I briskly took out our construction schedule to review it with him. At this, he snapped at me and told me in no uncertain terms that all construction schedules at this facility were crap. He was both more, uhm, "verbose" and more graphic in his terminology in relaying this to me. He went on and said, accurately so, that the primary purpose of our schedules was to publish something so we could later revise it. There was no real expectation by anyone, including the author, that the target dates would be met. Although no one had the courage to admit it, he was right, as usual. Yet great effort was expended to make these schedules. They were carefully and painstakingly made and published.

At least now I knew what he did not like; however, I still did not know what he wanted. So, with a certain amount of trepidation, I asked him. With a distinct passion in his voice, he told me that the main problem the company had with schedules was that there were not enough *engineering leaders* who would put together plans and then stick with them. He referred to us and our managers using a term that was less than "manly."

Our problem, he went on, "We bow to the god of company politics and it's the squeaky wheel that gets the grease not the wheel that needs to be greased." He continued, "The Division Production Superintendents look at the schedules, see their project, and complain that it takes too long. They then go to your bosses who crater immediately, and your bosses then come to you and you revise the schedule and republish it. Then the next superintendent complains, and the cycle starts all over again." I countered with, "But we get no guidance from them, only criticism." At which he said, "Your boss is not doing his job, and he never will. So quit whining. You need to do some things."

He told me to "put together a preliminary plan and highlight each commitment you will need from each superintendent. Then put together a serious plan that you *can* do, and then do it. Go directly to the superintendents, get those commitments from them, and don't let them off the hook. Badger those @#$%!s until you get the needed commitments. Make no mistake about it. They will turn on you like a dog when things go wrong, unless you get their fingerprints on your murder weapon." My mentor then went on, "Review it with all the superintendents and make them commit to the needed completion dates, and also commit to doing the work they need to do so your men can complete the projects."

As was normally the case, I took his advice and we did just that. I persisted, and persisted, and persisted, and we finally got all the needed commitments. We then put together a plan that included all the jobs of all the engineers, all on one schedule. It was reviewed and we put it in service. It worked, and very soon our group was significantly outperforming the other design groups.

The division goal was to have the cost of engineering and drafting less than 15 percent of the project total—a goal that was more often missed than met. After we got ourselves organized, we averaged 8.6 percent and led all groups by a large margin. In short order, this was noticed. I'm not sure how it got so quickly noticed, but I believe that this man who had adopted me as his student had done some "behind the scenes politicking." Not only did we spend less engineering time and money, we routinely met the project startup dates, projects took less time from start to finish, and our schedule meant something. The schedules of others were still looked at skeptically, but ours now had credibility.

Some other unintended consequences arose from this effort. First, no good deed goes unpunished. I found that the size of my group grew with the resultant increase in responsibility. Most engineering groups had five to seven engineers; ours routinely had 11 to 15. Although this was more work, I took it as the supreme compliment: They trusted us to get things done. Second and most importantly, I learned the value and power of good planning. While others were busy revising schedules and making excuses, we were finishing projects at a record pace, gaining self-confidence, and earning respect—both of the latter borne from our success.

The lessons I was taught by my mentor—about leadership, working together, making and meeting commitments, and planning—I never forgot. In every management position I have held, we have used goal development and deployment, and it has offered immeasurable success in our ability to meet our objectives.

Yet I see so many businesses slight this powerful tool, which further baffles me because it's relatively easy to use. I find time and again that poor planning is a critical weakness in plant performance and plant improvement. Consequently, I have expanded this chapter in the hopes that it will not only help in the implementation of your lean initiative, but that it will carryover into the entire management of the business as well. So I ask you, as my mentor asked me: "What's your plan, man?"—and I have included some materials that should help you.

Hoshin–Kanri Planning

Few things are as powerful for the manager as *Hoshin-Kanri* (H-K) planning. Any manager who has his pulse on the business can learn H-K planning and execute the basics of it rather quickly. It is not very time consuming to develop the plant goals, nor is it very difficult to review the goals monthly. Even if the manager is locked in his office and does not have any—or any real—contact with the floor (which is the real-time consuming step). H-K's goal formation and review can be powerful techniques.

I can think of no management technique that will leverage a manager more than goal creation and review, yet it is so seldom done well. It is one of the most powerful management tools, and one of the most humane, yet it is not difficult to do. It baffles me.

Why Are Goals and Goal Deployment So Important?

What Is the Purpose of Goals?

To answer this question, first let's explore what the purpose of goals is. Let me be simple and straightforward. The primary purpose of goals is to guide behavior.

Often the most critical and important thing a leader or a manager must do is act with courage and conviction on the plan he has created. When he

> **P**oint of Clarity   The purpose of goals is to guide behavior.

does this, he not only acts in the best interest of the facility but also shows, by example, the appropriate way to lead. If he does not act in this fashion, everything else he will do will be compromised. If he has developed a good plan for the facility, he then wants all those in the organization to act in consort with that plan, trying to reach those goals. This is his leverage. This then is his ability to get the work done through others and have confidence that the right work is being done, by the right people, in the right way, to reach the right objectives.

- If his goals are not meaningful, then the organization will not get to the right place.

- If his goals are unclear, then the organization cannot proceed with confidence, undermining their ability to reach the goals.

- If the goals are not directed to—and understood by—the right people, then they fall on deaf ears and will not be reached.

The goals become the primary tool, used by the manager, so he can convey both the needs of, and the desired destination for, the facility. Good goals, well deployed, will not only leverage the manager's ability to get the right work done and improve the performance of the facility, they will actually be a primary motivating tool for the workforce. People will always act better and more decisively when they know where they are going. Weak, unclear, or poorly deployed goals will doom the facility to perform at an inferior level.

> **"I**f we don't know where we are going, we might end up somewhere else. **"**
>
> Yogi Berra

Think about It

Imagine an organization where:

- Every employee knows what he needs to do, when he needs to do it, and how to do it in order for the organization to run smoothly.
- Every employee manages by facts and knows how to analyze and correct problems.
- All the needed information flows smoothly and concisely to the people who need it.
- Managers can establish the few key goals to meet the needs of the customer and so best leverage the facility. And these managers will have the time to perform daily management at the plant.

This is all possible, with good policy deployment.

Policy Deployment

The Manager's Task

Policy deployment, is more than goal setting, it is management's way of:

- Communicating
- Guiding
- Following up on issues
- Changing the important aspects of plant operation

When the manager creates goals, he will:

- Convert business concepts into understandable performance metrics
- Be able to compare these performance metrics to a standard
- Implement Just In Time corrective actions

Many managers have found that *Hoshin-Kanri* Planning (see Fig. 9-1) has proven itself to be a superior model of policy deployment. I strongly support it and have found it creates both horizontal and vertical integration of activity. It is a detailed method to achieve:

- The implementation of vision
- The alignment of goals
- A self-diagnosis of progress
- The process management of a plant
- Targeted focus at all levels

Some Unique Strengths of H-K Planning

In this chapter, we will touch only briefly on H-K planning, but I strongly suggest you pick up a good book on it and learn it more fully. Planning has many strengths, but I

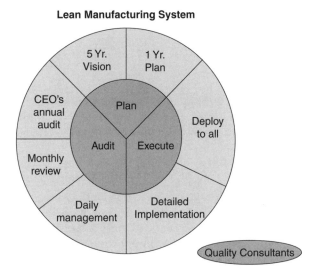

FIGURE 9-1 The H-K planning model.

wish to highlight two. First, H-K planning is an integrated continuous process that is graphically described in Fig. 9-1. It includes long-range planning such as the five-year vision and the one-year plan, and also requires periodic reviews as well as daily management. This is in contrast to most planning efforts, which have a huge influx of effort and management time at only the goal development and quarterly review. In Leanspeak, H-K planning is a continuous, rather than a batch, planning process.

However, beyond this the key uniqueness in H-K planning is the concept of "catchball," which is most evident at the goal development phase. Just what is this thing called "catchball"?

In "catchball," goals are first created by top management based on the needs of the business. These goals are "what" must be controlled to be successful. The manager then introduces this goal to the next level and asks "how" do we do that? The next level replies as to "how" it must be done, including help that may be needed from top management. *This creates a down-up-down-up-down process somewhat like negotiation, except it is not a negotiation. Its purpose is to make sure the goals are properly aligned and the means exist to execute the goals.* The next level turns their "how" into a "what" and asks their subordinates, "How will you do that?" and so forth...

For example. The plant manager decides we must improve OEE by 15 percent, that is the "what" that must be accomplished. This goal then goes to the production manager, and the plant manager asks him "how" he will do it. The production manager says he will improve OEE by reducing machinery downtime, but he needs another engineer... Maybe both agree, if so, the goal goes to the next level. Hence a down-up-down process that leads to agreement of the goals and the means to achieve the goals.

The process of "Catchball" has a number of extremely strong benefits, including:

- Goals are thoroughly deployed.
- Goals are mutually understood.

- Means are addressed.
- Ownership is clear.
- Measurement is clear.
- Priorities are clear.

Frequently, beneficial techniques have a downside as well. The downside of "Catchball" is that it is viewed as being time consuming. In a typical five-layer operation, it often takes two months to go through the creation of the one-year plan.

H-K Planning Success Rate

However, all my experiences with H-K planning have been positive and this early upfront time investment has proven to be an outstanding investment, without exception. In the end, I have found that the accomplishment of goals created using H-K planning has been far more successful than conventional goal setting. Once managers start using H-K planning, I find they continue with it, because it is an effective management tool that allows the facility to perform at a much higher level.

Additional Information on H-K Planning

A great deal of good literature exits on *Hoshin-Kanri* planning. I recommend the book by Bob King, *Hoshin Planning The Developmental Approach* (Goal/QPC, 1989), and *implementing A Lean Management System* (Productivity Press, 1996) (yes, the "i" in implementing is lowercase) by Thomas Jackson (which, by the way, is about H-K planning not Lean management), or *Hoshin Kanri: Policy Deployment for Successful TQM* (Productivity Press, 1991) by Yoji Akao.

Goal Development

At the heart of policy deployment is the development of goals that have:

- Purposes
- Characteristics
- Foundational concepts
- Deployment characteristics
- Owners

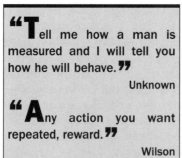

"**T**ell me how a man is measured and I will tell you how he will behave."

Unknown

"**A**ny action you want repeated, reward."

Wilson

The Purposes of Goals

The purposes of goals are multiple. The primary purpose is to guide behavior. Let me say that again, the primary purpose is to guide behavior. Goal development and deployment also provides a reason and means for rewarding the correct behaviors. The two guiding rules we use in directing human behavior are given in the quotes here.

The secondary purposes of goals are psychological/sociological in nature. One purpose is to convey where the company is at this moment and convey what it must accomplish. By conveying where the company must go in terms of performance, the goals create a future state. This future state is the biblical equivalent of the "Promised Land." In other words, once we get there (the future state), we will be in the land of milk and honey. Since the goals are, by definition, attainable, this has the effect of creating hope. Hope that the company will become "the best," "more competitive," or whatever is the desired vision of the company. It is a psychological fact that there are few positive motivators that are as strong as the concept of hope.

In addition to creating hope, properly deployed goals will have owners and clear objectives. This then has the highly desirable effect of instilling both confidence and a sense of ownership in those who must execute the goals. They will proceed with more confidence when they know they are the ones who are both responsible and accountable for the goals. Furthermore, they know that the goals are clearly an extension of the plant goals: important to the success of the facility. For those who achieve their goals, all these factors combine to develop a strong sense of accomplishment with the attendant self-satisfaction attached to accomplishment.

The Goals and the Goal Creation Process Are Strong Motivation Tools

Consequently, through the:

- Creation of hope
- The pride of ownership
- Renewed confidence
- A sense of accomplishment

the goals become a motivating factor, spurring the organization to new levels of performance.

Goal Characteristics

All goals must be:

- Written
- Challenging
- Believable
- Specific
- Measurable
- Have a deadline

Goal Deployment

Goals must be deployed in the right context, to the right owner, and with the right expectations. The context and expectation include the expected results in the expected time frame. In addition, good deployment will include the manager pointing out possible failure paths to avoid, and possible future conflicts that might naturally result from attaining the goal. It is also an aspect of good deployment to reach agreement on the available resources to allocate to the goal efforts and the consequences of achieving and failing to achieve the desired results.

The Owners of the Goals

Good deployment requires that the goals must have a clear owner who is responsible for the attainment of the goal. By responsible, I mean just that: "able to respond," and more than just being accountable, being "able to count." Hence, the owner must have:

- The awareness and the tools to determine if the process is performing properly. This is transparency.
- The imagination and values to determine what action is required.
- The desire to make a change when one is needed.
- The power to make it happen.
- The courage and character to accept the consequences of those actions.

Furthermore, the owner—and everyone for that matter—must recognize that we *cannot* live in a dependent world, and that total independence is neither real nor healthy in a society. The reality is that we live in a world of interdependence. Consequently, no one person can actually be totally responsible. In other words, we must work together and synergize for the common good, and therefore the owner does not have total control, but he does have *functional control*—that is, he can make things happen so that progress toward the goal can proceed.

Leadership in Goal Development, Deployment, and Determining What "Should Be"

In Chap. 6, we enumerated the three requisite skills of leadership as:

- The ability to develop a plan
- The ability to articulate this plan and engage others
- The ability to act on the plan

The first aspect of leadership is manifest when the goals are developed. The goals form the plans the manager will use. Consequently, the manager must have the skill to discern the few key metrics that will best guide the facility to success. I call these the "Plant Level" goals.

Plant level goals are almost always a subset of the three key customer needs of production:

- Quantities on time (on time performance)
- High quality (usually something like first-time yield)
- Fair priced (usually these are cost goals for the typical manufacturing plant)

There should be only a few—five to seven is ideal. Too often, where there are goals, there are too many. I frequently find 30 or more. Who can remember 30 goals? Furthermore, with 30 goals, the focus is being lost. Even if they can remember 30 goals, who can focus on 30 different areas?

In addition, often the wrong goals are chosen for plant level goals. A goal I see very often is the goal to reduce the cost of expedited freight. Not that the cost of expedited freight should not be reduced—that is not the point. The point is that these are just not the key metrics that should be used to guide behavior in the facility.

In fact, there is hardly a better red flag to indicate that a facility is in trouble than when you see that one of the plant metrics is expedited freight. Think about it. That means they have problems with on-time delivery, which signals a whole foray of production problems. Also, it means that expediting costs are a significant part of the plant's operating expenses. The choice of this metric sounds like a strong reason, in and of itself, to look into Lean.

Now, back to the development of metrics by the plant manager (PM).

Once the correct metrics are selected, the PM must select which levels these metrics "should be" by year's end, for example. This creation of a "should be," as you recall, has now just created a "problem" for his staff. It is an oddity of management that one of their key roles is problem creation. They do this by creating the possible future state of the facility—what "should be" attained.

When the PM selects the specific goals and the levels that need to be achieved, he is starting to make a very certain and definitive commitment. Actually, there are two major commitments. The first is about the facility. Based on his experience and abilities, he is saying that the attainment of these goals is how the plant needs to be "best" or "competitive," for example. Second, he is making the commitment about his future actions, including the rewards that are earned with successful goal attainment. Both of these place a great deal of pressure on the PM. But make no mistake about it, he must do both. He must select the best goals to guide the best behaviors, and he must reward those behaviors he wants repeated. Recall:

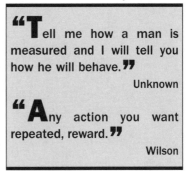

> **"T**ell me how a man is measured and I will tell you how he will behave. **"**
>
> Unknown
>
> **"A**ny action you want repeated, reward. **"**
>
> Wilson

If the correct metrics are not chosen with the appropriate performance levels, the policy deployment will not start properly. In short, the plan to improve the performance of the facility will not be a good plan. If the plan is not deployed well, it will not be understood and accepted by all, and execution of the plan will suffer. Finally, if the follow-up elements of H-K planning are not executed well, the leadership will not be acting upon the plan. All of these issues are just symptoms of weak leadership, which:

- Cause poor goal creation, which in turn…

- Cause people to pay attention to the wrong metric, which in turn…

- Causes them to act in a way that is not in the best interests of the plant, which in turn…

- Causes the plant to be less robust, which in turn…

- Is exactly *what we do NOT want!*

> **P**oint of Clarity   The key to good policy deployment is good leadership.

Chapter Summary

This chapter was added to not only assist the project as the goals for the Lean implementation are developed but to go further and assist the management in overall goal development and deployment. H-K planning has proven itself to be a superior management tool for the development and deployment of plantwide strategies and goals. Goals will not only guide the behaviors of the workforce, but goal development, deployment and execution will create the hoped-for future state, and create confidence and a sense of accomplishment. Properly developed and managed goals, as outlined in the H-K methodology, can be a large motivational tool, and a strong weapon for the manager as he tries to leverage his position power and maximize the potential of the plant.

CHAPTER 10

Sustaining the Gains

Sustaining the gains is a key foundational topic. We will discuss its importance and its application, especially the more powerful techniques of product and process simplification. Unfortunately, to most of us applying Lean tools today, the product is designed and the process in place, so we have to work with what we have. Seldom is it practical to change the product, so we are left with improving the process. Because of this, we will address how we can sustain gains in these circumstances.

Why Is It So Important?

It seems almost intuitive that to sustain gains is a key issue in any business. Why make the process improvements only to lose them over time? If the gains are sustained, over time the net effect is much larger, of course. The two techniques used to sustain gains include maintenance and standardization of improvements. Maintenance is the ability to restore equipment to the original condition so the status quo can be restored. Standardization, on the other hand, is the ability to get all people, machines, and methods to continue to do what has once been shown to be effective, be it the status quo or a process improvement. Hence, to sustain the gains from our process improvements, we need to at least standardize the gains achieved. (See Fig. 10-1.)

So why doesn't everyone do it? Sustaining the gains is something we consider to be of the greatest importance. This is what separates the companies that are prospering from all the others. Every company is encountering and solving problems—there is nothing particularly unique about that. Some do it better than others, some are more efficient, and some are more effective than others. However, how many of these companies are spending time solving problems that had been solved before? This cultural characteristic of spending lots of time on fixing problems, only to have them reappear and be re-solved, is incredibly common. It takes great discipline to check and double-check and audit to make sure the last problem was solved—and solved for good. It is simply a lot sexier to move on to the next problem and solve it. This pattern—the pattern of fixing problems without institutionalizing them—is the pattern of most cultures. It cannot be the pattern of a Lean culture—at least not one that wishes to survive.

> **"T**he only effective approach to quality is to make it part of your culture, that is our aim. **"**
> **The motto of Quality Consultants**

"Sustaining the gains," is an area in which our company, Quality Consultants, specializes. If the changes are not institutionalized, there will be natural degradation. Remember that the entropy of the world is increasing; all systems need maintenance;

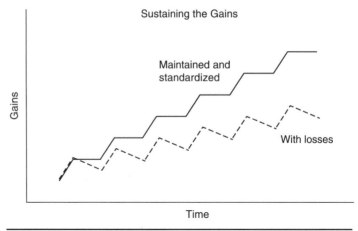

FIGURE 10-1 Sustaining the gains.

nothing is forever. This can be said many ways, but they all mean the same thing: If progress is your desire, then you need to fix it right, and fix it only once.

One of the real enemies of sustaining the gains is the rapidly changing world we live in, as outlined in the following list.

- First, to survive we must have a culture of continuous improvement, which means we need to change internally.

- Second, if our own internal efforts of improving OEE and reducing costs with those resultant changes are not enough, customers will in addition forever ramp up the required standards of their suppliers.

- Last but not least is the issue of employee turnover—the most difficult type of change of all.

When all three change factors—internal changes, external changes, and employee changes—act on a business, I cannot give a single example of even one firm that is excellent at sustaining the changes needed to prosper.

The first two enemies to sustaining the gains are not avoidable, so as a business we must manage the personnel in such a way as to minimize turnover. The issue of employee continuity is crucial to being able to sustain the gains. I know of several firms that are adequate at sustaining the gains, and they have one major item in common: a high level of employee continuity. In terms of numbers, they have developed a culture that holds on to the top management for an average of 16 years, supervisors an average of 12 years, and hands-on workers an average of seven years.

Now look at the example of the House of Lean, and you will find that "Sustaining the Gains" is a foundational issue. In fact, it is more than a foundational issue, it is *the* foundation of all foundational issues. It almost seems infantile to point out that we must continue to do the good things that allowed us to progress; however, it must not only be pointed out, there must be an entire system of activities in place to assure that we do continue to do those things that allowed us to progress. I often find it almost insulting to discuss this with top managers, but it is the most basic of problems. I can't give you one example of any company that does this well. I can give you some examples of several who are adequate, but not one that does it really well.

How do I measure that? It is simple: "Do the problems, once solved, stay solved?" Actually, this is a two-part issue. Can they solve problems? This is a major weakness with most companies (it was discussed in Chap. 6). So, problem solving is a precursor to sustaining the gains. Among those companies that have good problem-solving skills at all levels, they must now institutionalize those solutions. Causing these changes to be part of the normal fabric of doing business is not easily accomplished. Most managers recognize that sustaining the gains must be done, but put forth only a minimal effort, mostly because they do not make it a priority to ensure that the changes are institutionalized. Instead, they move on to the next issue.

> **"M**en stumble over the truth from time to time, but most pick themselves up and hurry off as if nothing happened. **"**
>
> Sir Winston Churchill

I said that most managers do not make it a priority to sustain the gains. In my practice, I have found many who will agree that institutionalizing the gains is important but a scarce few who can put it together in a simple flow chart that shows the actions they must take to sustain these gains. Most managers will readily agree that sustaining the gains is a process, but not many can make a process map that describes it.

> **"O**pportunity is missed by most people because it is dressed in overalls and looks like work. **"**
>
> Thomas Edison

Most readily agree that it is important, and just as rapidly they skip right over it. Yet others have an idea of what to do, but are not willing to put forth the effort—which is frequently substantial. Once again, the reluctance to sustain the gains was predicted by those men of yore when they said:

We are not a voice in the wilderness decrying the need to standardize and institutionalize improvements. This concept of sustaining the gains is deeply embedded in Lean and also in Deming's philosophy. Read Deming's landmark book, *Out of the Crisis* (MIT CAES, 1982), and spend a few minutes on his 14 Obligations of Management and his 7 Deadly Diseases. Deming is clear when he states the following three points:

- We must sustain the gains.
- We can not make progress without sustaining the gains.
- Ultimately, we cannot survive without sustaining the gains.

How Do We Know There Is a Loss?

Transparency

In regards to transparency, refer to Chap. 20 of this book, which shows that transparency is one of the foundational issues. It is the concept that lets you "see," in real time, right now, what is happening in your process, letting you make a determination as to whether anything has changed or needs attention. *Andons*, 5S tool outlines, and *heijunka* boards are all examples of transparency. For example, by just glancing at a *heijunka* board, we can tell if production is ahead, behind, or on schedule. We can make that evaluation right now in real time. There is no need to check the information center. It is not necessary to check with the storehouse or some computer database; the *heijunka* board will allow us to "see" it directly, at this moment. Today, it is more common to hear the phrases, visual management or visual system used in place of transparency. However, I prefer transparency because it better captures the concept, as you will "see" in the example later.

Transparency Misunderstood

A great deal of the original concept of transparency has been lost. Most of what is commonly displayed as transparency is really visual management and it is very long on the "visual" and all too often short on "management." Unfortunately, most of what is posted is not consistent with the original concept of being able to "see" what is happening in the process. A common example of "visual management" is display boards that show the status of such things as monthly production volumes, the status of problem solving, and preventive maintenance, for example. Frequently, this information is kept in one location, known as an information center. The information center often has process information such as standard work combination tables, standard work charts, and other engineering intensive information that generally I find are not valuable on the floor, but make for nice wallpaper at the information center. All of this information is needed, at some time, at some place, and for some reason. What should be kept at the line is what will be *used* at the line. Keep that information there, the rest should be kept where it is needed.

The Ultimate Purpose of Transparency

> **"N**ot everything that is faced can be changed, but nothing can be changed until it is faced. **"**
>
> James Baldwin

However, the ultimate purpose of transparency, starting with situation assessment is Rapid Response PDCA. Recall that PDCA stands for "Plan-Do-Check-Act." This is the iterative process improvement cycle that is inherent within the *kaizen* improvement process. As part of PDCA, first it is necessary to determine something has changed, such as a production rate flagging. Next, it is necessary to:

- Plan a corrective action or countermeasure (plan).
- Implement it (do).
- Confirm whether it is successful or not (check).
- Determine if additional actions or thoughtful inaction is appropriate (act).

The cycle then starts all over again; hence, it is an iterative process. Thus, it is critically necessary to have some information to be able to:

- Discern something has changed.
- Confirm that the countermeasure was or was not successful.

Furthermore, to perform PDCA in a "Rapid Response" fashion, or in Leanspeak, to do that in a JIT (Just In Time) fashion, requires JIT information. This availability of JIT information is what transparency is all about. In a transparent system, these data are available JIT.

An Example of the Implementation of Transparency

For example, in a high-speed electrical manufacturing plant with multiple lines operating at 6.65-second cycle times, it was hard to tell, except after several hours of operation, if the lines were producing to the plan. Even once we got a grasp on the production, if there *were* problems, we didn't have any idea where the problems might be, so JIT (Just In Time) problem solving was something that wasn't possible.

In addition, the technicians, line leader, and area supervisors were always busy. They were always correcting something; however, the stories of what was done and why they were done were filled with generalities and terms like "It was not operating quite right."

These are clear signs of a number of things. Let me point out three:

- Fortunately, the workers, as well as the management, were both motivated and engaged. They were trying to do the right things and they were doing work that resembled their job description.

- Unfortunately, they were doing many wrong things.

- Yet more unfortunately, even when they did the right things, there was inadequate feedback of information to confirm that something positive had been done.

At this time, the relevant information to assess the production rate, which was readily available on the floor, included *andon*s, rejected product segregated into collection bins and production boards that covered a full day's operation.

The *andon*s had no recording, only warning features. Hence, they were not useful for problem solving unless we were right there when they were activated or cleared. The scrap data was sorted and recorded hourly, but actual scrap was very low and not a factor in low production.

The production boards had hourly and cumulative production goals with actual production numbers entered at the top of each hour. These goals were calculated based on the hourly goal of 600 units per hour, taking into account lunch and rest breaks. There were 21.5 available hours per day, so the daily production goal was 12,900 units. This hourly goal was met nearly 50 percent of the time, but the daily goal was met less than 3 percent of the time. A review of the last month's data showed production of 9,330 units/day, or 434 units per hour: a full 27 percent below goal.

As part of their management system, there was a daily production meeting on the floor, run by the line supervisor. It lasted about 15 minutes and had a good agenda, but when the production shortage was discussed, which was a topic nearly every day, the answers were—well, amazing is the best word I could use. First, they were very general, and in almost every case where a specific problem was discussed, it was decreed to be solved. Unfortunately, these same problems would reappear later, and it did not seem odd to anyone that, although these problems had already been solved, they still reappeared. Quite frankly, during this part of the meeting everyone was on "autopilot." It was apparent they had accepted these "amazing" explanations for so long that they sort of believed them themselves. Yet day after day the production remained below goal and they were forced into working seven days per week to meet a plan for five days. In this case we can conclude that the production data were inadequate— there was no "transparency" of the production data. We could neither understand if we were on plan, nor could we solve problems when we *did* understand them.

After a meeting with the top management, we decided that taking on the low production was our number one goal. It was the key reason this product was highly unprofitable. The first thing we did was try to break down the problem into solvable pieces. We asked three questions. Is the production shortage due to:

- Quality losses
- Availability losses
- Cycle-time losses

We had the quality information, and at this step of the process, quality losses were insignificant. Our segregation bins gave us all the real-time information we needed. We could "see" that quality losses were not the answer to our low production problem. Or in Leanspeak, our transparency regarding quality yield was adequate for this issue.

Now we had to determine if availability was an issue. A quick check showed that material stock outs were virtually nonexistent but the technicians were working on the machines seemingly all the time. Significant downtime occurred, but we neither knew how much of it there was nor did we know what was causing the downtime. We had serious concerns here but had no information at all about the availability losses. Our transparency in this instance was not only inadequate, it was nonexistent, but we did have the *andon*s.

Next, we looked into cycle time. Other than a few time studies done by the engineers, no data on cycle time was available. The measured cycle time was advertised to be 6.0 seconds. If the process performed at this cycle time, production should be 600 units per hour—the hourly goal referred to earlier. No measurement of cycle time—of any kind—was done on the floor. The transparency about cycle time was similar to availability information. It was inadequate and practically nonexistent.

With this review, we decided to implement an OEE (Overall Equipment Effectiveness) program. (See the description of OEE in Chap. 4.) OEE information would allow us to begin the understanding of production losses and segregate the information into quality losses, availability losses, and cycle time losses. Forms were made, training was done on gathering and entering data, and the information was set up to manage these data with feedback after each shift.

We soon found out that the losses were about 20 percent on cycle time and almost 10 percent of availability losses. Although OEE is always lagging information, it was valuable information and told us we needed to work on both cycle time and availability. However, when it came to understanding and improving the process in real time, we were no closer than a shift away from good information, so we needed to improve that.

Nonetheless, we set up some goals, created an improvement plant, and the first thing we attacked was cycle time. We did a controlled study and found the cycle time was really 6.65 seconds, which surprised everyone. It meant the goal of 600 units per hour was not even attainable. The 6.65-second bottleneck was a manual operation in the welding process. This welding machine was operated by a robot that had a microprocessor with a small display screen. Using some great imagination and innovation, the engineering supervisor found a way to program the microprocessor and display the cycle time for the manual operation. The operator now had real-time information regarding cycle time. Immediately, the cycle time began to drop and stabilize, and likewise production increased. The drop in cycle time was amazing. In less then two weeks, it had improved to 5.5 seconds. We implemented several *kaizen* activities to improve the work station even further. One of these activities was to program the microprocessor to display the average cycle time. It told what the average was, how many units were produced with that average, and when the averaging had started. We retrofitted the station with a reset button and—*voilà!*—we had an excellent piece of transparency. Ultimately, after a few *kaizen* events, we were able to achieve cycle times of 4.6 seconds, consistently. Furthermore, now the operator, supervisor, and anyone interested could look at the display and see how the process was performing. Literally, we could find the instantaneous production rate. This was an imaginative solution that added significantly to the transparency of the system.

However, remember that availability losses were about 10 percent. The OEE information was valuable in helping us understand the losses, but not very helpful in solving many of the problems due to the lagging nature of the information. Nonetheless, we were able to reduce some losses. Check sheet information logs on white boards were installed on the *andon*s to gather information at the time the *andon* was activated. This helped distill the information further and made it available for future use. We now had a recording function for the *andon*s. However, the review of the OEE information showed that 90 percent of the availability losses were due to machine adjustments. A quick review showed that most adjustments could not be explained and were just apparently "tinkering" by well-meaning technicians. We did some training, clarified some work procedures, and these availability losses immediately dropped like a rock—as did the workload of the technicians. They were not comfortable with this at all, however, since they were highly motivated, very energetic and weren't used to having any spare time.

It might be asked how we ever met the hourly goal of 600 units when the process cycle time would not support it and our average production was only 433 units per hour? Well, the answer is both simple and revealing.

The simple part is this. The way production was measured was to count the production of trays that each held 120 units and multiply by the number of trays. No partial trays were counted. So, production was in multiples of 120. A tray of 120 units was the transfer batch. Also, since the line leaders knew that the managers wanted the production goal to be met, it was better to meet the goal some of the time than none of the time. Consequently, at the top of the hour when production was counted, if a tray was nearly filled, the line leader might wait to enter the data. Of course, the next hour the production was even shorter than it would have been if the data had been reported accurately. This 120-unit transfer batch created some problems beyond accounting for production, so we later cut the transfer batch size to 48 units. This not only helped the accounting but reduced the processing lead time, as you might expect.

So how is this so revealing? It was a symptom of the entire facility. Everyone wanted to do well and be seen doing well. As I said earlier, they were a motivated group. And this motivation served us well as we implemented other process changes and improvements.

We could go on, but let's briefly review what we just discussed.

- First, our production system improved. The production rate increased by 42 percent with virtually no invested capital.

- Second, once the improvements had been made, we could now "see" the system status, and if there were problems, we could rapidly find and implement the needed countermeasures. We now had the information to do Rapid Response PDCA. Our system transparency had improved immensely, allowing our problem solving to improve immensely, too.

A Story within the Story

Within this story of production improvements through improved transparency are numerous stories, one of which we will elaborate on. It is the story of engaged workers and a dedicated, motivated management team that was hampered by a certain type of blindness. A type of blindness we all share—to some degree.

In this case, this product was not profitable at all. After 18 months of production, the company was losing 30 percent on this product! The primary problem was the insufficient production rates. Under design conditions, this product needed to produce 330,000 units per month to meet the business plan. The management was absolutely frustrated because, as they said, "We have done everything we can to improve the production." They then listed the things they had done to get to the 6.65-second cycle time. The list was long and impressive. It included a great deal of training and efforts to motivate the workers, as well as a long list of changes made to the equipment.

Although they were both sincere and passionate about the statement: "We have done everything we can to improve the production"—they were very wrong. They *had not* done everything they were able to do, as you will "see" in this story.

Specifically, when we measured and posted the cycle time, it dropped from 6.65 seconds to 5.5 seconds, with no other process improvements. Two conclusions are inescapable.

- Once the workers "knew" what management wanted and once the management allowed them to "see" what was really happening, the rate improved by more than 18 percent. In rough terms, that meant we could now produce the same volume in one day per week less. Our fixed costs had now just dropped 18 percent per unit. Isn't that incredible? Consequently, we could conclude, with certainty, that the workers were both capable and motivated to improve the production.

> **"T**ell me how a man is measured and I will tell you how he will behave. **"**
>
> Unknown
> (from Chap. 9)

> **"I** often say that when you can measure what you are speaking about, and express it in numbers, you know something about it; but when you cannot express it in numbers, your knowledge is of a meager and unsatisfactory kind ... **"**
>
> Lord Kelvin
> (from Chap. 5)

- The second conclusion is a bit more uncomfortable to those in management. That is, the management had not "done all they could do," actually they had done all they knew how to do, but still, they were the obstacle that prevented the improvements.

Let me refresh your memory of some information contained in other chapters. A review and deeper understanding of these comments may assist some managers in their efforts to remove roadblocks that are hindering the ability of others to perform.

Now let me put these two quotes into context with this story.

First, the problem of low production could only be measured in a meaningful way, after a full day of production—a full 24 hours late. Then, they could discern, as was normally the case, that the production rate was low. But how did they numerically describe the problem of low production. They could not tell if the losses were from quality problems, materials delivery problems, machine availability issues, or poor cycle-time performance. So their knowledge of "low production" was, as Lord Kelvin would say, "… of a meager and unsatisfactory kind." Their knowledge *was* meager, to say the least, and it was unsatisfactory to solve these problems, for certain. The same could be said about their "knowledge" of cycle times.

Second, they had failed to explain to the operators what the cycle time goals really were. They had told them to improve production—that is, "work better"; and improve the cycle time—that is, "work harder." They had done this and expected the operators and the process to improve, without giving them any meaningful way to determine if they had done either. They had no meaningful way to measure the performance of the system and they had no meaningful way to measure the performance of the individual.

Behind each of these requests—"work better" and "work harder"—was some management belief. These beliefs, respectively, were:

- Since we have low production, the workers must not be performing well enough, so we need them to "work better."
- Since cycle time is clearly too slow, operators obviously must apply themselves better, thus they need to "work harder."

Both of these beliefs were wrong, as has been shown. The two exhortations of management were heard, but the operators were virtually powerless to make the corrections they so vigorously sought. What was needed was not more effort by the worker but rather more effort by management. What management needed was a *cold*, *hard*, *dispassionate*, *honest*, *introspective* review of the system, and especially their role in the system. To their credit, they did this. They did it well and they did it rapidly. Once this was done, the beast of Rapid Response PDCA was unleashed and progress followed immediately. Furthermore, the management learned, at this time, and only at this time, that they had not "done all they could do to improve the production."

This was the case here, and nearly 90 percent of the time, this is the case for everyone.

In a nutshell, there was no lacking on the part of the rank and file workers. The system was deficient. The system that management created was deficient. Hence, to improve this system, the leadership and involvement of management was needed. But, let me not be too critical of this particular management since, quite frankly, I found them to be open, honest, very hard working and sincere in their efforts to improve not only the plant, but all who worked there. They worked long hours and applied themselves fully. There was no lack of trying on their part.

Their weakness was inadequate awareness.

We are all blind to certain things, at certain times and in certain places. Often it takes some outside influence to help us "see" those things we are not aware of; those things that are our blind spots. Once this management group was able to "see," we were able to make great progress.

Deming spoke of this in his writings, so let me quote him here using his 2nd and 12th Obligations of management, both taken from *The Deming Route to Productivity and Quality* (CEEPress Books, 1988) by William Scherkenbach.

> **"A**dopt the new philosophy. We are in a new economic age, created by Japan. Western management must *awaken to the challenge*, must learn their responsibilities, and take on *leadership for change.* **"**
> —(Point 2)
> W. Edwards Deming

> **"R**emove the barriers that rob the hourly worker of this right to pride of workmanship. The responsibility of management must be changed..."
>
> —(Point 12)

> **"I**nstitute leadership. The aim of leadership should be to help people, machines and gadgets to do a better job. Supervision of management is in needs of overhaul, as well as supervision of production workers."
>
> —(Point 10)
> **W. Edwards Deming**

In this case, as is always the case, leadership was needed to improve this situation. Again, I cannot say it better than Deming, so I wrap up this section with some words of his also from *The Deming Route to Productivity and Quality*.

This story-within-a-story shows the manifest power of transparency. Once the workers could "see" what was wanted through clearer goals, and once they could "see" what they were doing through better metrics, they simply performed better and the system improved. The workers were already motivated; they simply needed the proper tools and then they could make the system perform at a higher level.

This is their story.

A story that is repeated way too often.

At some level, this is your story as well. Don't forget it.

Back to Transparency

Just how did the greater level of transparency work for the future activities? First, now the hourly production was more accurate and we could determine if we were performing as planned. At the top of the hour, the supervisor would enter the production rate, which was now within 48 units of being exact; or accurate to within four minutes of the schedule. Second, if we were off schedule, we could look at:

- The rejected product segregation bins to see if we had a quality problem, or...
- Look at the *Andon* log to see if downtime had been a problem, or...
- Look at the cycle-time information to see if the process was performing to cycle time

Generally, once we had this information, it was easy to focus our attention on the specific problem machine or issue, for Rapid Response PDCA.

We were not done improving the transparency of this system, but we had been able to improve significantly. Earlier, we were trying to solve problems 24 hours after they happened and having very low success rates. Now we could find out nearly all the data we needed to solve most problems in a 15-minute window. We had made significant progress and it was largely due to the transparency we had built into the system.

Transparency and Imagination

Transparency is one of those concepts that is truly ripe for development as part of Lean manufacturing—we are only limited by our imagination. With transparency, along with the SMED (Single Minute Exchange of Dies; quick changeovers) and *poka-yoke* technologies, there simply is no end to what we can develop to fully exploit these tactics. Just keep two concepts in mind.

- We want "transparency" to be able to distinguish and tell us that something has changed.
- We want "transparency" to show us the necessary information, immediately at hand, in order to implement Rapid Response PDCA.

Transparency is that way of showing the information such that these two aspects of process diagnosis and process improvements can be accomplished.

> **"I**f you see in any given situation only what everybody else can see, you can be said to be so much a representative of your culture that you are a victim of it. **"**
>
> S. I. Hayakawa

What Is Process Gain?

Most process gains are achieved by either reducing the variation in the product, the process, or both. This variation reduction will then reduce waste and produce gains that are manifest as higher yields, shorter cycle times, or greater uptime, to name a few of the typical manufacturing plant gains.

So, to achieve the gains we need to reduce the variation, and there is a specific approach that can be made to reduce variation. We can work with:

- The product
- The raw materials
- The process equipment
- *Poka-yokes*
- Process procedures

Simplify the Product

The greatest leverage in reducing variation is to simplify the product. In one case I can recall, we were working on improving yield on an electronic control unit (ECU) that had 13 functions. This ECU had over 300 components populating an 8" by 8" printed circuit board (PCB). The next generation of ECU, although it had 42 functions, required only 46 components and the PCB was 4" by 5".

Technology had allowed this design simplification and, of course, the processing equipment was both dramatically reduced and simplified as well. For example, in the solder application process, done with a screen printer, the PCBs were initially printed in a pair—that is, two PCBs per panel were printed. After the design simplification, six PCBs per panel were produced. Before the redesign, there were five placement machines in series. After the redesign, the new version required only three placement machines.

This is just part of the impact, but it is easy to see that more units were produced in shorter times using less machinery with less investment. In addition, space requirements were reduced and future maintenance was also reduced. Also, as expected, when the new process started up, initial process yields were greatly improved over the earlier design. This type of product simplification is the most powerful, but in the typical manufacturing environment, unfortunately, it frequently is not possible. More often than not, you have to deal with the product you now have.

Reduce Raw Materials Variation

Often the incoming raw materials have significant variation and this is a problem of varying degree to everyone. However, most manufacturing firms work with the supplier's at the process level to improve their processes. They use tools such as 8Ds, Supplier Certification Programs, and some of the most forward-looking producers have good Supplier Support organizations. Supplier development was a key tool used by Ohno to improve his manufacturing system, but he really did not implement this until nearly 20 years after he began his quest for quantity control. Very likely, this will not be a large issue in your facility for a few years. When it is a priority, luckily, there are a series of good books on supply chain management, so we will not go into detail here on this topic.

Simplify the Process

The second most powerful technique is to simplify the process. Most often, the effort is focused on reducing human interaction. Frequently, this means you will use robotics or other types of automation. Often, and it is especially true for a tier-1 automobile supplier, this type of simplification is not practical. First, there is the cost issue of the capital improvements, and beyond that suppliers are required to jump through a whole series of hoops that tend to discourage these type of improvements. These hoops, known as process validation, which are mandated by the typical automobile customer, are so costly and time consuming that even if the change is warranted, it is often not done due to the rigors involved. In addition, these changes also open the door for the topic of price reductions. All in all, process simplifications of any magnitude, such as converting manual operations to robotics, are not done very often.

Poka-Yokes

Poka-yokes (see Chap. 4) are powerful tools and should be fully exploited before other procedural changes are implemented. I find it curious that *poka-yoke* technology is not used more often. I attribute this underutilization to the low levels of imagination demanded of most manufacturing support engineers. It does not help that I find very little in-house training being done with *poka-yoke* technology, although there are several good references in the literature. Other than a little periodic maintenance required of most *poka-yokes*, I see very little justification for their underutilization. The beauty of *poka-yokes* is in their simplicity and effectiveness. We can give numerous examples of *poka-yokes* that saved $10,000 per dollar invested. That kind of investment is hard to beat.

Standardize the Process Procedures

Unfortunately, the most common type of improvement comes about by improving process procedures. What this generally entails is an attempt to reduce variation through the use of better work descriptions. But in the end we still rely on the human element to perform well. I used the word "unfortunately" in the first sentence for two reasons.

- First, the human element is often the greatest source of variation and we are simply trying to do it better, not really differently.
- Second, it is unfortunate because, often times, even with the most imaginative engineers and managers, that is the best we can do. In fact, this is the most typical approach taken in the manufacturing world to reduce variation.

Short-term improvements are often realized but the larger problem is *"how to sustain these gains?"* That is the subject matter of the rest of this chapter, so we can realistically sustain the most common of our improvement techniques—improving the processes done by humans.

The Prescription

A Five Step Prescription exists on how to sustain the gains. It is not amazing, but once implemented, it is always effective. It includes:

1. Good work procedures
2. Sound training in the work procedures
3. Simple visual management of the process
4. Hourly and daily process checks by management
5. Routine audits by management

Now let's look at the prescription, one element at a time.

Good Work Procedures and Standards, Reflective of the Facility Goals

First, we must have good work standards and procedures, including checkoff lists, startup procedures, maintenance procedures, and standard work to name a few. In short, all the procedures necessary to run the business and do it efficiently and effectively. There is a lot of good literature on writing good job instructions so I will not belabor this point, except for two points of major concern. The work instructions and standards:

- Must be written in behavioral terms
- Must be auditable

People Trained to the Standards, Including Being Reality Tested on Performance to the Standard_

Next, we need to train the employees on these standards and procedures. Again, there is a lot of good material in the literature, so I will not elaborate further except for one point. The training, once completed, must be evaluated using a practical test done at manufacturing conditions. Let us say, for example, we have trained operators to visually inspect parts for attribute characteristics, prior to final packaging of the parts. These operators need to be tested—in this case, with an attribute Gauge R&R study—and it must be done at line conditions, at *takt* time. Written tests and video training are okay, but in the final analysis the purpose of this training is behavioral modification, and that modification must be tested; *there is no substitute for this.*

Simple Visual Checks to See if the Standards Are Met…Transparency

At the jobsite, we now need simple visual checks to see if the objectives are being met. This is a key element of transparency. If the operator is to meet a specific cycle time, how can he tell? Is there a visual readout that shows his performance? If not, how does he know? If routine maintenance needs to be done, can that be audited by the manager walking by?

Routine Management Evaluation, Daily If Not Hourly

In my experience, most companies do the first three steps with some degree of efficiency, and here in steps 4 and 5 is where the system to sustain the gains often breaks down. A key element of *hoshin-kanri* (H-K) planning is daily management review. More and more, I find it less common for even midlevel managers to go to the floor daily. These trips to the floor are absolutely necessary if continuity is the concern. In today's

> **P**oint of Clarity Managers do not get what they EXPECT, they get what they are willing to INSPECT.

rapidly changing world, management must have their thumb on the pulse of the plant; there is no way to get this understanding from the office. It has several names, including management by walking around, or going to the *Gemba*, but the managers, at all levels, must spend time on the floor, daily, if not hourly.

Managers must be aware of actual operations and operating conditions. Also being on the floor keeps them in touch with the people. It helps the manager evaluate not only what is happening on the floor but helps him evaluate the information he gets from others, which is generally the bulk of his information. And finally it helps the manager learn: about the people, about the process, and about the product. It is a critical part of Deming's 14 Obligations of Management and Toyota thought it was so important they gave it a name, *genchi genbutsu*, which means, "go and see for yourself, thoroughly understand the situation." In fact, to Toyota it is so important that it is Principle 12 of the "Toyota Way," their guiding document to management. Sometimes the manager may have a specific agenda, sometimes he may just want to observe to see what happens—but there is no substitute for his or her presence.

The most common management practice is to show up on the floor only when there is a problem. Consequently, when the workers see the manager they know that something is amiss. There is nothing like this to create an atmosphere of concern which then evolves into a culture of fear and secrecy. The manager must be on the floor because it is part of his normal job—not only to investigate when things go wrong, but also to investigate when work is going well. He can then tell that the methods to standardize are working.

In addition, I have found the very best at this go to the floor for two other reasons. Really good managers have an ability to go to the floor and just listen to the rank and file workers and be able to learn about the process—directly from the horse's mouth. It is the "How is it going? And then saying nothing more, simply listening skill" that only a few managers possess. Another technique that pays high dividends is to have on your calendar the birthdates, company service dates, and wedding anniversaries of all your employees. These provide reminders to go visit the workers on a more personal basis. Then the manager cannot only get to know these people better, he can stop by and discuss their first day at work, for example. This type of touch needs to be sincere, but if it is not your style, don't do it, you will only come across as phony and, hence, gain nothing. However, if it is your style, it pays huge dividends. In the future, these people will be far more willing to "tell it like it is" rather than the "what I think you want to hear" mode so common in most plants.

Routine Management Audits...to Teach the Managers, to Check the System

Again, the H-K planning model includes the concept of management audits. The general paradigm of audits is to check the system to see if it is working, and most auditors are not really happy unless they can find a few things done wrong. It seems to quench their thirst and convince them that they have done a good job. In my experience, nearly all auditors have that desire: to catch people doing things wrong. Those audits are not really helpful, much like the standard itself. On the other hand, the objectives of these Routine Management Audits are much different and are twofold:

- To teach management
- To check the ability of the system to meet the policy

It is my experience that the major opportunity for improvement in manufacturing lies not in improving the people; rather, it lies in improving the systems. Let me say that again, it is my experience that the major opportunity for improvement in manufacturing lies not in improving the people; rather, it lies in improving the systems. In fact, in controlled studies we have done, we routinely find that 85 to 90 percent of all variation is system created. More simply said, most of the variation is because people are:

- Using the raw materials they are supplied
- Running the machines they are supplied
- Following the instructions they are supplied
- Working in the environment they are given

Consequently, most of the time, when we wish to make progress and when the analysis of the variation is complete, it is the raw materials, machine operating conditions, work instructions, and work environment that must change, not the people. Keep in mind that the selection of raw materials and machines, the writing of work instructions, and the creation of the work environment are all done by management. These four things largely define "the system." (Recall the definition of variation in Chap. 3: *the inevitable differences in the individual outputs of a system.*) Unfortunately, and all too often when problems appear, the managers do not have the necessary understanding of variation to respond properly, and all too often they inappropriately focus the attention on the workforce when it is these systems that must change.

The managers, particularly the middle managers, have a lot of emotional investment in these four aspects of the system and do not really want to change them—after all, they created this system. To change this system is then an admission that they had created a defective system. The truth is, no system is free from deficiencies—all can be improved.

All kinds of cultural forces protect the status quo, yet the status quo must change, and frequently the middle managers do not see or do not want to see the changes necessary. For this reason and others, I have found that there is no substitute for top management presence on the production floor. These audits provide just such an opportunity and provide the teaching of the managers and the middle managers as well.

There is a third benefit that is achieved when the managers perform these audits. Here I do not mean, "make sure these audits get done," I mean to do them. I mean:

- To review the standard
- To compare the actions to the standard
- To draw conclusions and develop corrective actions

- To follow up that the work was completed

- To document the audit

These audits are, by definition, nondelegable. The benefits are tremendous here. Let me state just a couple, but the list goes on.

- It provides a real-life connection of the manager to the floor and the workers know the manager is committed, not just involved. The managers will be seen as real workers, not just shiny pants looking at spreadsheets in their office.

- It provides an understanding, at the worker level, that their problems are understood by top management, and that they, the workers, are not totally insulated by middle management. The workers can easily see that they have a conduit of communication to the top managers.

Through these audits, system deficiencies can be found and corrected. They complete the PDCA cycle for the system. Few companies do management audits, and even fewer have an annual audit policy. This lack of an auditing system—and the lack of management audits in particular—hinders system improvements and is a major impediment to sustaining the gains.

Chapter Summary

One of the key foundational issues is the ability to sustain the process gains, once they are achieved. This is a key issue to maximize process gains over time, but strangely enough few companies do this well. The first problem to sustaining the gains is to make sure there is, in place, a system to assure that losses become obvious, when they occur. The technique used to make the losses obvious is called transparency. The two strongest techniques to sustain the gains, once they are found, are through product and process simplification. Although these are strong techniques, for the typical Lean initiative, they are not practical. The next best method is to standardize through the use of *poka-yokes*. This is also a very strong technique, but this is also often underutilized. Finally, the most common but least robust technique to sustain the gains is to standardize the process procedure. A Five Step Prescription exists on how to standardize the process. You will not only find this prescription amazing, but once implemented you will find it to be effective as well. It includes:

1. Good work procedures and standards, reflective of the facility goals

2. Sound training in the work procedures, including being reality tested on performance to the standard

3. Simple visual checks to see if the standards are met—transparency

4. Routine management audits, daily if not hourly

5. Routine audits by management … to teach the managers, to check the system

CHAPTER 11

Cultures

Ⓘn this chapter, we offer a brief introduction into cultures—business cultures in particular. We will just skim the surface, but our objective is to give you enough information that the complexity and depth of the Toyota culture can be appreciated, as well as the depth of effort needed to truly attain a Lean culture. We also want to explain why a concerted effort into changing the culture can wait for a while—at least until the major foundational issues have been implemented.

This chapter was included because the defining aspects of the TPS are its cultural elements. As an extension of that thought, some of the truly unique aspects of the TPS were consciously and painstakingly developed over an extended period of time; hence, they are not easily imitated by others.

Background Information on Cultures

What Is a Culture?

We define a culture as "the combined actions, thoughts, beliefs, artifacts, and language of any group of people." It could be the culture of the Catholic Church, the culture of AARP, the culture of the South, the culture of Toyota, the culture of the New York Yankees, the culture of your plant, or any group of people. The people within these groups think, talk, and behave within predictable patterns of behavior. These thoughts, language, and behaviors then identify them to be a member of the culture. Often, these cultures have specific artifacts that help identify them as part of a culture. These artifacts may include such things as symbolic necklaces, or uniforms. However, simply put, a culture is "how we do things around here."

For example, when I started as an engineer in the oil industry in 1970, all engineers wore a dress shirt and tie, at a minimum. Generally, even first-line supervisors wore a suit or at least a sport coat. I worked in Southern California, the heart of the "take-it-easy-and-let's-go-to-the-beach" culture, yet the dress was very formal. When we asked why, they would say, "That's just how we do it around here." A second example at my employer was that there was no formal program to indoctrinate engineers. For the most part, we were given work and expected to find out what we needed to do to perform. A middle manager once said, "Our engineer training program is like asking the engineers to put on ankle weights, throwing them out into the middle of a pond, and telling them to swim to shore. Along the way we lose a lot, but the ones who make it to shore are real strong." That too was an aspect of their culture—that is, engineer training was not very highly valued, yet we were still expected to perform. A less obvious, although equally strong, aspect of that culture was the intolerance for failures. Failures were not easily forgotten. If an engineer completed ten projects, for example, and nine were very successful, with one having some problems, he was often chastised for the failure and it was seldom forgotten. Should another engineer complete maybe only three projects and all three were successful, frequently he would be viewed as a superior engineer. It was the "It only takes one ah-shit to cancel 100 attaboys" aspect of that culture. It was joked about, at least behind closed doors. But in the end it caused engineers to hesitate before taking even the slightest chances. Consequently, qualities like imagination, creativity, and innovation were repressed. As you might expect, when these qualities became repressed, other qualities would rise as being important. In this culture, company politics became a dominant quality that helped in salary increases and promotions.

These qualities—dress, training, and development—are all important cultural aspects. They need to be understood and managed just as costs, profits, and customer satisfaction are understood and managed.

How Are the Cultural Rules Set?

Cultures are created and sustained by way of two major factors: the history of the group, and the actions of the few top people. These few top people are the ones who set the cultural rules. Whether the rules are stated or silent, they are made and enforced by the top few people. Most cultural rules are not stated, and when we are unaware of them, this makes them potentially very dangerous.

> **"I**n studying the history of the human mind, one is impressed again and again by the fact that the growth of the mind is the widening of the range of consciousness, and that each step forward has been a most painful and laborious achievement....Ask those who have tried to introduce a new idea!**"**
>
> Carl Jung

The Most Common Rules: The Silent Rules

This seemingly odd aspect of cultures, the silent rules, creates a major cultural problem. It is often the reason why those within the culture, especially the rule-makers, simply do not see what is happening within their own culture. For example, in the culture of my early engineering job, it was not unusual for some high-level manager to comment and even criticize the organization for being so "close to the vest"—for not being willing to step out and be innovative. He might say something like, "Why is it that the new ideas always come from our competition?" Or maybe, "Where is the rugged individualism we Americans are so proud of?" In fact, he might say this just minutes after having reprimanded someone for some mistake that was made. Once you become aware of what is happening within your culture, some of these comments are almost laughable—that is, if they were not so tragic.

The silent rules would not be so damaging, except (more often than not) we are not consciously aware of them. And if we are not conscious of them, they control us; we do not control them. This is not only dangerous, it leads us into all kinds of aberrations.

Healthy Cultures

For a business culture to be "healthy," it must have two basic qualities. It must be strong and it must be a culture that is appropriate for the business.

To be strong, a culture must have two characteristics.

- Its thoughts, beliefs, and actions must be widely accepted, acknowledged, and practiced across all levels and functions of the group.

- Its thoughts, beliefs, and actions must be in harmony with one another. There needs to be a logical consistency. For example, for a business that needs a highly structured environment, like that necessary for firefighters, it is not possible to have an "everyone can do it his own way" attitude and still expect things to get done. Those behaviors are not consistent.

Simply put, to have a strong culture, the thoughts, beliefs, and actions must be consistent with each other, as well as vertically and horizontally integrated throughout the group.

The second quality of a healthy culture is that it must be appropriate. Appropriate for the needs of the group and, in this case, the needs of the business. For example, a culture that is appropriate for a professional football team would not be appropriate for a business such as Disney, and may not be appropriate for a manufacturing facility. This concept of an appropriate culture is missed by most. Often, a culture that has demonstrated success is one that many will want to copy. Take the successful football team that demonstrates their competence by winning the Super Bowl, for example. The coach is often next seen on the motivational speaker's circuit explaining the "Road to Success" or some such thing. Altogether too often, many business managers flock to these meetings trying to get the most recent success formula, believing fully that if they could embody the principles of the football team, they too could be successful. Well, it just doesn't work that way.

Appropriate Cultures: An Example

A key cultural characteristic is leadership style. This characteristic then dictates a whole set of behaviors by both the leaders and the followers within the culture. Now back to our football metaphor. For example, on game day, the quarterback is the offensive leader. It is his responsibility to call the plays and audibles. In his role as the "on-field leader," he will perform based on his judgment alone. In so doing, he does not consult with those affected. His actions are very autocratic. In fact, he acts in a very dictatorial fashion. In Super Bowl I, Bart Starr was the quarterback of the winning Green Bay Packers, and Forrest Gregg was his all-pro offensive tackle. While he was in the huddle, calling a play, do you think he might have bent over to Forrest and said something like:

> "Well Forrest, I know you have blocked that big guy across from you for some time now, and I know you could use some rest. I really appreciate your efforts and am pleased to play with you. But would it be all right if we ran over your position just one more time? We really do need the yardage to get the first down, and I firmly believe this will be the best thing for the team. But before we do anything that might affect you, we wanted to solicit your opinion. So do you think you could support that and knock that big guy on his butt one more time?"

Well, for a quarterback to say such a thing in a football game would be patently ridiculous. The quarterback, the team leader on the field, is a dictator—there is no better word to describe his style. He is not seeking input or agreement, nor is he trying to create relationships. He tells people exactly what to do, when to do it, and how to do it, and if they do not do it, he has them replaced—they have no options. Worker freedom, on this occasion, is nonexistent. The quarterback is under extreme time pressure and his necessary style of leadership is dictatorial. Yet no one finds this odd. It is the way it has to be for that business. It is *appropriate* for that business, at that time.

So when the plant manager looks for the answers to his cultural problems in the football team, he is often looking within the wrong type of a culture. He may not be able to learn much from the football coach and apply it directly to his culture. What is appropriate for a football team is not necessarily appropriate for a manufacturing plant.

Cultures and Dependence, Independence and Interdependence

When we are little children, we are totally dependent upon our parents. This dependence, coupled with our survival instincts, goes a long way towards shaping our personality. But as we grow and get older, we are expected to become far more independent: being able to dress ourselves, later keep our rooms clean, and still later do some work around the house. Becoming independent is also a sign of maturity, and is often equated with maturity. We are so enamored by independence, we coined the phrase, "rugged independence" to somehow capture the American spirit. Well some of us just don't buy into this as the highest of ideals.

In fact, I for one, do not even think it is an accurate description of reality. A far more accurate representation of reality is the concept of *inter*dependence. It is the concept that gives recognition to the reality that all things are intertwined, and if one aspect of an entity is changed, almost without fail, sympathetic changes occur in other aspects. It is a key aspect of "systems thinking." For example, at the human level, if you change your exercise habits, your patterns of eating and sleeping will naturally change because your body is an interdependent system. At the family level, if one person gets sick, frequently all are affected, because the family is an interdependent system. At the business level, production lead times cannot change without a resultant change in inventory, overtime, or delivery performance. Thus, your business is an interdependent system as well.

Some call this "systems theory," as I mentioned, and parts of it are taught in engineering, business, and medical schools. Systems theory implies that whenever one part of a system changes, other parts must also be able to adjust or the entire system will break down. This adjustment requires several characteristics.

- The system must be able to recognize that a change is occurring, and it must have a conscious awareness of its state.
- It must be flexible enough to make the change.
- The system must be responsive.

This is true of all systems, including human systems.

I still find a large number of managers who either do not understand this or do not believe in it. For example, I still see managers deciding that to improve bottom-line profit, all they need to do is cut labor. As if there is no impact other than to reduce the overall costs.

- What happens to the overall skill level? Is it affected?
- What about the morale? Will it cause a reduction in productivity when they see the layoffs?
- What about the effect of working as a team now that some members are gone? Certainly this has an effect.

But it is easier for the manager—*not better, just easier*—to ignore the impacts and do some simple straight line mathematics, as if that represented reality.

We see the same thing with expedited freight, for example. I have seen many managers who decided that they needed to reduce this cost (they must think their personnel incompetent and so expedited the shipments, even though they didn't have to). When the manager imposes limits on their ability to expedite the shipments, what happens to on-time delivery or overtime? This will impact the system somewhere, but where? Or maybe their people *are* incompetent or poorly trained, or maybe they don't understand the goals and objectives. If they are any of these, then these managers have some failing in their management system, which also needs changing. Yes, it is very much intertwined, this systems concept.

> "**T**he key to the Toyota Way and what makes Toyota stand out is not any of the individual elements....But what is important is having all the elements together as a system. It must be practiced every day in a very consistent manner....not in spurts. **"**
>
> Taiichi Ohno

The beauty of Lean is that it recognizes these concepts of systems and interdependence. For example, the concept of transparency is prevalent, so we can understand and become conscious of the workings of the system. These concepts were understood early on during the formative phases of the TPS (Toyota Production System). Ohno is quoted as saying:

It is clear Ohno recognized the systems issues and the concept of interdependence.

How Are Cultures Developed?

Most cultures are developed unconsciously. The level of awareness is more about the behavior, about the actions, than it is about the underlying culture. It is this behavior that then begins to create the culture. Take a business—a restaurant, for example. In our restaurant, we are in business to make money. We decide we want to cater to upscale patrons with the upscale prices they are willing to pay. However, we must also develop a group of chefs and waiters who are compatible with our upscale business. Let's discuss the waiters, for example. In our upscale restaurant, we will need waiters with significant social skills, like the ability to carry on a conversation—something we would not need, nor even want, if our restaurant was a fast-food business. At our upscale restaurant, we might even want them to get some training in handling customers, or we ourselves might even give it to them. As soon as we start to define the behaviors we want, we start to define our culture. Now, as we proceed to develop our business, we further define the behaviors—the skills—we need. This then goes further toward developing the culture we will have. The more aware we are of the behaviors we seek and the impacts of these behaviors, the more we understand our culture. Hence, the culture is developed based on the required action of the personnel, which is dictated by the needs of the business. Pretty simple, huh? Well, *not* so simple, because there are always conflicts. But more on that later.

How Do We Plan a Cultural Change?

In most cases, when some top manager decides his culture needs to be changed, he is aware of some weaknesses that he would like to see corrected. Maybe he thinks he needs to modify his culture to be like that of Toyota, for instance. That being the case, it is then necessary to do a cultural evaluation. This is a tool that allows an evaluation of the present situation of the culture—how it is structured, right now. We do this using a variety of techniques, including a formal written survey, individual and group interviews, and observations of the culture in action. These data are compiled and compared on our Cultural Matrix. This will give a good picture of what the culture is like now.

Next, it must be determined what type of culture your business needs. This can be done in a facilitated workshop, and will determine what type and structure of a culture will be needed in the future. With these data in hand, it is now possible to define and begin the changes needed.

Following the cultural evaluations of the present state and the possible future state, improvement and growth then focuses on defining and solving problems in three major areas of opportunity. These are:

- Are the thoughts, beliefs, and actions appropriate for the business and meeting the needs of the customers? (Appropriateness of the Culture)

- Are the thoughts, beliefs, and actions in harmony, one with the other? (Harmony of the Culture)

- Are the thoughts, beliefs, and actions disintegrated either vertically or laterally in the facility? (Integration of the Culture)

This becomes a project and is managed like most large projects with one huge exception. The desired cultural changes must be prioritized and the largest cultural changes done first. Recall that every cultural change will affect all other aspects of the culture, so once the first major change is accomplished; a minor reevaluation is in order. It sounds complicated, but in the hands of a cultural expert, many of the dependent changes are predictable.

How Do We Modify the Culture?

At this point, it is worth discussing how we might proceed to make the changes. Let's say we want to implement our Lean initiative. There are two schools of thought on how to go about changing the culture. They center on the question, "Should we change a man's attitude so his behavior will change (first point), or should we change their behavior and expect their attitude to follow (second point)?"

Once I did a great deal of research on this topic and was able to find statistically validated studies that scientifically proved the first point, to the exclusion of the second. I also found studies that proved the second point, to the exclusion of the first. Go figure. I guess they both work. However, I am a bigger fan of the second premise. I have found that either will work in the short term, but the issues that tend to be sustained are based in action, which yields results, and that in turn creates a sense of satisfaction, which I have learned is a natural motivator. I have seen much of this "rah-rah" positive mental approach stuff (the first point), which in the end usually fails. It fails because it is not supported by the needed actions, especially the actions of management. Consequently, it dies of its own weight and does not sustain itself. So how do we proceed to change the culture?

Simple, we put together a plan that is action-based. We do this with a lot of help; remember most of the culture is unconscious to those within the culture, so it is almost guaranteed that to properly modify your culture, help is needed from someone who has not been "contaminated" by your culture.

The Toyota Production System and Its Culture

After having studied the culture of Toyota, and having worked with many Toyota facilities and suppliers, I have come to not only appreciate but quite frankly admire the Toyota culture. It is the healthiest culture of any business I have ever studied. It is a culture that is both strong and appropriate for their business.

An Appropriate Culture

Their culture is absolutely appropriate for their business. First, while using their culture as a strong tool to improve their business, they have grown from a small manufacturing firm making a few thousand small trucks per year to a manufacturing giant. Second, Toyota has not only grown, their production system has been copied all over the world in every type of manufacturing environment. Finally, they have used their culture as a strong weapon in the battle for survival and prosperity. In today's environment, while Chrysler and General Motors are struggling to survive, Toyota continues to prosper. All of this is largely due to the culture within Toyota.

A Healthy Culture

The culture is healthy when the beliefs, thoughts, and actions are consistent throughout the business. These attributes are vertically integrated as well as laterally integrated throughout the entire Toyota business, suppliers included. They are so consistent it is almost boring. Furthermore, it is such a strong culture that it has developed a language of its own with new words like *"kanban"* and "autonomation," and old words like "leveling" that have an entirely new meaning.

A Culture of Management Responsibility

A facet of Toyota that has always stood out to me was their unrelenting respect for people. Even more so is their thorough acceptance of responsibility. Especially the responsibilities of management. Think about their policy of no layoffs. This has been their policy since before the days of Kiichiro Toyoda, who resigned as President in disgrace in 1948. His responsibility was manifest when he resigned because forced layoffs were required to avoid bankruptcy—all to save the company. Toyota has maintained this policy of no layoffs, even until today. It is clearly a part of their culture. A part that is neither well understood nor appreciated by Western managers.

Today, a few companies have agreements like this with employees. These companies are not the norm, but some do exist. However, I can provide no examples where, when the company suffered financially, the President stood up, took responsibility for the problems, and resigned. Quite the opposite is the norm. The stories are legion of those CEOs whose companies failed but they dropped out to safe landings with their golden parachutes. But back to Mr. Toyoda for a second. Can you imagine what impact it has—on the culture, on the entire organization, and on the managers and workers alike—to know that this level of responsibility is not only expected of all, but is also practiced by the highest echelon of management? Well, it is huge, to say the least.

What Toyota is saying through their actions is what any responsible company would say, that is: "If the company fails, it is because management has failed." Across the business world, unfortunately, what is seen differs greatly from what Toyota says. Rather, almost without exception, management will take credit for the successes, but not for the failures. As soon as a failure appears, it is amazing how management will work to find both circumstances and "others" to blame. Yet at the first blush of success, they will stand in line to not only collect the accolades but the bonuses as well.

What kind of logic supports that? It is said in a folksy way that, "Success has many fathers, but failure is an orphan."

Well, some management teams will take responsibility for their failures … or sort of. They will talk about it and say nice things like, "We take full responsibility" or "the buck stops here," or some nice-sounding phrase—and they might even feel bad about it. But that is not the measure of what true responsibility is really about—it is about RESPONSE-ability. Responsibility is simply the ability to respond. So, just how do they respond? Clearly Kiichiro Toyoda not only felt bad, not only did he say he took responsibility, but his actions were unmistakably supportive of what he said. This man, and hence this company, "walks the talk." He set an example for what is expected of each Toyota Manager, what is expected of each Toyota Supervisor, and what is expected of each Toyota employee. Consequently, they have a culture that expects responsibility, and they work hard to maintain that. They are responsible.

A Culture of Worker Responsibility: *Jidoka* and Line Shutdowns

The second example, also focusing on responsibility, but closer to the shop floor, is the *jidoka* principle that, upon finding a defect, the line is shut down and not started until the problem has been resolved. This line stoppage is done by the worker or anyone who finds a defect. It is not only their right to do this, it is their responsibility. Toyota considers this a good, if not mandatory, business practice.

Most American plants would consider this like turning the asylum over to the inmates. At the typical American plant, this idea is close to the truth. Not because the workers are insane, but often because they do not have what I call "the context of the problem." Consequently, when they find a defect, they cannot make a good business decision to shut down or keep it producing. They do not understand enough about the problems and the consequences of shutting down the line, or the consequences of *not* shutting down the line.

What does this say for the Toyota culture? Let me just mention three aspects.

- It shows a great respect for the decision-making ability of everyone. What this means culturally is that they demand—and expect—respect from their people. It is a culture of individual respect.

- They have confidence in the training of these people. They not only train their people but expect them to use this training. It is a culture that values knowledge and training.

- It shows, beyond a doubt, that the worker is fully capable of producing 100 percent good product. It also says that if we cannot produce good product, we will produce no product. It is the ultimate statement of the importance of quality.

Again, how does your manufacturing system and your management measure up to this?

- Are your line workers shown this degree of respect? If not, why and what are the consequences of this?

- Is your training treated as a process with a high level of importance manifest through the actions of the training group? Or is training the first item to be compromised in a conflict, or actually cut when costs need to be trimmed?

- Are your workers trained and supported by a system that will allow them to produce 100 percent salable product? If not, why not, and what are the consequences of that?

What Should We Do with Our Lean Culture?

We can now see that the culture cannot be orchestrated directly; rather, the culture is the result of what we do. Consequently, it is not wise to develop a culture until we decide precisely what we must do, and it may take a while for us to decide just what we want to do in our Lean initiative. Let's say, for example, we want to implement line shutdowns by the line operator when a quality defect is found. What if we make one defect an hour? Are the workers trained to make the evaluations? Are the problem solvers ready to jump into action every hour? Are we ready to suffer the production impacts of a line stoppage each hour? These are but a few of the questions that need to be answered. If we do not have the right answers to these questions, we should not try to implement this Lean technique. Worse than not starting is to start, fail, and then regress to an earlier state.

This is why I recommend not worrying too much about the depth of your culture at this point. First, every change made will create a change in the culture. So, without trying to directly change the culture, it will change anyway. Second, the large cultural impacts— such as empowering workers to shut down production lines—did not come early, nor did they come easily to Toyota. Now it is commonplace in the TPS, but they did not start this way, nor should you. Finally, many people talk about cultures and changing cultures, but a scarce few really know what they are doing. At some point, it will be necessary to seek help in this aspect—even your *sensei* is probably not an expert in cultures.

Make the Minimum Necessary Changes

We have consistently avoided the topic of cultural change as a separate entity. Not because it will not be beneficial, but because there is plenty of technical and organizational stuff to do. Focused projects of cultural change are very time consuming and management-intensive, and often the changes obtained do not yield immediate financial benefits. We have found it is better to modify the manufacturing system first, much as Ohno did, and then focus on the culture later.

However, two aspects of the TPS culture should be addressed at the assessment. Two of The Five Precursors to Implementing a Lean Initiative are major cultural change issues. They are:

- A continuous improvement policy
- A policy to sustain the gains

Both of these should be included in the initial Lean Implementation Plan until both are at level 3, per the assessment.

What If We Still Want to Modify Our Culture?

If this is still your desire, I laud your attitude. There is no technique with more power to leverage your management skills than to work on your culture. Unfortunately, this is a very delicate area and it is like doing psychotherapy. Unless you are both very skilled and very careful, it is easy to do more damage than good. I strongly recommend contacting an expert and getting an assessment before proceeding with direct cultural change.

Next, make sure to have the time to take on the needed issues uncovered in the assessment. Simply implementing a Lean initiative is a huge time-consuming effort, so be careful. To proceed with a programmed effort to modify the culture, generally the first topic undertaken is, "How do we get the culture to not only change but to accept change and, furthermore, to embrace that change because it is not only needed but is also healthy for our culture and our business?" In so doing, I strongly recommend very heavy reliance on your *sensei* for advice. So strongly that what he says … *goes.*

Chapter Summary

A culture is the "actions, thoughts, beliefs, artifacts, and language of a group," and it could be the group at your facility. It is "how we do things around here." The culture is created and perpetuated by two factors. The history largely creates it, and the few people at the top tend to perpetuate it or change the cultural rules, most of which are silent rules. The silent rules often create a lot of problems in the culture. Healthy cultures are those that are appropriate for the group and also are widely shared by the group. Cultures can be classified, evaluated, and modified to a desired state. They can be corrected just as any problem can be corrected.

The Toyota management was acutely aware of its culture and created a healthy culture that is both strong and appropriate, and is a culture of responsibility at all levels. Regarding our Lean culture, my advice is to not try to orchestrate the culture directly, at least not initially. Work on the implementation of a Lean system emphasizing the necessary behaviors. Getting these behaviors in place will be a major undertaking by itself and in so doing the culture will begin to make major changes interdependently. Then, as your production system grows, with advice from your *sensei*, properly evaluate and develop the remaining portions of your culture, as they are needed.

CHAPTER **12**

Constraint Management

The management of bottlenecks is often overlooked in Lean applications because it is not a large part of the Toyota Production System. It, however, is often a strong tool in the Lean Tool Box since it can be used very powerfully for process improvements. In this book, I show numerous examples of bottleneck reduction which then led to huge process gains. Often, bottleneck reduction provided the "low-hanging fruit" for the early process gains. For this reason, it is included herein.

Bottleneck Theory

What Is a Bottleneck?

Bottlenecks are the limiting aspects of a process, much like how the neck of a wine bottle limits the flow of its contents as you pour. Often, instead of "bottleneck," we use the more sophisticated term "constraint." Every process has a constraint. There are constraints to our manufacturing processes, our engineering processes, and constraints to our business process. Every process that has an objective has a constraint, unless the objective is fully met. In a typical manufacturing process, the constraint is usually the step in the process that has the longest cycle time. If so, it can normally be identified with a standard time study, as described in Chap. 7.

Moving Constraints

In Lean Manufacturing, we are always striving to eliminate wastes. Often, this entails using the tool of line balancing. When we balance a line, we try to design the process steps so they all have the same cycle time. This brings about another phenomenon. When the cycle times are very similar, a small variation in one step can cause a cycle-time increase, making it the constraint at that moment. The variation disappears and yet another process step incurs a minor variation that causes *it* to now be the long cycle-time step, and thus the new process bottleneck. In this manner, the constraint moves from one process step to another. We call this a moving constraint. (Refer to Chap. 19 for an example of this.) If the variation of the individual process steps is not too large, we are often not concerned with this moving constraint and do little about it. If, however, the variation of the individual process steps is significant, this variation will measurably affect the process performance and thus the process will not produce to the design cycle time. In this case, there is a process loss. There is a cycle time loss. This loss can be quantified by the Lean metric of Overall Equipment Effectiveness (OEE) discussed in Chap. 4, and once it is quantified we can decide if we wish to work on this constraint to improve the process.

Some Constraints That Are Often Not Called Constraints

For the moment, think about a product you make, produced by some process. What if the process is performing well? For example, the quality yield is 100 percent, as is OEE and on-time delivery. The process is producing to *takt*, making good margins, and meeting customer demand in every respect. Do we still have a constraint? Well, if your company is in business to make money and you do not have 100 percent of the market for that product, then the answer is a resounding yes. In this case, the constraint is likely your sales department. Why sales? That sounds odd! The logic goes like this:

1. The business objective is to make money.
2. There is more market share to be had and more money to be made.
3. What is our limit?
4. Answer: The constraint is probably sales.

The constraint is not always a step in your process. It can be any aspect that limits your ability to meet your objective. The constraint can be the process itself; it may be a raw materials supplier; it may be a resource; or it may be another aspect of your business.

Point of Clarity All businesses have constraints, and these constraints limit the business's ability to make money!

Policy Constraints

The most disturbing constraints are often policy constraints. This can occur when the company makes policies that turn out to limit the facility. Seldom are these policies designed to be limits. Rather, they are often created with the best of intentions but without a good understanding of the intended or unintended consequences of the policy. Two types of policies drive me absolutely crazy.

- The first type is "We don't know what we are doing so we will create a policy to cover it." For example, one client explained to me that their policy on inventory was to have 30 days on hand. It was a corporate-wide policy. And no one could give even a rough explanation of why this 30 day policy existed.

- The second type is the "I don't trust you, so we will create a policy to limit your authority." An example is described next.

On one occasion, I was called in to make an evaluation of a production line. The plant manager needed to increase production by 38 percent and knew that the capacity constraint was their electrical testers. In an evaluation that took less than one hour (he wanted to discuss what I found over lunch), I was able to spot potential capacity increases of over 22 percent with no capital investments. The recommendations consisted of staffing the test station during breaks and lunch and moving one test station that had significant scrap. This test station was after the bottleneck, and by placing it in front of the bottleneck it would improve throughput instantly. I was feeling pretty good about myself and just figured he would jump at these ideas.

We met for lunch and although the plant manager was intrigued, he flatly rejected both ideas. These ideas would add about $200 per day in labor costs to the 24-hour operating expenses but would increase revenues by over $26,000 per day. They were making about 22 percent on sales, so this was extremely profitable, to say the least. The plant manager was interested, but it was the policy of the company (that ugly word, *policy*) that they would not increase manpower for any reason, above their current levels. Hence, he no longer had the authority to add these people. Well, as you might expect, there was a lot more to this experience ... but this was not the first, nor the last, time I encountered a business decision where "The Policy" was the system constraint. These constraints are usually very costly and frequently the management is blind to them and does not see them as constraints to the business.

The Economics of Constraints

The system constraint will limit the ability of the business to make money. However, in most cases, if the constraint is broken, the resultant increase in production is often the most profitable production the company has. Let's look at a simple example of a 5 step process, shown in Fig. 12-1, which shows the process cycle times for each process step.

It is clear the 1.0-hour process step is the constraint and will limit production to 24 units per day and, as shown in Table 12-1, profits will be $20/unit.

Let's say we have added possible sales and we want to increase production. We could design and build a complete new line, just like our 5 step process, but someone notices that the line is not well balanced and suggests we break the system constraint. The constraint is the one-hour cycle time at step 3. It is easy to see that if we wished to double production, we could duplicate the third step and place it in parallel with the third process step. We do this and have a new 5 step process, as shown in Fig. 12-2.

The third step would now produce two units in one hour. We have broken the process constraint and the new process constraint would be any one of the 0.5-h process steps—either steps 1, 3, or 5. Very likely, we could now produce up to 48 units per day, and on a good day our sales department could sell them and everyone would

| 0.5 hrs | 0.1 hrs | 1.0 hrs | 0.1 hrs | 0.5 hrs |

FIGURE 12-1 The 5 step process.

Cost Category	$/Unit
Sales Price	200
Variable Costs	20
Fixed Costs	60
Raw Materials Cost	100
Profits	20

TABLE 12-1 Economic Profile, One-Hour Constraint

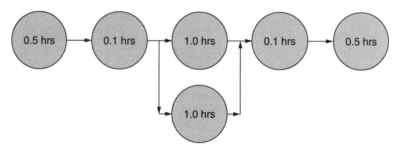

FIGURE 12-2   The new 5 step process, with added capacity at step 3.

be happy. But just how happy would they be? The economics of breaking the constraint are very important to understand. In the case with the 1.0-hr constraint, our profits were $20/unit.

How does breaking the constraint affect the profits? Our production unit will still need $100 of raw materials, and we will likely have most of the $20/unit variables costs as we use the consumables that it takes to make the unit. But the fixed costs are driven to almost zero. For example, our building and all the property taxes and all those fixed costs are already covered and will not be affected by the rate increase. Neither will such things as home office burden or staff and management costs, which are often very large. Simply put, the incremental fixed costs are very low, and for the sake of argument we will say they are zero. So let's look at Table 12-2 to see the economics for the incremental production volume after the constraint is broken.

In this case, with a $200 sales price, our profit per unit has risen from $20 for the first 24 units produced (up to the constraint) and $80 for all units produced after the first 24. Our profit margins have *quadrupled,* for all incremental production after the constraint was broken. If our sales department can sell, say 12 more units per day, our gross profits will go from $480 per day to $1440 ($20/unit × 24 units + $80/unit × 12) per day. With this you can see the power of "constraint economics."

> **P**oint of Clarity   The most profitable production occurs after a system constraint has been broken.

System constraints need to be understood and aggressively attacked because this is one of the most powerful tools available for improving the economics of a business. This concept can not be understated. When you read the stories of the Bravo Line in Chap. 15 and Larana Manufacturing in Chap. 16, pay particular attention to the huge

Cost Category	$/Unit
Sales Price	200
Variable Costs	20
Fixed Costs	0
Raw Materials Cost	100
Profits	80

TABLE 12-2   Economic Profile, One-Hour Constraint Broken

early gains that were achieved through breaking the system constraints. To a large extent, this was how they "harvested so much low-hanging fruit."

When Is It Best to Address Constraints?

Very often, during the review of the foundational issues, constraints will be found and highlighted. It is also, in this timeframe, that the constraints should be aggressively attacked and removed. First, it will make money for the facility and you will be able to accrue huge early gains. And second, it is more efficient to achieve the appropriate process flow rate, stabilize the flow, and then work on quantity control techniques like *kanban*. However, everyone should constantly have an eye out for constraints and should be attuned to removing them. After all, once the system constraint is broken, the system's profitability is vastly improved.

How Do We Spot the System Constraint?

This is the simplest of questions. Look for the inventory build-up. It will always be in front of the constraint.

What if there is inventory between all process steps? It will then be the largest pile of inventory. What if there is inventory galore, as is so often the case? Then do a process study to calculate the time for each step. The step with the longest cycle time is usually the constraint. I say usually, not always. It is possible that a given process step may not have the longest average cycle time, but if it has huge variations in the cycle time, this step may be the system constraint.

Often when a multistep process has a lot of WIP (work in process), it may experience a broad array of problems. One of which is that it may operate at a cycle time that is well above its potential, thus producing at a rate well below its design. This is especially true if it is dominated by manual operations. In this case, unless there is some clear way to measure and control the cycle time, the system constraint may bounce from one station to the next and the whole line will under-produce. This is commonplace (see Chap. 16). Consequently, at the end of the day when the production quota was not met, no one can point to any specific problems that the line experienced. Often, what follows is some of the most elaborate, but incorrect, rationales for why the production goal was not met.

The answer, of course, is first understand and reduce the variation and then get rid of the WIP inventory so you can "see" the process. This will allow you to find the constraint and work on it—if necessary.

Almost always, once the inventory is removed, the entire process, often every step, will speed up. To most people's amazement, the quality will improve as well. It is not logical to most that this should occur—that is, a simultaneous improvement in both quality and rate. However, it is a metaphysical truth, and also a very common occurrence, that production will speed up and quality will improve once the inventory is eliminated.

> **P**oint of Clarity When we rationalize, we generally just create "rational lies."

Why Ohno Does Not Even Talk about Constraints

If this constraint stuff is so important, why is there nothing of it in either Ohno's writings or those of Shingo? There are two explanations. First, Ohno and Shingo are so good at process design and management that achieving minimum cycle times is second nature to them, they take it for granted—we seldom fret about those things we take for granted. Second, their approach gets them to a minimum cycle time anyway. They break constraints by first removing the waste and then rebalancing the work.

Chapter Summary

Every process has a bottleneck, a constraint that limits its ability to produce. The constraint can be a specific step that has the longest cycle time, or if the steps have large variations in cycle time, the constraint can be a moving one. Generally speaking, the worst constraint is a policy constraint because they are often not recognized as things that can limit the system. The beauty of breaking through system constraints can be seen in the economics of constraint removal, because the incremental product produced, once the constraint is broken, is much more profitable. Constraints are generally easy to find since the inventory will accumulate in front of the constraint.

CHAPTER 13

Cellular Manufacturing

Cellular manufacturing is a key element of Lean manufacturing—this is common knowledge. What is not so commonly understood is how powerful a variation reduction tool cells can be by virtue of their design. This is a well-kept secret. In addition to explaining the benefits of cells, and how they fit into the battle of waste reduction, I will guide you through an actual redesign of a complex production line that was converted from a long flow line into cellular manufacturing. Grab a pencil and pocket calculator and follow me through the calculations as we redesign the line and *improve the production rate by 63 percent, with virtually no capital expenditures—and less manpower!*

You will be amazed!

Cellular Manufacturing

The Definition of a Cell

A cell is a combination of people, equipment, and workstations organized in the order of process flow, to manufacture all or part of a production unit. I make little distinction between a cell and what is sometimes called a flow line. However, the implication of a cell is that it:

- Has one-piece, or very small lot, flow
- Is often used for a family of products
- Has equipment that is right-sized and very specific for this cell
- Is usually arranged in a C or U shape so the incoming raw materials and outgoing finished goods are easily monitored
- Has cross-trained people for flexibility

Cells are advertised as taking up less space than "island" production, but I find this is not always the case.

The Advantages of Cells

Cells are an integral part of Lean manufacturing. The use of cells is so basic that in the TPS (Toyota Production System) it is not even questioned. For Toyota, that works. Unfortunately, for others, cells may not work in quite the same way, so it is worth understanding the benefits and drawbacks of cells before we embark on a full-blown effort to convert everything to cells.

The primary purpose of a cell, is to reduce wastes in the manufacturing system. The seven wastes again are

- Transportation
- Waiting
- Overproduction
- Defects
- Inventory
- Movement
- Excess processing

It is easy to see that cells reduce transportation due to the close coupled nature of the workstations. They do nothing directly to reduce waiting since that is a function of work balancing and variation. However, cells make the balancing much easier to manage. Cells, in and of themselves, also do nothing directly to prevent overproduction or defects. They do minimize the inventory when one-piece flow is achieved, which is their basic design. As for movement, they actually promote movement, more efficient movement, depending upon the cell design and do nothing to reduce excess processing.

Thus, cellular production is designed to reduce the wastes of transportation and inventory. Consequently, it is also designed to speed up the process and make it flow. Cells almost always do this, resulting in production advantages such as:

- The reduction of first-piece lead time
- The reduction of lot lead time

So with the reduced lead times, we have greater flexibility and responsiveness.

However, both of these benefits can be achieved with a flow line. A flow line is a linear, rather than a U- or C-shaped, closed arrangement. To achieve these benefits in a flow line, there needs to be the same low inventory approach as well as making sure the stations are close coupled as in a cell. So why are cells so popular? There are other benefits to cells over flow lines that may be a bit harder to quantify, but are possibly even more powerful reasons to choose cells over flow lines in many manufacturing circumstances.

Cells or Flow Lines?

Just what are these other benefits in choosing cells over flow lines?

The first and often the largest reason to select cells over flow lines is the production rate flexibility possible with cells. For example, let's say we have a typical balanced cell staffed with six work stations and six workers. If we use three workers instead of six, we can have each worker do the work of two stations and this will double the cycle time or halve the production rate. If the workers are designed to move from station to station, it is possible to use either one, two, three, four, five, or six workers and get 16 percent, 33 percent, 50 percent, 67 percent, 83 percent, or 100 percent of capacity with no increase in labor unit costs. This allows the cell to operate at different rates as the customer's demand changes. This rate modulation is much more effective than starting up, running the cell at full rate, and then shutting down when the month's demand has been met. The cycling up and shutting down is nothing more than creating batches. The TPS is a batch destruction system, not a batch creation system. This rate modulation by modifying staffing is not practical to do with a flow line.

When cells are arranged in a C or U shape, worker communication is facilitated. For example, all workers are in proximity to one another, so worker interaction is encouraged. Worker interaction to assist in cross training and worker interaction to assist in problem solving are just two such benefits. This proximity just makes communication much simpler. They are also not only able to communicate better but assist each other as well.

The typical U-cell situates the first and last work stations near each other. This makes cell supervision much easier and gives everyone a better sense of work completion. Also, in the typical U-cell, workers usually sit or walk in the center of the cell. This frees up the exterior of the cell to supply materials to the cell more easily.

Two Hidden Benefits of Cells

All of the items listed earlier are benefits, but in my experience I have found there are two major benefits of cellular manufacturing that are seldom mentioned, but that are very real, very positive, and very powerful.

First, the very nature of a cell creates a team with a sense of flow and synchronization not seen in flow lines. In the flow line, you have two neighbors; in the cell, everyone is in close proximity. The personal dynamics are changed considerably, leading to a feeling of a group, of a team. The team concept is very powerful and there is a real sense of assisting each other. In the cell, since the process is all around the worker, there is a sense of flow and a sense of synchronization that is not present in the flow line. We have documented cases that show this sense of flow and synchronization actually creating a faster pace in the cell with reduced cycle times. We have found that it is not uncommon for cells to reduce cycle time by 10 and even 20 percent as they mature. I have often witnessed this cycle-time improvement in cells, yet I hear many engineers attribute it to training and worker maturity. These same engineers, however, cannot explain why we do not see the same benefits in a flow line as it matures.

Nevertheless, the greatest benefit of cells is a well-kept secret. Cells are a tremendous tool to assist in reduction of variation.

I will later describe a case study of a high-volume production flow line that was converted to cellular production. The plant achieved the rate modulating benefits earlier mentioned, but in addition—with less staff and the same equipment—the cellular option, compared to the flow line, was able to improve production by over 63 percent.

The Gamma Line Redesign to Cellular Manufacturing

The Background

The Gamma Line was a 21-station flow line with a 16-second cycle time. The first 16 stations were all manual assembly, followed by two tests and three packaging stations, which utilized some expensive test and packaging equipment. Material was delivered to one side of the 200′ long assembly line, the same side on which the workers were stationed. Each station was staffed with one worker. The first 16 stations had almost no automation; the most sophisticated tools were some ergonomic screwdriver stations. It was a highly labor intensive production line.

Since most of the skills were very basic, the operators would learn all 21 stations in less than six months with little effort or inconvenience to the work schedule. They had an aggressive operator cross-training program.

For this corporation, new production lines were designed by the home office engineering staff and then sent to the facility to debug them and bring everything up to speed. Consequently, this facility did not have a full engineering staff.

This production line was a straight flow line with a conveyor. The conveyor would advance and then remain stopped for the 14-second work cycle time; the transportation time between work stations was two seconds. The appliances were mounted on rotating tables to facilitate construction, and the tables were fixed to the conveyor. There was no new technology in this line, but the design cycle time was considerably shorter than prior designs. It stood out in the facility as being different and placed significant pressure on everyone, especially the materials delivery staff. But the demand for this product was high and the management wanted to only invest in one set of equipment—hence, one high-velocity line.

Earlier, we had done significant training in statistical problem solving at the facility and conducted two waves of Greenbelt training, as well as one wave of Blackbelt training. Following this, Greg, the general manager asked us to review the operation of the Gamma Line. He said he was in a hurry and wanted us to evaluate why the labor efficiency was so low. Labor efficiency was one of their most important plant metrics.

Greg explained the situation. The line had been placed in service over three months before and had never achieved design rates. As the demand ramped up, they had to schedule Saturdays and even some Sundays to meet demand. That was why the labor efficiency was low. In addition, the union was becoming a significant obstacle. From the beginning, the union was against the design. The shop steward claimed the short cycle times placed too much pressure on the workers. Greg had the same concern.

We were not familiar with the metric of labor efficiency, so we asked about this metric and the 0.85 standard. It was explained thusly: It is the allotted labor, which is based on the cycle time and design line staffing, divided by the actual hours worked for all the hourly workers. Each line is calculated based on only their direct labor headcount for those on the line, including those not working. The 0.85 factor was to account for labor that was scheduled and working but not producing due to machinery failures or anything that would cause production to be reduced. We asked if the production losses included quality dropout, stock outs, machine downtime, and not performing to cycle time. They said that was the concept and the minimum standard was 0.85. Anything over 0.85 (or 85 percent) was gravy. On this line, the best they had done was just recently when they struggled to get to 70 percent.

Without knowing it, their labor efficiency was a type of OEE. Not as powerful and not as usable as OEE, but a very similar concept. The unfortunate part of this metric was twofold. First, their objective was to reach 85 percent. If they got there, they would be happy. There was no strategy to go beyond the 85 percent. Second, this labor efficiency was affected by all the aspects that affect OEE, such as quality, machine availability, material supply, and cycle-time stability—to name the major issues. Yet they framed it as a labor issue. Quite frankly, it was everything *but* a labor issue.

A problem exists when a facility is managed as only a cost center. Clearly their only effort was to make sure they did not use too much labor. But as we will see, labor was not their problem—they had neither begun to understand, nor begun to attack their real problems. It shall be shown that their true problems were waste and variation—which is just waste by another name.

Greg went on to explain some more history of the line. He said that upon startup, unlike prior startups, tremendous scrap was generated, over 30 percent. When they slowed down the line for a while, scrap dropped, but when the line was sped up again, scrap increased to unacceptable levels. The workers claimed they did not have enough time to complete the tasks and that the short cycle times caused a lot of stress.

To remove the stress from the workers, they tried the Red-Tag Procedure. The company told the workers that if they could not finish a unit, they should leave it on the line and simply place a red tag on the unit. When the unit progressed down the production line, no one would work on it, and just prior to the test station all red-tagged units would be removed and sent to the rework station for completion. Unfortunately, this did not work at all. The first rework station quickly became overloaded and when the second station became overloaded also, the company decided a change was necessary. To make product, they again slowed down the line while they worked on a solution.

After some thought and interaction with the union, it was decided to give the operators a little control over the line. At each work station, they placed a delay button so the operator could delay the advance of the conveyor and hold it for another five seconds. If the operator was afraid, he/she would not finish the unit in the allotted 14 seconds; the operator would hit the delay button and the conveyor would be held for five additional seconds. Greg said it had been the plan to institute line stoppages for quality issues, their attempt at *jidoka*, but since the line was not yet stable, they had not implemented this feature. The company thought this might be a good transition into *jidoka*, as well as solve the current production problem.

Immediately, the number of red-tagged units dropped to practically zero and the line produced with only a minor quality issue. There data showed there was 1.2 percent rejected product at functional test and electrical test combined. This could all be reworked. In addition, there was another 2.9 percent rejected from final inspection, mostly cosmetic defects associated with handling. This, too, could be reworked. Unfortunately, the line rate, since installing the operator-controlled line delays, had not achieved rate. They struggled merely to reach 70 percent. (The design rate, with no losses should be 225 units per hour, [60 × 60/16] but were only able to produce 155 units per hour, or about 70 percent).

Greg mentioned he was impressed by our training and asked us to review the line operation and see if a redesign was in order. We had previously spoken with him about some of the problems inherent in a long flow line—especially one with very short cycle times. He asked for our recommendations and requested we look at converting the flow line to cellular manufacturing with the caveat that there was very little capital available for new equipment. His guidance was to maximize the production rate. Although customer demand was 1,000,000 units per year, demand changes were common, and if the production could be increased and costs reduced, the company would consider dropping the price in an effort to increase market share.

Strategy 1: Synchronizing Supply to the Customer, Externally

Customer demand was produced by running 24 hours per day, five days per week, less 11 holidays. They had 250 scheduled days per year to produce the 1,000,000 units, hence daily production should be 4000 salable units. Each of the three operating shifts had 50 minutes for lunch, along with their breaks, so the available time was 21.5 hours per day with a design rate of 186 units needed per hour. Hence, *takt* should be about 19.4 seconds (21.5 × 3600/4000), and with an OEE of 0.85 the necessary cycle time should be around 16.5 seconds. Their design of 16 seconds should have worked nicely… But it didn't. In fact, it wasn't even close. But why wasn't it?

Our first step was to look at 24 hours of production. The data culled from this are shown in Table 13-1.

The average for this day was 163 units/hr (3516/21.5), which was well below the design rate of 186 units/hr, although our data were slightly above the recent average of 155 units/hr. In addition to the low rate, there was large hour-to-hour variation. Hourly production rates varied from a high of 204 to a low of 156 (ignoring the first hour of each shift, which is naturally erratic due to the shift change). This range of 48 units is plus or minus 15 percent of the average. No one could give any rational explanation for the rate variation beyond the obvious explanation that the operators were hitting the

Shift	Hour	Production	Cumulative per Shift	Cum Total
Shift No 1	7am	156	156	156
	8am	180	336	336
	9am*	132	468	468
	10am	180	648	648
	11am**	84	732	732
	12n	192	924	924
	1pm*	132	1056	1056
	2pm	204	1260	1260
Shift No 2	3pm	156	156	1416
	4pm	168	324	1584
	5pm*	132	456	1716
	6pm	156	612	1872
	7pm**	72	684	1944
	8pm	156	840	2100
	9pm*	120	960	2220
	10pm	156	1116	2376
Shift No 3	11pm	144	144	2520
	12m	156	300	2676
	1am*	132	432	2808
	2am	180	612	2988
	3am**	84	696	3072
	4am	156	852	3228
	5am*	132	984	3360
	6am	156	1140	3516
	Total	3516	3516	3516

*30-min. lunch

**10-min. break

TABLE 13-1 Gamma Line Production, Base Case

delay button. At first glance, the production numbers do not look too bad until we recall that the design of 186 was based on an OEE of 85 percent, or 15 percent losses due to quality, availability, and cycle-time losses. The data we analyzed showed:

- Quality losses to be 3.1 percent
- Zero availability losses
- All other losses were cycle-time losses

So what were the cycle-time losses? The design hourly rate at 100 percent efficiency was 225 units but we produced 163 units so losses were 62 units or 27.6 percent. If quality and availability losses total 3.1 percent this means cycle time losses total 27.6 percent –3.1 percent or 24.5 percent … which is huge! With all these losses the effective cycle time is 22 seconds ($21.5 \times 3600/3516$).

Our design is synchronized externally since we need a 16.4-second cycle time and have a design of 16 seconds, which supposedly should work nicely. It does not, however, and the reason it does not work is because the production cycle time is not producing anywhere near *takt* × OEE.

So, just what can we do about it?

Well, we need to redesign the production process so it will produce to *takt*, and to do that, we will proceed to Strategy 2: Synchronize Production, Internally.

Strategy 2: Synchronize Production, Internally

A New Approach, Multiple Work Cells

We will attack this problem with cellular design, which will directly reduce the variation and better balance the flow, using multiple cells in our new design.

The questions we must first answer are:

- How many cells?
- How many persons per cell?
- What is the cell layout?

The Time Study and Line Balance Review

A time study was performed and a line balance chart was constructed (see Fig. 13-1).

As we evaluate the line balance chart, we find the following.

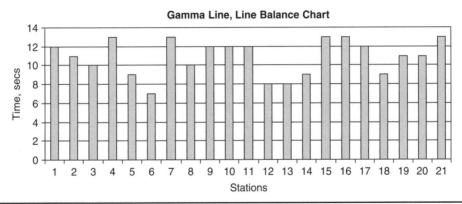

FIGURE 13-1 Gamma Line balance chart, base case.

The first evaluation to make on a balancing chart (see Appendix C in Chap. 7), is to evaluate the waiting time. Recall that the waiting time is the distance from design time to the station cycle time, plus the time for transportation. In this case, the waiting time for station 1 is two seconds, plus two seconds for transportation for a total of four seconds, station 2 waiting time is three seconds plus two for transportation for a total of five … and so on, rendering the total wait time as 108 seconds per unit produced. This is the *design* waiting time, not the actual waiting time. Quite frankly, that is both huge and amazing. The actual work time is 228 seconds per unit for a total labor time of 336 seconds per unit, this means the designers planned for labor losses, due to waiting alone, of 32 percent [$100 \times (108/336)$]. However, since the process is not able to perform as designed, our labor used is 462 seconds per unit (22 seconds × 21 stations). The *actual* waiting losses are 234 seconds per unit [$(22 \times 21) - 228$]. And now it is obvious that the time losses exceed the actual work time. To put this in simple terms, "we need a lot more time and a lot more labor than we planned for." So, in summary:

- We have 228 seconds of planned work in each unit.
- We are paying for 462 seconds of labor per unit.
- Our actual wastes due to waiting are 234 seconds per unit.
- The planned waste was 108 seconds per unit.

These represent huge losses that we may be able to exploit in the redesign.

Second, we evaluate the balance, qualitatively. In this case, the balance is very bad. There are five stations with a 13-second cycle time, and seven with cycle times of ten or less. That is a 30 percent variation, which then translates into a lot of wasted labor. Clearly, rebalancing will benefit this process.

Third, we find the bottleneck. In this case, we only evaluated to the nearest second, so it appears there are five bottlenecks. There is, by definition, one bottleneck in any system.

Although we do not know which station is the bottleneck, the evaluation is extremely revealing. First, it shows the average times are all below *takt*, yet we know the process cannot sustain this average rate. The graph also shows that there are five stations right near the 14-second maximum time. Consequently, if a slight variation in cycle time occurs, the operator would likely hit the delay button and this would delay the entire process for five seconds. Since on every unit of production, this opportunity is presented seven times, I was actually amazed we were able to produce as much as we did. Recall that our real cycle time was 22 seconds. If we add the designed cycle time of 16 seconds (14 seconds of work plus 2 seconds of transportation) to the time of one delay, 5 seconds, and compare this total of 21 seconds to the actual cycle time of 22 seconds, it means, that on average the delay button is struck once per production unit. It is clear that a redesign is needed.

How Many Cells? How Many Stations per Cell?

With the time study and balancing study completed and evaluated, we needed to come up with a reasonable cell design. As you recall, little capital was available to spend on this project, so we were not able to change the end of the line, which had some expensive machinery. Stations 17 through 21 consisted of two tests and three packaging stations, which culminated in pallets of 12 units being shrink-wrapped and ready to load on a semi. These five stations all produced at cycle times below *takt* and had a total of

56 seconds of work. Consequently, we decided we would design multiple cells using the work from the first 16 stations, (which totaled 172 seconds of work) and these cells would then feed into our final cell, which would be testing and packaging. Now, what would those multiple cells that feed test and packaging look like?

We knew we wanted to increase the cycle time. The short cycle times were a root cause for much of the waiting time, and the increase in cycle time was causing the low production. We have found that, for repetitive small work of this nature, cycle times of 30 to 90 seconds seem to work best. So we did some rough calculations.

Each cell will now have 172 seconds of work, if we use four work stations and can balance perfectly, each station will have $172/4 = 43$ seconds of work plus 2 seconds for transportation. With three cells, operating in parallel, that means we would produce 3 units per 45 seconds, or a theoretical cycle time for overall production of 15 seconds compared to our design cycle time of 16.5 seconds. On the surface that looks good, but we know we would not be able to balance the work stations perfectly and since we were not really worried about overproduction, we decided upon four cells. At this point, Greg chimed in and mentioned they had already approved some design improvements in final packaging and test, which would reduce the required work on the fifth cell.

So our final design was four cells of four stations, each working in parallel, and feeding a fifth cell. We decided to lay out all four new cells in the U design layout and leave cell 5 in a straight flow line with five work stations.

We now need to calculate the approximate cycle time. We have 43 seconds of work for the cell, and if we allow two seconds to pass the unit to the next station in the cell, we would have a cycle time of 45 seconds. If we use four cells in parallel, we could produce four units in 45 seconds, or about an 11-second cycle time. We are well below *takt*, were producing more with the same staff and so we knew we were on the right track.

Wow! Were we happy with that! And we had just begun!

Why Not More Cells?

First, when the restraint of no available capital was placed on us that set the lower limit of the cycle time at about 12 seconds. This is the current cycle time for the two expensive testers at the end of the line, hence it made them the de facto bottleneck and the limit on our rate. Now, in all designs of this nature—single station in series—the theoretical number of work stations will be the work time divided by the cycle time, or in this case $172/12 = 14$. If we add a little for OEE losses, it is easy to see we need 15 or 16 stations.

Next, at this point it becomes a matter of style if we want three five-station cells, four four-station cells or five three-station cells. This calculation can be refined in additional ways, but we did not feel our data were accurate enough to draw these conclusions. Somewhat arbitrarily, we selected four, four-station cells, which allows production modulation in 25 percent increments yet are easier for material supply than five cells of three each. Quite frankly, any of the three would have worked well in the beginning.

Balancing the Work within a Cell

Now, just how do we balance the work in the four stations of any cell? In our new cell, we will have four work stations, which we will call stations A, B, C, and D. As a first pass, we will try to combine the work from the existing stations 1 through 4 into the new workstation A in each of the four new cells. Likewise, existing work stations 5 through 8 were combined into new work station B; 9 through 12 into C; and 13 through 16 into D, respectively.

Work Station	Work in Secs	Trans in Secs	Total in Secs
A	46	2	48
B	39	2	41
C	44	2	46
D	43	2	45

TABLE 13-2   Gamma Line Balance Chart, New Cellular Design

Using the data from our time study, we now have:

- New station A is 12 + 11 + 10 + 13 = 46 seconds
- New station B is 9 + 7 + 13 + 10 = 39 seconds
- New station C is 12 + 12 + 12 + 8 = 44 seconds
- New station D is 8 + 9 + 13 + 13 = 43 seconds
- Flow Line E is 12 + 9 + 11 + 11 + 13 = 56 seconds, less 3 seconds from the modifications

We will then need to look at the balance. The results are shown in Table 13-2.

In checking the balance for workstations A thru D, we find about seven seconds difference, or about 15 percent. Although we would not normally consider this good enough, we accepted it for now. We had made so many changes we wanted to get the cells in service and check it out. In addition, we were under severe time pressures. The risk here is that any changes we now make in the cells will need to be done four times. Even considering this, we were comfortable and decided to proceed with this design, in addition, this would make for a neat transition. For example, we could reuse almost all of the work procedures.

We Plan the Modifications and the Test Run

With this design in hand, and plans to implement these changes, Greg placed yet another restriction on us. We would need to prove this design before he was willing to convert the entire line. Although we were very confident, his stipulation was reasonable, so we agreed to construct one cell (as if we had any choice at all) and direct its product to station 17, while the original line continued to produce normally. We would operate this cell for one or two shifts per day, measure the performance of the cell and then proceed from there. Meanwhile, we implemented the changes on the final test. They were simple and successful and reduced three seconds of work.

So far we have designed a cell, operating at a lower cycle time (remember that *takt* is not really of interest to them, they want more, faster), and balanced so it will flow nicely with one-piece flow. Our practical limit was the inability to add capital. Practically, this meant we wanted to increase the flow rate up to the ability of stations 17 – 21, which was about a 12-second cycle time.

The Test Run

Over the weekend, during the construction of the first cell, we had problems. We planned to convert the automatic conveyor to a track so the platforms could be manually advanced. The conveyor track would not fit in the U cell as designed. We either needed to enlarge the size of the cell or change its shape. We immediately converted this cell to an L-shaped cell and Sunday night tested it. It worked just fine.

Monday morning came and we trained the operators assigned to our new cell, arranged for a new materials handler and by morning break we had the cell in production, although nowhere near the design cycle time. Materials delivery was a problem, but that was quickly ironed out. By the end of the shift, the bugs had been worked out and the cell was producing, with no quality losses, at a 55-second cycle time. Although we had hoped for 45 seconds, we were still pleased. The workers were pleased as well. They responded extremely well to the longer cycle times. They simply said that they felt more comfortable and it wasn't as stressful. The looks on their faces were unmistakable—they were grateful. We made some minor changes and prepared for the next day.

Tuesday was a better day. Many of the efficiencies we had hoped for were realized. The cell increased in speed and the cycle times for each station had improved per Table 13-3.

We were really pleased, but let me tell you: Greg was ecstatic. Not only was our experiment clearly showing success, but for the first time in a very long time the shop steward visited him with good news. The steward had gotten unsolicited comments from the cell operators. Uniformly, they liked the new cellular arrangement. They particularly liked the longer cycle times. Their stress was reduced significantly since they could now advance the unit when their work was done. They no longer had to worry about the conveyor taking their work away before they were ready.

We spent the rest of that week using only one cell while we planned for installation of the other three cells. Meanwhile, the original line was in operation as usual. The cell gave the line added production and this gave the plant a chance to catch up on production. While this was underway, each day we would rotate a new crew into the L cell, and after a little training they would begin production. By the end of the week, nearly all operators had been trained on the new work layout. It was received very well.

Over the weekend, we dismantled much of the old line and constructed the three additional cells, per Fig. 13-2. The construction was simple and did not take long except for the raw materials supply which needed to be arranged for each cell. On Monday morning we started up and there were a few problems, but by the end of the third day the cells were stable and producing at a record rate with all four L cells producing. We completed a time study and the new cycle time was 12.5 seconds.

Work Station	A	B	C	D
Design—Original Work, Secs	46	39	44	43
Design—Trans Time, Secs	2	2	2	2
Design—Total, Secs	48	41	46	45
Current Cycle Time, Including Transportation Time	45	40	44	42

TABLE 13-3 Cycle Times for Work Stations, Cellular Design

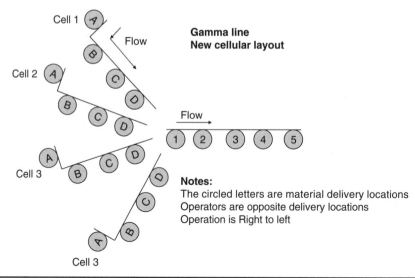

Cell 1

Flow

**Gamma line
New cellular layout**

Cell 2

Flow

① ② ③ ④ ⑤

Cell 3

Notes:
The circled letters are material delivery locations
Operators are opposite delivery locations
Operation is Right to left

Cell 3

Figure 13-2   Gamma line, new cellular layout.

The Results

Actual production was as shown in Table 13-4.

- Production had increased from 163 units/hr to 266 units/hr!
- Cycle time had reduced from 22 seconds to 12.5 seconds!
- This was a 63 percent increase in production using the same people, on the same machines, with the same raw materials ... and all we changed was the work environment by the conversion to cellular manufacturing.

Where All These Production Gains Came From

Just to refresh, see Chap. 5 again on how to reduce lead time and improve flow. Review the seven ways, which are:

- Reduce production time.
- Reduce piece wait time.
- Reduce lot wait time.
- Reduce process delays.
- Manage the process to absorb deviations, solve problems.
- Reduce transportation delays.
- Reduce changeover times.

We did nothing to reduce the production—that is, work time—and since it was already one-piece flow with standard WIP of one, lot wait time was already zero. We

Shift	Hour	Production	Cumulative per Shift	Cum. Total
Shift No 1	7am	264	264	264
	8am	276	540	540
	9am*	216	756	756
	10am	288	1044	1044
	11am*	144	1188	1188
	12n	264	1452	1452
	1pm*	204	1656	1656
	2pm	276	1932	1932
Shift No 2	3pm	252	252	2184
	4pm	276	528	2460
	5pm*	204	732	2664
	6pm	276	1008	2940
	7pm**	144	1152	3084
	8pm	276	1428	3360
	9pm*	204	1632	3564
	10pm	264	1896	3828
Shift No 3	11pm	240	240	4068
	12m	276	516	4344
	1am*	216	732	4560
	2am	264	996	4824
	3am*	132	1128	4956
	4am	264	1392	5220
	5am*	216	1608	5436
	6am	276	1884	5712
	Total	5712	5712	5712

**30-min. lunch
*10-min. break

TABLE 13-4   Gamma Line Production, New Cellular Design

had no process delays and there were no changeovers, so from this list we see that three of the tactics were employed. They were:

- Reduce piece wait time.
- Manage the process to absorb deviations, solve problems.
- Reduce transportation delays.

Recall that the design had 42 seconds of *transportation time* per unit. We effectively reduced that to 18 seconds (2 seconds per station × 9 stations) by the use of cells. That directly translates into a faster cycle time by about 1.5 seconds [(42 – 18)/16]. This does not sound like much, but on a 16-second work cycle, it is a 9 percent improvement in production rate. Or viewed from a cost context, we just converted 9 percent more raw materials into finished goods at no increase in operating expense. (See Chap. 11 for more on this.)

Next, we eliminated the *wait time* of the workers. For the first 16 stations, there were 52 seconds of wait time—we turned this into productive time. That accounted for about three equivalent seconds of cycle time (52/16 = 3.25).

Turning those wastes—the waste of transportation and the waste of waiting; actually, in this case it was all waiting—into productive time effectively reduced our cycle time by over four seconds.

But, hey, not so quick! That doesn't fully explain all the gains.

If the line had been actually producing at a true cycle time of 16 seconds originally, even with all the waste of waiting, it would have been making 225 units/hr (3600/16 = 225) at 100 percent OEE. The real OEE was not 100 percent, rather the line had 4 percent scrap and less than 1 percent availability losses, so OEE was really about 95 percent. Consequently, we should have had about 214 units/hr (225 × 0.95 = 214). We did not have 214 units/hr; rather, we had only 163 units/hr.

So how do we account for this missing 51 units/hr (214 – 163 = 51)? The answer has to do with the fact that the line could not perform at the design cycle time of 16 seconds.

And why was that?

You got it! The answer is *variation and dependent events*! (See Chap. 18 for more information on this.)

Yes, this effect is huge, and in this case it is easy to understand. Whether the process is in lock-step when the process is fully synchronized with a conveyor, as this one was, or if there is no inventory, the effect is the same. Any time one station performs at a time above the cycle time, the effect is felt in all stations. This effect accounted for a huge loss of production, 51 units/h, over 20 percent of the design rate of 225 units/h! Most people find this interaction of variation and dependent events amazing! Well, amazing it is, but it is also true, and it is also often an overlooked phenomenon.

Think about this concept of variation and dependent events for just a second. Since all 21 process steps were synchronized by a conveyor in this case, any time one station would perform at a time above the design cycle time, there was a time loss for the whole line—for all 21 stations. It was exacerbated by the short cycle times, but the basic problem was that 21 people had to be totally synchronized to make this work. In this case, each of the 21 work stations were dependent upon the other 20, otherwise no station could maintain its cycle time. That is the nature of variation and dependent events and it must be understood. Conversely, in the four-person cells there are now only three levels of dependency, so even if one person slows down, they now only affect three others, not 20, and with 4 cells that one person only slows down 25 percent of the production, not all of it.

Point of Clarity There is always a loss associated with the variation in the system... Always!!

In this example is explained one of the beauties of cellular design. They are a natural variation reduction device and they help us to execute the flow improvement tactic of, managing the process to absorb deviations.

Summary of Results

So, summarizing, just how did the conversion to cells improve our production?

- With minor losses, the 21-station flow line could only produce 163 units/hr.

- By using cellular design we reduced the waste of waiting, caused by the interaction of variation and dependent events and we should have been able to produce about 214 units/hr.

- However, we produced 266 units/hr, and this increase above the 214 units/hr was due to the reduction of the wastes of waiting and transportation, which was also made possible by the cellular design concept.

> **P**oint of Clarity Cells are a natural variation reduction device.

Make no mistake about it. Properly designed cells are

- a variation reduction device.
- a waste reduction device.
- a productivity improvement tool!

Chapter Summary

So what happened to Greg's labor efficiency? Well, it skyrocketed and is now well over 100 percent, which of course cannot be. Needless to say, he was very very happy.

The Gamma Line is a story often repeated in manufacturing and differs from the situation that Ohno and Toyota had to deal with in the automobile business. The Gamma Line produced a household appliance where the life of any given product is seldom three years. In addition, the monthly demand for household appliances has huge swings due to the sales and promotions by their customers. In short, the concept of *takt* does not really apply to these products, and overproduction is handled by running the line in less time—seldom is the production line slowed down. Nonetheless, we were able to apply the concepts of Lean and achieve huge process gains. To achieve these gains, exactly which tools did we utilize?

First, we looked at the waste in the system; our focus was waste removal. Then, utilizing cells, we were able to eliminate most of the waste. They already had a pull production system in place, using one-piece flow, but we utilized line balancing and conversion to cellular manufacturing. This allowed us to reduce the wastes of waiting and transportation and more effectively utilize our manpower. Cycle-time reductions were a large part of our gains, as was an understanding of how variation and dependent events were affecting our production rate. Changing the design to cellular production allowed longer cycle times so that small variations in techniques did not affect production nearly as much as with the shorter cycle times. These were the Lean techniques employed to achieve the 63 percent increase in the hourly production, as well as achieve a more level hourly production.

However, these gains were not achieved in a vacuum. A number of Lean techniques had already been implemented and were functioning. First, to achieve flow, the original

> **"E**veryone is somewhere on the journey to become Lean; no one has yet arrived. **"**
>
> Wilson

line already had one-piece flow. Consequently, their first-piece lead time and lot lead time were relatively good. To supply the line, they used a two-container *kanban* system for the small items, which represented 75 percent of all parts. The large parts were delivered on a scheduled route by the materials handlers. In addition, they obviously had multiskilled workers, which made this conversion possible. All of this helped as we embarked on an effort to improve "labor efficiency."

CHAPTER 14

The Story of the Alpha Line

Many lessons can be learned from the story of the Alpha Line, but the best one, by far, is how management stepped up to the challenge, recognized their responsibilities, and provided the leadership to guide the company through the needed cultural changes. In so doing, they managed the Three Fundamental Issues of Cultural Change extremely well. The Three Fundamental Issues of Cultural Change are covered in detail in Chap. 6.

How I Got Involved

The Alpha Line was the first of many production lines at this Mexican maquiladora, Bueno Electronics. It was the first plant in Mexico for this European-based manufacturer, who moved to Mexico to take advantage of low-cost labor and the proximity to the U.S. auto industry. Since this was their initial interaction with the U.S. auto industry, and didn't know how to deal with their customer, nor had the required skills to meet their customer's demands, they retained us to assist them.

At Bueno Electronics they started up the Alpha Line, and after their initial customer audit, in which they achieved a failing score of 41 (75 was the minimum acceptable), we were contacted to assist them. Their primary weakness was in statistical techniques, where they scored 0 out of a possible 10 in four different areas of statistical applications. Of particular concern to them was the need to implement statistical process control (SPC) for all critical product and process characteristics—a skill they were completely lacking. Here is where the story of leadership and management commitment unfolds.

For several years, as Bueno Electronics grew, we helped them implement a number of lean systems, including pull production systems, cellular manufacturing, and an OEE initiative, to name but a few. In addition, they made great strides in improving both their flexibility and responsiveness through lead-time reductions. However, the best part of the Bueno Electronics story occurred when we were initially asked to assist them on the Alpha Line. Of particular importance was how their management responded when they first started up this production line. That set the stage for all the success that followed, including their journey to becoming a lean facility. This is that story.

Initial Efforts to Implement Cultural Change

We were retained by the plant manager and reported directly to him. Our first assignment was to teach the required skills of statistical process controls (SPC), measurement system analysis (MSA), correlation and regression (C&R), and designs of experiments (DOE).

Of particular significance was the way in which the training was done. This sent a clear message to the facility and set the stage for later successes.

The training started with top management in the various departments, including Production, Engineering, Purchasing, Maintenance, and Human Resources—all of which attended every class. The initial training was an SPC class: 36 hours of training that covered SPC and focused on the topics of attribute and variables control charting techniques. In addition to the classroom work, they needed to complete a project. At the initial class, the plant manager addressed his "students" and explained that: "... management has a distinct role in the success of this and there are no shortcuts. They must be directly involved ..." More importantly, the plant manager attended the entire training, and like all the others—managers and nonmanagers alike—he completed the project and passed the final exam. The plant manager set an unmistakable example. Following this management group, other supervisors, engineers, and technicians were trained in SPC. Following the SPC training, other techniques such as MSA, C&R, and DOE were taught. In each case, the management team was the first group trained. In addition, classes were given in Kepner-Tregoe problem solving, and later we were retained to assist in the implementation and support so often needed in these efforts.

> "**P**ut everyone in the company to work to accomplish the transformation. The transformation is everyone's job."
>
> —(Point 14)
> W. E. Deming

> "**A**dopt a new philosophy. We are in a new economic age. Western management must awaken to the challenge. They must learn their responsibilities and take on leadership for change."
>
> —(Point 2)
> W. E. Deming

It was not surprising that the entire effort had great traction and very rapidly the process improvements became obvious.

Around this time, the plant manager approached me with a specific concern. He was given an appropriations request to construct the second rework facility. It disturbed him. He had been assured that all work stations (they had a 28 station line) were at 98 percent effectiveness or higher, except for one that was struggling at 88 percent. Rework for this product was permitted, but he knew something was amiss. We did a quick analysis of the line and found that its first time yield (FTY) was less than 50 percent. This means that less than 50 percent of the total product went through the production process the first time, with no rework. Over 50 percent of the product needed to be reworked at least once, with some units getting reworked more than once. When this was explained to the plant manager, he was amazed but immediately approved the rework station. Being the good manager he was, he called together the managers of engineering and production, and with us also in attendance, issued the following instructions:

- Start using the metric of FTY as the plant's measure of internal quality. He wanted it to be calculated and posted by next Monday.

- Develop and execute a training course in FTY, that not only taught engineers and managerial personnel the concept of the Poisson distribution, but also its detailed calculations. Until this was done, the production manager would calculate and post the FTY daily.

- Create a specific plan to improve the FTY.
- As part of the plan to improve the FTY, one of the action items would need to be the dismantling of the second rework station.

The metric of FTY became the facility's measure of internal process quality, and the key tool to improving the FTY became the use of SPC.

The plant manager had shown excellent leadership in two key areas. First, he had required that the entire management team know SPC, and then he had created the key metric for the plant in FTY. Next, he set about supplying all the needed training. Operators were taught how to take data, make Xbar-R chart calculations, and plot the data. They were also taught how to read the charts for special causes (1 point beyond +/–3 sigma, runs and trends, for example) and then the operators would solicit assistance for out-of-control conditions from their supervisor. Supervisors and engineers were taught the necessary problem solving skills, plus how and when to recalculate limits and general chart management. In addition, one engineer and one full-time assistant were dedicated to this effort as the number of charts skyrocketed (see Fig. 14-1) with the success of the effort (see Fig. 14-2).

The leadership shown by the plant manager went well beyond the training. He made sure that:

- The metric of FTY was clearly posted at the front of all production lines.
- FTY was discussed at each morning meeting.
- FTY was a key topic in the weekly production planning meeting.
- Once a month, chaired by the plant manager, there was a meeting to address FTY and FTY alone. This meeting was attended by all plant management, and the minutes were distributed to everyone in a supervisory position.

There was no question that FTY was important, and there was no question that the plant manager was leading this effort. This effort could be heard, seen, and felt in all

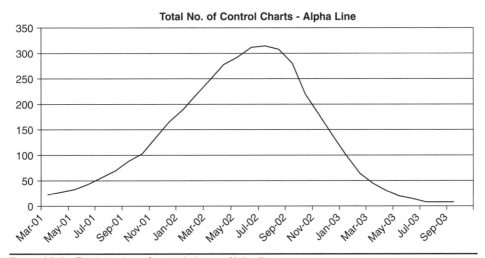

FIGURE 14-1 Total number of control charts: Alpha line.

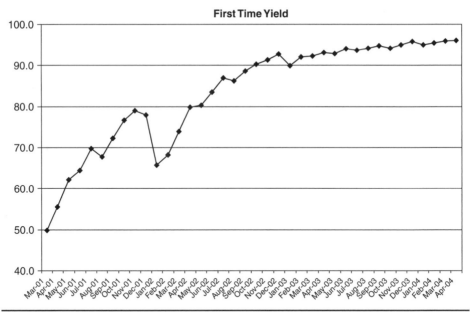

Figure 14-2 Alpha line, FTY.

aspects of the business. And judging by the results as shown in Fig. 14-2, it is obvious that this effort and leadership not only persisted, it succeeded.

Some of the Results

As you can see in Fig. 14-2, yield climbed significantly and the entire effort was a huge success. Though it was a success because the yield climbed, the effort's success went beyond that.

As it turned out, demand for the plant increased dramatically, largely due to the process improvement efforts. The customer of the plant had three alternate suppliers (each of the four now had 25 percent of the demand) of this product, but the customer's stated desire was to reduce that to only two suppliers. One would be the primary supplier; the other a backup supplier. The purpose was clear: The customer wanted to reduce his variation—that meant fewer suppliers for each product, which meant the plant needed to be the best supplier or lose some or all of the business.

> "**C**reate constancy of purpose toward improvement of product and service with the aim to become competitive and to stay in business and to provide jobs. **"**
>
> —(Point 1)
> W. E. Deming

Because of its efforts, by the end of the first year this plant had outperformed all others. At the annual "Give me your quality improvement ideas" meeting called by the customer, we were able to finesse the following information from the customer. This plant was the only plant of the four that was meeting the customer's goal of less than 500 ppm field failures—the plant was at 220 ppm. The other three suppliers were at 700, 1100, and 3300, respectively. It was no surprise that by the end of year one, the plant

	Mar-01	Jan-02	Jan-03	Jan-04	Apr-04
First Time Yield (%)	49.9	65.7	89.9	95	96.1
Assembly line returns (PPM)	1200	780	320	155	62
Field failures (PPM)	220	120	47	22	8
Cost of scrap (M$/Mo.)	$150	$120	$46	$4.2	$0.5
Production rate (units/day)	2800	3300	5200	8300	10,500

TABLE 14-1 Alpha Line, Quality History

now had 45 percent of the demand, while the fourth supplier had been eliminated entirely. The remaining two backup suppliers each had 27.5 percent of the business. In subsequent years, the Alpha Line increased in capacity, and after three years it was supplying 70 percent of the demand, with one backup supplier taking on the remaining 30 percent. A look at the progress made by the plant, as shown in the Quality History table in Table 14-1, shows how well the plant performed.

Continuous Improvement, as It Should Be

There is yet another interesting aspect of this story. Take a look at the FTY graph in Fig. 14-2. It started at about 50 percent and rose almost steadily to over 96 percent, over a three-year period. It should be noted that this was an old design with lots of manual operations, and with 28 stations that meant each station on average was over 99.8 percent effective. During this period of continual process improvement, the key tools used were the statistical tools of MSA, DOE and SPC. In addition, small group problem solving was implemented, and other than basic good work practices, nothing of note was implemented—yet the line continued to prosper. It prospered month after month and year after year. This was a classic, almost textbook, example of good leadership, good process management, and process improvement. It was continuous improvement as it should be.

The Cool Story of SPC: SPC Done Right!

There is another fabulous subplot in the story of the Alpha Line. It is the story of SPC and how to implement it.

In our SPC trainings, we tell our students that one of the purposes of SPC is "to cease doing SPC." This sounds odd, but recall that SPC is not a value-added process. It is a waste. Maybe a necessary waste, but a waste nonetheless. So just how do we get out of the business of SPC once it is started? That was the beauty of this effort. The Alpha Line did it right and this is how it was done.

The primary purpose of SPC is to gain intimate process knowledge about a process. Then, using this knowledge, it is possible to stabilize and improve the process. Once the process is improved to a certain level and process parameters have been standardized, the SPC should show that the process is stable and capable (a high Cpk [Process Capability Index], for example). Once this is accomplished, there may be no need for the SPC. Let's discuss just one specific example on the Alpha Line.

In this process line, there was a solder "touch-up" station—actually, the plant had several, including the rework stations. Here, an operator would make minor repairs using a solder iron. This station had a high fallout rate. To better understand the process, a DOE (Designs of Experiments) was performed and it was found that the solder quality and solder iron temperature were the two key process characteristics. The proper solder was purchased and the solder iron temperature was monitored and placed on SPC. An Xbar-R chart was used. Subgroups of three were gathered every hour and plotted. The chart quickly showed the appearance of a special cause after about 20 hours of operation. A quick analysis determined that flux and residue had been built up on the iron and additional mechanical cleaning was instituted. It was made part of the standard operating procedure and at the start of each shift the operators all cleaned their solder irons. Also, at this time the sampling frequency was changed from hourly to every two hours.

Some minor problems were encountered, but FTY increased by about 4 percent at this station. (Please recall that one station was at 88 percent FTY—this was it.) Over the next two months, sampling was changed to every four hours and yield climbed another 2 percent, though we are not exactly sure why—I have always suspected it was due to the Hawthorne Effect.

The Hawthorne Effect

At the Western Electric plant in Cicero, Illinois, from 1927 to 1932, studies were done on workplace conditions and worker performance. In one series of tests, it was hypothesized that worker productivity would improve if workplace lighting was improved. They improved the lighting and, not surprisingly, worker productivity rose. Someone questioned the validity of this testing, so they performed another test. In this one, they told the workers that the test was such a success that they were going to increase the lighting even further. They made their changes and worker productivity rose once again. However, instead of raising the lighting levels they actually had lowered them; leading the researchers to conclude that there were other dynamics to consider. This and other experiments led them to develop what has become known as the "Hawthorne Effect." One definition of the Hawthorne Effect is:

> People singled out for a study of any kind may improve their performance or behavior, not because of any specific condition being tested, but simply because of all the attention they receive.

> **"C**ease dependence on inspection to achieve quality. Eliminate the need for inspection on a mass basis by building quality into the product in the first place.**"**
>
> —(Point 3)
> W. E. Deming

But then we noticed a common occurrence on many of the SPC charts. Somewhere between 45 and 90 days of operation, the process would go out of control. Upon investigation, we found that the solder irons just wore out. It seemed the tips would change metallurgically and not be able to hold the temperature. So we implemented a program to change out the tips. This was done to all solder irons during the monthly plant preventive maintenance (PM). The tips were only a few dollars a piece and even though we changed them out prematurely, we did not want to scrap even one electronic control unit (ECU) since each was worth about $65. Shortly after this process change, the solder iron SPC changed sampling to once per day and we used only one data point, transferring the information to an XmR chart for individuals. This continued for several months, and the process showed remarkable stability, with Cpks exceeding 3.0. At this point, we discontinued the SPC entirely and the process continued free of defects at the solder iron stations.

So just what happened? Exactly how did we execute this purpose of SPC, which is to cease doing SPC?

Well, we used the information from SPC to make the process more robust. First we implemented a simple form of maintenance: tip cleaning. Since the process was stable, we reduced the sampling frequency to reduce the cost of the SPC. Then, through careful monitoring of the SPC, we gained additional information about the process. We used this information to make process changes—in this case, a monthly PM, which increased the robustness of the process. All the while, we were reducing the cost of sampling and analysis until we were able to do away with the SPC entirely.

So you see, here we were successful. By implementing SPC, we were able to cease doing SPC and in the interim we made the process robust. Neat, huh?

Just How Committed Was Management of the Alpha Line?

The Commitment Test

Let us test their commitment by grading it against :

The Five Tests of Management Commitment to Lean Manufacturing

1. Are you actively studying about and working at making your facility leaner and hence more flexible, more responsive and more competitive? (All must continue to learn and must be actively engaged; no spectators allowed!)

2. Are you willing to listen to critiques of your facility and then understand and change the areas, in your facility, which are not lean? (We must be intellectually open.)

3. Do you honestly and accurately assess your responsiveness and competitiveness…. on a global basis? (We must be intellectually honest.)

4. Are you totally engaged in the Lean transition with your

   • Time

   • Presence

   • Management attention

   • Support, (including manpower, capital and emotional support)

     (We must be doing it; we must be on the floor, observing talking to people and imagining how to do it better; Lean implementation is not a spectator sport.)

5. Are you willing to ask, answer to and act on, "How can I make this facility more flexible, more responsive and more competitive? (We must be inquisitive, willing to listen to all including peers, superiors and subordinates alike, no matter how painful it may be and then be willing and able to make the needed changes.)

The evaluation below gives a test by test critique of how the management performed as they made the huge progress on the Alpha Line.

How Did They Score?

1. By attending all the training—and completing a project, too—they showed their commitment to learning. So on this point, they receive the top score.

2. As for point two, they showed intellectual openness throughout. First, when we recommended they be the first ones trained, the plant manager readily agreed. When the request for the second repair station was made and we gave them our recommendation, they listened. In addition, not only did they make a clear assessment of their skills—admittedly, this was highlighted by the low score on their initial customer audit—but as problems arose they listened to others, accepted their responsibilities, and provided the leadership to succeed.

3. Regarding intellectual honesty, it was their hallmark, so they scored well in this category, too.

4. Their firm was the poster child for management engagement. I am not sure how they could have become more engaged as managers. They did the easy managerial tasks of hiring the right people, creating the goals, and finding the finances to pay for the efforts. They also did the more managerially difficult tasks of attending trainings, leading meetings, and spending time on the floor and with the people. These last three items cut deeply into a manager's schedule and so are often slighted—but in this instance, no such thing occurred, even though they were already working long hours.

5. They were inquisitive and open to changing whatever was needed to attain the goals they had established.

Quite frankly, on all Five Tests, they scored extremely well, which is a sign of why they were able to make such huge gains over a long and sustained period. Results like this do not come about without a committed management team.

How Did the Alpha Line Management Team Handle the Fundamentals of Cultural Change?

Let's look at the actions of Alpha Line management in relation to the implementation of any cultural change initiative (see Chap. 6). How did the plant manager handle the three fundamental issues?

Did They Have the Motivation to Make It Work?

Absolutely, they were staring at an audit score of 41 and the thought of losing business if they did not improve. This was communicated to and understood by everyone in the plant. In addition, to management this was a grass-roots effort at getting a foothold in the American automobile business as well as moving into the low-cost Mexican labor market. No one wanted this to fail. Everyone worked long and hard to make it succeed. Overtime and weekend work was commonplace.

To further motivate the employees, plant metrics were posted and discussed often. This was not done to create a punitive or intimidating attitude, but to clearly and simply communicate to the employees in hopes of succeeding. Since the necessary work was done, the effort was a success. This early success fueled even greater motivation. The management team had created an environment of success—an environment that clearly said, "We will do what is necessary to succeed and we expect the best from you." Everyone responded accordingly, as people normally do.

Did They Have the Necessary Problem Solvers in Place?

At first, the necessary problem solvers were not in place. It was key that they recognized this and responded openly and honestly to it. In my experience, this honest evaluation and admission is not common. Most try to, instead, "just make do," and usually fail. However, the plant manager showed the commitment to not only start at the management level with the training, but as the effort progressed, frequent assessments were made and the necessary training and staffing was supplied. This included the addition of both engineering and support staff, along with the addition of a consulting firm— ours—to provide ongoing support. Within three months, all the necessary problem solvers with the necessary skills were in place.

Did They Have the Necessary Leadership?

So did they have the necessary leadership? The answer is a resounding, "Yes!" The company's leadership was evident in:

- The direct involvement and commitment of management
- Their open and honest evaluation
- Their acquisition of the necessary resources, especially of people
- Their implementation of a plan to train
- Their plan to improve
- Their creation of a metric: FTY
- Their creation of a whole series of "problems" by establishing goals for FTY
- How clearly management communicated all of this to the entire workforce. Everyone knew the objectives and the direction of the quality initiative.
- The fact that every time there was a problem, management acted—doing so thoughtfully, quickly, and decisively.

Leadership was especially evident in the actions of the plant manager who spent time on all that was required to make this effort a success. At every step, he were not only involved, he was committed. He set an excellent example. Actions by the plant manager were swift and effective. Thus, no one at the plant ever doubted the plant's direction and its goals.

> **P**oint of Clarity A good leader will make it so clear through his/her actions that he/she need not say much.

And the results were obvious, as we have seen!

Chapter Summary

As you can see, the story of the Alpha Line is one of management commitment, leadership, and support above all else. They followed the basics of cultural change (as outlined in Chap. 6) and in turn created a culture that was producing a better product for their customer. The customer satisfaction was manifest in additional business. Not only did the actions of management allow them to better satisfy their customers, but they simultaneously made the plant a better money-making machine, with increased employment and job security for their employees.

It is not just a wonderful story of cultural change managed well, it is a wonderful success story as well.

CHAPTER 15

The Story of the Bravo Line: A Tale of Reduced Lead Times and Lots of Early Gains

If process improvement and the elimination of waste is your objective the Bravo Line improvements, show in dramatic fashion how important lead-time reductions are. They also show just how important *jidoka* is to these process improvements. Lead-time reductions serve many worthwhile functions, the most important being:

- Quick feedback for problem solving
- Quick turnover of product, requiring less inventory with its attendant losses of excessive space, excessive equipment to manage it, and excessive manpower to handle it—to name just a few
- Improved cash flow
- Generating future business

Finally, we explain how the seven techniques to reduce lead time and improve flow were applied at the Bravo Line and how these changes resulted in a reduction of first-piece lead time by 97 percent, a reduction of lot lead time by 81 percent, and a greater than 60 percent reduction in manpower consumption.

Background Information

My company was called into "use Lean technology to improve the performance" of the Bravo Line. The problem was stated as: "The line cannot meet the demand of 10,000 units per week. It is well laid out, has a cycle time that should easily meet demand in a five-day work week, but we consistently need to work overtime. In fact, we are working Saturday and Sunday even though the production plan says we should not have to. It has gotten so bad that we have even had to put the utility line into service on occasion, effectively working an eighth day, just to meet regular demand. Quality is extremely good, a process with Six Sigma yield, but we just cannot meet demand."

We inquired about the OEE (Overall Equipment Effectiveness) for this line, and they did not know, but we were able to discover the following information.

- The production process consisted of two cells, operating in series.
- Nearly all steps were simple manual assembly, with no major equipment until final test.
- They believed that line availability was well over 95 percent, close to 100 percent.
- Operators were highly cross trained.
- They had not had a stock out in over a month.
- They had recently conducted line time studies, and had both line balancing charts and standard combination work tables posted at the line for the product.
- The entire plant had begun a lean initiative with some withdrawal *kanbans* and had begun utilizing U-shaped manufacturing cells.
- On this line, which produced 11 different models, they had also made an effort to go to small lot production.
- Within each cell they had a pull system in place, using *kanban* spaces, and the operators would not forward the small lots unless the *kanban* spaces were empty.
- All models were very similar, with over 90 percent of the component parts being the same in all 11 models.
- They described a small lot as 50 units, which was a tray.
- Four trays were stacked in a small box and five small boxes were packed in a larger box, for a total of 1000 units per box, and a typical shipping lot was 2000 units, or two large boxes on one pallet.
- The customer demand was 10,000 units per week of a model mix, but since changeover time was minimal, product mix was not an issue. Their problems were threefold:
    - They could not produce to schedule and frequently missed shipments.
    - It took nearly twice the time.
    - It took nearly 70 percent extra labor to make a batch.

Implementing the Prescription

We did some preliminary calculations and could easily understand some of the problems with production times and missed shipments. We could not yet explain the magnitude of the extra labor required, however. Nor could we explain the need to work eight days to produce five days of product.

Our approach was simple: We used the prescription outlined in Chap. 7 and we decided to:

- Synchronize supply to the customer, externally
- Synchronize production, internally

- Create flow, including
  - Establishing *jidoka*
  - Working to destroy batches
- Establish a pull demand system

Synchronizing Supply to the Customer, Externally

The *Takt* Calculation

Demand was 10,000 units per week and the normal workweek was five days. Available time was 24 hours per day, less a 30-minute lunch and two ten-minute breaks during each of the three shifts, so we needed to produce 2000 units in 21.5 hours or generate a 39-second *takt* ($21.5 \times 3600 / 2000 = 38.7$).

We checked the standard work combination table and it listed the cycle time as 28 seconds, but the line balancing studies appeared to be balanced to 25 seconds, so we were baffled. First, why have two different cycle times? And second, if the cycle time design is way less than *takt*, where are their problems? None of these questions could be answered by the production supervisor or the process engineer—or anyone, for that matter.

Synchronizing Production, Internally

The Basic Time and Balancing Studies

Even though they had done time studies, we redid them and found the following (as shown in Fig. 15-1 and Fig. 15-2):

It is obvious they had made an effort to balance the cycle times to 25 seconds. We could not find a basis for the 25 seconds, which was also the basis used in their planning program. Ever since the original balancing was done six months earlier, problems had occurred at both cells: at Station 1 in Cell 1, and Station 3 in Cell 2. As a result of these problems, the work procedures had been modified but the balance chart and planning

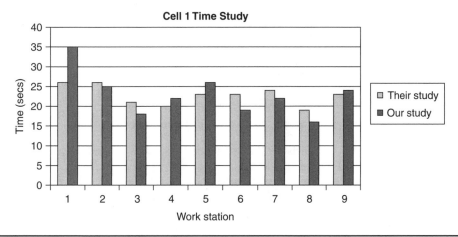

FIGURE 15-1   Cell 1 balance chart, base case.

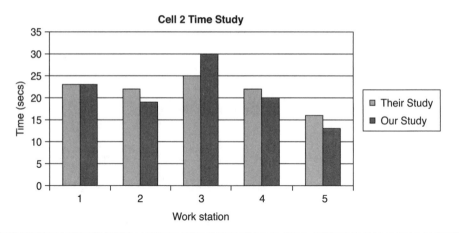

FIGURE 15-2 Cell 2, balance chart, base case.

numbers had never been updated. This, of course, was part of their problem, but only a very small part.

We completed the three-part evaluation—waiting time, station-to-station balance check, and bottleneck analysis—of the balancing chart, but it was not at all revealing as to why the system could not meet demand. *Takt* was 39 seconds, above the line bottleneck of 35 seconds. Yet they say it takes twice as long to produce an order, which means, they have an effective 78-second cycle time. It is clear that the design cycle time was not an issue. This evaluation was not particularly revealing, but it did show that the line balance was simply terrible. Cycle times varied from 35 to 13 seconds—this needed to be corrected, but as I said, it does not explain our problems. At some level, that explains why they had extra staffing. But why did the jobs routinely take the extra lead time?

To unravel this question, we needed to do a line balance chart with our data. First, we needed to find a reasonable cycle time. Normally, this is the *takt* time multiplied by the OEE (as a decimal fraction). In this case, we chose to take management at their word. Recall, they said, that quality losses are less than 3.4 PPM, and availability losses are practically zero. In that case, all the OEE losses were due to cycle-time variations, and so the OEE calculation basically lost its worth. We now knew where to direct our attention. Consequently, we calculated the rest of the redesign based on an OEE of 1.00, so *takt* time effectively became cycle time.

The work (from the time study) for Cell 1 was 207 seconds and the work for Cell 2 was 105 seconds. Consequently, for a 39-second *takt*:

- At Cell 1 we needed 207/39 = 5.3, or six stations
- At Cell 2 we needed 105/39 = 2.7, or three stations

The line balance charts for the redesigned flow now looked like those in Figs. 15-3 and 15-4.

We now needed to redesign the work stations. It turned out to be remarkably simple. We created a "Time Stack of Work Elements" (see Chap. 16 for an example) and were able to redistribute the work very nicely at both cells. This was easy to do since the work elements were almost all manual and almost all short-duration process steps. In addition, each work station had several process steps and the staff was highly cross-trained.

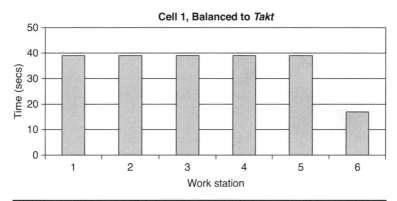

FIGURE 15-3   Cell 1, balance chart, redesigned flow.

Creating Flow, Including

- Working to destroy batches
- Establishing *jidoka*

Working to Destroy Batches

To destroy batches, we wanted to implement one-piece flow in place of the 50 unit batches. This was relatively simple since the line previously had a conveyor and all we needed to do was move it into place to connect the processing stations. The conveyor eliminated the need to pass 50 unit trays from station to station, and in this case, no *kanban* system would have been better. The conveyor had gates to divert the product to the work stations. Since we knew little about the process stability, we modified the *kanban* spaces to hold a maximum of five pieces.

To better connect the cells, we decided to use the *kanban* cart operator to transfer the small boxes from Cell 1 to a makeshift storehouse we created in front of Cell 2. We changed the transfer lot from 1000 units to 50 units.

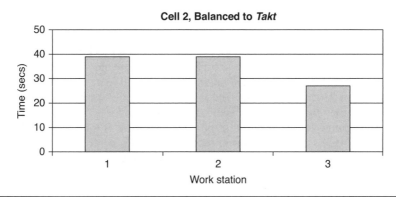

FIGURE 15-4   Cell 2, balance chart, redesigned flow.

Establishing *Jidoka*

We made one other agreement with management, which we knew would be critical. The agreement was that if we had a quality problem, we would stop the line and not start it up until the problem was fixed. We knew it would be crucial to have this type of *jidoka* concept in place. At first, they baulked at this proposal. We did not even try to explain logically how important this was, but when we reminded them that they had characterized the line as having few quality problems, they reluctantly agreed.

Establishing Pull Demand Systems

Our *kanban* system provided a good pull system within each cell, but we had no pull signal from the storehouse. We had enough information to design a good *heijunka* board with a make-to-stock system. We inquired about implementing a *heijunka* board with a production *kanban* system but they were opposed to taking the time to do it. They said "Maybe later," which usually means, "No." It was clear what they wanted: to increase the production rate and reduce the labor it took to produce a unit. We obliged.

The Production Run, with Problems Galore

Our *Jidoka* Concept Works Like a Charm

With these preliminary steps in place, we set up for the first production run. It took less than one hour for us to see one of the major issues. Station 6 of Cell 1 was now only used for final visual inspection and packing. The very first unit it received had an incorrect O-ring installed. Following our *jidoka* guidelines, we shut down the line, gathered the cell workers, and began to investigate. It was simple: The changeover had missed that this model needed both a different shaft and a different O-ring. The shafts had been changed but the O-rings had not. Fortunately, due to the *kanban* system, we had less than 25 units to rework. We reworked the work-in-progress, modified the error in the changeover procedure, acquired the correct O-rings, and started again. On this run, we had to stop four more times for quality problems in the first four hours of production alone. It was clear at this point that process stability was nonexistent. They needed a good dose of Lean foundational issues—that is, quality control (see Chap. 20).

Back to the test run and the finding of defective products. In each case, with small lot production, the problem was easily corrected and we only had a few units to rework. If we had maintained the lots of 50 as before, when the incorrect O-ring was found we would have had 50 units per work station to rework. That would have been over 400 units on the old nine-station arrangement for each of the five quality problems we encountered. Combined, that would have been 2000 units reworked with the old cell design in the first four hours alone—recall that daily production was only 2000 units! It was becoming obvious to everyone how important the small lot and *jidoka* concepts were. And it was equally clear that we had found at least one major source of the extra time it took to make an order. The rework, in the old arrangement, could easily have taken as much time and labor as the design workload.

Suspicious that quality might be a larger problem than previously portrayed, we did a little questioning and found that almost every run had one or two, or sometimes more "start-up burps," as they called them, and rework like this was common. In fact, that was one of the reasons they had everyone so cross-trained: so they could do the rework. There were several, huge, absolutely unmistakable messages in what we had just uncovered.

- Their concerns about quality were focused on the product only. They had a very low interest in, and understanding of, process capability and process stability.

- Management never once mentioned that rework was a concern. They literally had to be blind not to see it. It was a culturally acceptable norm to overlook rework. Once we started the process, and observed its operation, the defects stood out. Our *jidoka* concept made it procedurally unacceptable to proceed without fixing the problems.

- Their view of downtime was odd, to say the least. They did not consider the line to be out of service while all this rework was going on, even though they were able to produce precisely nothing. That was "just how we do things around here," they said.

And the list could go on. At this point, we "began to see" that the required rework both extended the time to produce a batch and also consumed much more labor than was designed.

Production Improves Even Further

With these problems behind us, the line speed picked up and right before the shift change we increased the conveyor speed so there were only two pieces between stations. Production went along without a hitch. After about 16 hours, we completed the first batch of 1000 units in front of Cell 2. There was already a queue of 46 hours in front of Cell 2. The next day, we returned and were able to break into Cell 2's production schedule, which had operated flawlessly with only three people. When our model got there, it went through without event. In the ensuing days, we were able to work off the materials in the Cell 2 queue and get it more in balance with Cell 1. After one week, we had the flows essentially balanced with a small buffer (two–ten hours) in front of Cell 2. We again tried to push the concept of a *heijunka* board. Had we done that, we felt we could have effectively eliminated the queue in front of Cell 2. They again decided not to change and continued to plan the two cells independently.

At this time, something very interesting, although not uncommon, happened. By the end of the first week, the whole line production had sped up—some work stations were under 30 seconds, and it looked like we could rebalance and thus need only five people in Cell 1 and only two in Cell 2. No one could really explain it, but this phenomenon is known as the "Hawthorne Effect" (see the explanation in Chap. 14). After a week, the line was producing smoothly, making over 2000 units per day and meeting the model mix. Furthermore, they were doing it with less people, less space, and most importantly, with less chaos.

The Results

Let's now look at the impact of our efforts on the performance of the line. We will review several aspects including:

- The impact on lead time, which was our original intent
- Other typical gains achieved in a lean initiative
- How the gains were achieved
- What should be next for the Bravo Line improvements

Cell No.		Base Case—14 Work Stations	Redesigned Flow— 9 Work Stations	% Reduction
1	Lead time, first piece	3.9 hrs[1]	6.5 min[2]	97%
1	Lead time 1000 unit box	13.6 hrs[3]	—	
	Queue time for Cell 2	46–52 hrs[4]	2–10 hrs[4]	88%
2	Lead time, first piece	1.7 hrs[5]	5.2 min[6]	99.5%
2	Lead time, 1000 units box	10.0 hrs[7]		
	Total shipment lead time	145 hrs[8]	28.4 hrs[9]	81%

[1] $([35*50]/60)*8/60 = $ **3.9 hrs** (the line fill time for eight stations, first piece appears at Station 9 for inspection; total lead time would be over nine stations)

[2] $(39*2*5)/60 = $ **6.5 min** (at two-piece flow fill time for the five stations, appears at No. 6)

[3] $([([35*50]/60)*9] + [([35*50]/60)*19])/60 = $ **13.6 hrs** (the line fill time for all nine stations, plus the time to produce 19 more trays to make the 1000 unit transfer batch)

[4] By observation

[5] $([30*50]/60)*4/60 = $ **1.7 hrs** (line fill time for four stations; Station 5 also was an inspection station)

[6] $(39*4*2)/60 = $ **5.2 minutes** (line fill time for four stations; Station 5 also was an inspection station)

[7] $([([30*50]/60)*5] + [([30*50]/60)*19])/60 = $ **10.0 hrs** (line fill time for five stations, plus the time to produce 19 more trays)

[8] $(13.6*2) + 46 + 52 + (10*2) = $ **145.2 hrs**, or 6.05 days (lead time to make two 1000-unit batches in cell 1, plus two queue times, plus the lead time to make two batches in cell 2)

[9] $(6.5 + [(39*50)/60] + 360 + 5.2 + [(2000*39)/60])/60 = $ **28.4 hrs** (this is the 1st pc lead time in cell 1, plus the time to fill one tray, the transfer batch; plus average queue time, plus cell 2 1st pc time, plus the time to finish the batch

TABLE 15-1   Lead Time Results

The Lead Time Results

As a result of the *jidoka* and the batch destruction techniques, the process improved dramatically and the lead time results in Table 15-1 display just some of the benefits achieved.

Other Benefits Achieved

When we started redesigning the Bravo Line, they had already begun a Lean initiative. But, as you can see, they had not made much progress. In addition to the process lead time improvements in Table 15.1, other benefits are summarized in Table 15-2. All these changes were accomplished with virtually no capital investment.

Some Points Worth Repeating

The following are some points worth repeating, a few with a slightly different twist:

- The most important quantity control tactic for process improvement is often first piece lead-time reduction. It is accomplished mainly by a reduction in lot size, ideally going to one-piece flow. Once achieved, this will allow the problem-solving process to be responsive, and in this case, rework can be reduced on both a batch and permanent basis.

Other Impacts	Original Case	After Leaning Out the Process	% Reduction
Space Impacts	8 line-days per week, plus 600 sq. ft. of buffer inventory	Less than 5 line-days per week, with less than 100 sq. ft. of buffer inventory	~70%
Manpower Impacts	14-person line running 3 shifts for 8 days per week = 336 md/ 10,000 units	9-person line running 3 shifts for 5 days per week = 135 md/10,000 units	60%

TABLE 15-2   Other Benefits

- By implementing *jidoka* as we did, using line shutdowns, we literally equated quality with production. If quality was bad, production went to zero until quality was reestablished as good. This is the beginning of a huge cultural change from an attitude of "production trumps all" to an attitude of "we will only allow the production of product that is satisfactory to our customer."

- Overall lead time is a very important metric. It is the sign of both flexibility and responsiveness. The reduction of overall lead time is basically done by reducing lot wait times and process delays, or "queue time" as some call it. In this case, process delays, due to the imbalance situation also contributed greatly to the overall lead time.

- By converting to one-piece flow (actually two in this case), we allowed the quality problems to be found and solved. Finding and eliminating quality problems are not basic Lean elements. Finding and eliminating quality problems is not *quantity* control, it is *quality* control. It is a precursor to Lean. This allowed us to reduce the manpower needs from 336 md to 210 md per 10,000 units. That is a huge savings, or better yet a huge loss reduction. (See Table 15.2 earlier: 210 md is calculated by 14 persons, over five days for three shifts to produce the 10,000 units—in other words, their planned staffing.)

- The second reduction, from 210 md to 135 md per 10,000 units came about by balancing line production to *takt*. This is not a new Lean technique. It is an age-old industrial engineering tool known to every manufacturing firm that has significant labor costs.

- After the line was balanced and flowing, it improved itself! Recall that production rates increased even after we had completed our changes.

In Summary

- The company was now able to produce to schedule, and missed shipments dropped to zero.

- First piece lead time was reduced from 3.9 hours to 6.5 minutes.

- They had now reduced the lot lead time from over six days to less than 30 hours.

- They found they could achieve these numbers with much less manpower. They reduced from 336 md/10,000 units ... to the plan of 210 md/10,000 units ... and then to a new low of 135 md/10,000 units. All with virtually no extra investment. And they were just getting started!

How the Gains Were Achieved

The story of the Bravo Line is like many Lean stories. It was not just an effort at making a process Lean through quantity control efforts. It was both an effort at taking care of the foundational issues of quality and utilizing quantity control tools to make further waste reduction improvements. Note that the majority of the huge gains came from the implementation of the foundational issues.

The Three Fundamental Issues of Cultural Change

Just how did the company score on the three fundamental issues of cultural change? A good argument can be made that the reason they got into these problems was due to a weakness in their culture.

However, for the time being let's focus on the culture that surrounded this change in the process. They affected a successful change in the process because, at least for this project, they had good leadership. A plan existed that was understood by those involved and they acted upon that plan. Second, they clearly had the motivation to proceed. The escalating labor costs for the line were making the product unprofitable, and the situation was deteriorating. Finally, they recognized they did not have the necessary problem solvers in place, so they acquired them. It remains to be seen if they will institutionalize these cultural issues, but for the moment they properly addressed the three fundamental issues of cultural change and so this effort was a huge success.

Foundational Issues Addressed

Before the *quantity*-control issues could be implemented, foundational *quality*-control issues needed to be implemented. Review the House of Lean and notice all the foundational issues that were addressed, including:

- Problem solving by all
- Understanding variation
- Process stability (Cp, Cpk)
- Cycle times reductions
- Standard work
- Availability

Quantity Control Techniques Applied

While these were being addressed, we were able to also implement the *quantity* control techniques, including:

- *Jidoka*
- One-piece flow, small lot flow
- Balancing to *takt*
- Minimizing lot sizes
- Reducing both WIP and lead times, which caused a significant improvement in the flow

The quantity control tools would have been much less effective if the quality control issues had not been addressed. Taken all together, these tools then improved the performance and the following occurred:

- Quality was improved.
- Lead times were reduced.
- The cost to produce was reduced.

These three are the key objectives of Lean (again, refer to Chap. 20).

Application of "The Seven Techniques to Improve Flow"

We can employ seven basic techniques to reduce lead time and improve flow (see Chap. 5). The following describes how they were applied to the Bravo Line.

1. **Reduce processing time** Here the major gain was achieved through the reduction of defects and especially the reduction of rework, which had become a part of the normal process. All the rework was simply non-value-added time. Regarding the necessary work, here we did nothing to intentionally reduce the processing time, although the time decreased as the operators became more engaged in the process.

2. **Reduce piece wait time** This is the time a single piece is waiting to be processed. Here the wait time is reduced by balancing, so the flow is synchronized. In the original case, the cycle time was controlled by station one at 35 seconds, so it took 35 seconds per station, times nine stations in Cell 1, to go through the line, or 315 seconds, but only 207 seconds of work were performed. There were 108 seconds of wait time per piece caused by poor synchronization at Cell 1 alone. Again, in this case, the wait time was reduced, but it did not have a large effect on lead time; however, it certainly made the process flow better. Following the rebalancing of the work the gains achieved by reducing piece wait time were found in manpower reductions.

3. **Reduce lot wait time** This is the time that a piece, within a lot, is waiting to be processed. In this case, it is substantial and adds to the overall lead time, but more importantly it adds to the first-piece lead time. This time is often overlooked but is incredibly important. There will always be quality and production issues and these issues must be uncovered quickly so they can be solved. If the lot wait time is like the original case of the Bravo Line, it takes 3.9 hours for the first pieces to get to final inspection. If a problem is found at final inspection, we now have over 400 pieces to inspect and rework—for each problem we find. If the lead time is shorter, the problems surface more quickly and we can react more quickly. In our first run, we stopped five times in the first batch to correct problems. This impact alone could have accounted for the huge excess in labor expenses to run this line in the base case. To reduce lot wait times, shrink lot sizes as we did here. The goal of minimum lot sizes is one-piece flow.

4. **Reduce process delays** This is the time an entire lot is waiting to be processed. Often it is called queue time. Here we were able to reduce it from an average of 49 hours to 6 hours. This is caused by lack of synchronization and also by transportation delays. Close coupling processes, or making them in the same cell, mitigates this problem. *Kanban* can help this, but *kanban* is only a way to

make the best of a lousy situation. Better ways to reduce processing delays are to synchronize operations, close couple processes and production rate leveling for both rate and the model-mix.

5. **Manage the process to absorb deviations, solve problems**   For the Bravo Line, we did not specifically highlight any applications of this technique. There are many sources of deviation that increase production lead times such as machinery breakdowns, stock outs, and stoppages for quality problems, to name just a few. All these deviations cause inventories to rise, and inventories are the nemesis in Lean manufacturing—we want to go to zero inventory if we can. So whenever we have variation in the system, do not add inventory … instead, attack the variation.

6. **Reduce transportation delays**   One-piece flow, synchronization, and product leveling all place emphasis on transportation, which is also a waste. To reduce this waste, several strategies can be employed:

   • *Kanban* is the first thing most people think of, but *kanban* has inventory, and kanban creates a second delay, the delay of information transfer, so it is a double waste in itself. Try to avoid *kanban* systems by instead using close coupled operations such as those in a cell and the use conveyors, as we did here.

7. **Reduce changeover times; employ SMED (Single Minute Exchange of Dies) technology**   The Bravo Line had very short changeover times, so we made no applications of SMED technology here .

This is clearly a Lean success story!

What Could Be Next for the Bravo Line?

Further Analysis

Now it is time to do another time study and rebalance the line. We did paper *kaizen* and determined we could:

- Standardize the improvements to date
- Make minor processing changes
- Rebalance the workload
- Create better flow by moving Cell 2 to be next to Cell 1, and eliminate the queue, then convert to a single seven-station U-cell arrangement
- Eliminate the conveyors and go to one-piece flow

Potential Results

The paper *kaizen* improvements showed:

- The company could reduce space by another 55 percent.
- The company could operate at a cycle time slightly less than *takt* and run both cells with only 6.4 people—really 7.
- The labor consumed would be less than 105 man days per 10,000 units.

And the Story Goes On

In this case, we should make these changes and then proceed to implement the prescription and utilize even more *quantity* control techniques. Each improvement brings about a new present state that will have another set of opportunities—and the story goes on, forever.

Chapter Summary

The story of the Bravo Line starts with a simple request. The plant manager wants to make more product, faster, using less labor. Something all managers want. To reach that end, earlier, the company embarked on a Lean initiative, but it can now be seen that they had some serious flaws in their approach. First, it was apparent they wanted more to *appear* to be Lean, rather than actually *be* Lean. However, the biggest flaw in their approach was that they had not evaluated the foundational issues very well, if at all. Consequently, once we backtracked and took care of the quality control issues, progress was possible. After implementing the quantity control techniques on this stronger foundation, even greater gains were achieved. In all, we were able to reduce first-piece lead time by 97 percent, reduce lot lead time by 81 percent, and in the end reduce the required manpower by 60 percent. As a result, not only was the Bravo Line leaner and its process more robust, but the gains made were huge and achieved early.

CHAPTER 16

Using the Prescription— Three Case Studies

In this chapter, we will cover three case studies that follow the prescriptions. In each case, the prescriptions are applied somewhat differently. These examples are included so you might see the variety of applications for the prescriptions. It is not, "my way or the highway," nor is it "all or nothing." Nevertheless, the prescriptions will take you to the huge early gains that most find so beneficial.

Why These Case Studies?

In my experience, especially since starting my consulting practice in 1990, I have found that *applying* techniques, principles, and theories for most people is very difficult. More difficult, for example, than *understanding* the techniques, principles, and theories. I have found there is an "applications gap" in industry. In addition, especially with Lean, I find there is frequently a mentality that to do Lean is an "all or nothing" proposition. I do not ascribe to that principle at all. I do, however, believe it must be done well or not at all. It is very practical to implement just parts of the entire House of Lean without danger of regressing as long as those parts are done well. Of course, the Lean principles work together, and to get the full benefit, you must apply the entire complement of strategies, tactics, and skills, but you can apply *some* of them and get *some* of the benefits. And almost always, some of the benefits are better than none of the benefits.

These three examples will show that there are different ways to apply these prescriptive methods and still achieve significant results. These examples will also show that there are different degrees or depths of application and you can still achieve huge benefits.

Larana Manufacturing

The first study, which concerns Larana Manufacturing, is the story of a facility that wanted to implement a Lean initiative. This company is a good example of the second prescription of How to Implement Lean-The Prescription for the Lean Project, without really applying the first prescription, The Four Strategies to Becoming Lean. In fact, they had so many foundational issues (quality control issues) to work on that their initiative did not get very deep into real quantity control efforts. In addition, this facility was under severe financial pressures, and had an inexperienced staff with myriad obstacles to overcome. Nonetheless, they applied the second prescription very well, took on the foundational issues and made monstrous early gains in spite of the obstacles. It is an amazing and energizing success story.

The Zeta Cell

Next is an application to a single processing cell, the story of the Zeta Cell. I have included it since they were able to apply the First Prescription, The Four Strategies to Becoming Lean, and make huge early gains with it alone. I have also included this since it is often the type of project undertaken as part of a Lean effort. That is, conversion of a single process or a single cell to Lean manufacturing. Again, here, the gains made in this production cell were truly outstanding. They, too, were made under some very adverse circumstances.

QED Motors

In addition, there is included an example of a much more complicated processing arrangement for a larger value stream in the case of QED Motors. In this case, they applied the entire, Second Prescription: How to Implement Lean. They addressed the fundamental issues to cultural change, completed an entire systemwide evaluation and thoroughly addressed the First Prescription—The Four Strategies to Becoming Lean for one of their production lines, the Motors Line. So they addressed the quality control and the quantity control issues including doing a full value stream analysis. Like the previous two examples, their gains were also huge and rapid.

Maybe You Can't Fix It All ... But

My hope is that this book and these techniques will reach a broad range of people who want to make their facilities leaner, making those facilities securer and better money-making machines. I often find an engineer for a value stream who would like to implement Lean techniques but tells me, "You know I would like to apply Lean principles but the facility is just not interested." That may prove to be an obstacle, but don't let that stop you. The following real-life examples show why you must not stop:

- That was almost the case at Larana Manufacturing. Some of the local management wanted to change but there was literally no support at the corporate level.

- That is precisely what we did in the case of the Zeta Cell. In fact, the plant owners DID NOT WANT to change and we found a way to make them change.

> **"Do** not let those things you can not do, prevent you from doing those things you can ..."
>
> John Wooden

Just because we can not change the whole plant does not mean we cannot change our value stream. And just because we may not be able to change the entire value stream does not mean we cannot change things in our area of influence.

Lean Preparation Done Well: The Story of Larana Manufacturing

Background Information

The story of Larana Manufacturing is a story of real success, not only of a Lean implementation but how to work through the precursors to Lean and achieve success all the way. It is the story of getting huge early gains on the way to becoming Lean.

- Larana Manufacturing makes electrical parts. They are a tier-two supplier to the automobile industry.

- The facility was hampered by a large list of what most people would call weaknesses. First the plant was old. Only three of the nine production lines were less than three years old and none of the lines had the newest technologies to produce their products. In addition, the process flow was poor. Most lines were poorly balanced, set up in islands, and saddled with mismatched production capacity. Larana Manufacturing was burdened with a dangerously small technical staff and had only the basic quality skills in place with no Lean experience at all. The one staff person who was earmarked to be the Lean implementation manager quit, taking another position outside the company. Their employee turnover was about 8 percent per month at the line operator level. To make matters even worse, the plant was under severe financial pressures. I will spare you the details, but the bottom line was this: If the plant was to survive, they had to reduce operating costs. If that was not enough, their home office exhibited a rather heavy hand in the business, sending representatives there almost constantly. With all this visibility, the "home office help" was, more often than not, the "home office burden."

- To top all of this off, the most severe problem was that senior management was using old-world paradigms and were not Lean-thinking at all. Unfortunately, they actually thought they were Lean thinkers. That proved to be a significant burden to everyone as we embarked on this initiative.

Even with all these issues and obstacles, Larana Manufacturing did an exceptional job of working on the precursors to Lean, generating huge financial gains for the facility. It is truly a story of "Lean done well … under some very grim conditions."

How They Handled the Three Fundamental Issues of Cultural Change

Leadership

First, the movement toward cultural change was lead by the plant manager, Kermit, who was in his first major manufacturing leadership position. Kermit had a degree in engineering and was one of those natural problem-solvers that all companies look for but few can find. In addition, even though he was quite young, 30 at the time, he had a maturity beyond his years. This maturity was manifest both in his excellent judgment and his wonderful rapport with the entire workforce. He turned out to be an outstanding leader and directed the effort extremely well.

Together, we created a plan. He made sure all the training was delivered on time, that all follow-up was done, and that he was actively involved in all aspects of the initiative. Leadership was a strong point for this effort, and it was seen in the results.

Motivation

The motivation of the local workforce was not an issue. All while I was at this facility, I can say with certainty that the people were not only engaged in the effort, they put in extra time to *make* it work. With all the distractions mentioned earlier, I was both surprised and pleased with the level of employee motivation.

> **"W**e are told that talent creates its own opportunities. But it sometimes seems that intense desire creates not only its own opportunities, but its own talents. **"**
>
> Eric Hoffer

Adequate Problem Solvers

As for adequate problem solvers, that was an issue both in numbers and in the quality of their problem-solving skills. To enhance the abilities of the problem solvers, we provided training, which helped immeasurably. Regarding the number, we were always very lean in that regard. Consequently, we made some changes to augment the problem-solving staff.

The Systemwide Analysis of the Present State

The Commitment Evaluation

We discussed the commitment evaluation briefly and at the senior management level there were serious problems. Nonetheless, we decided to proceed, knowing full well, that at some point we would run into some serious obstacles. Our hope was that we would have huge early gains, which we did, and thereby get a greater commitment from senior management, which unfortunately we did not.

The Evaluation of the Five Precursors to Become Lean

Kermit and I performed independent evaluations, the results of which are shown in Table 16-1.

Ten Reasons

On the Ten Reasons for failure, we had trepidations about nine of the ten reasons—reliable raw materials supply was the only area in which we had no major concerns.

Process Maturity

We completed a review of the production lines for process maturity and none reached level 2. Stability was a problem in every case.

Both rate and quality levels were unstable when evaluated on a Shewhart control chart. This was true of all three lines we evaluated. Based on this, we decided we needed to do more foundational work before we proceeded to a quantity control initiative. We then proceeded to implement "An Overall Equipment Effectiveness (OEE) Initiative." We also decided that Line 9 was the most stable and felt we could improve it rapidly, so we choose it to implement some appropriate Lean strategies, tactics, and skills—that is, quantity control techniques on this line only. We would address quantity control on the other lines after they had taken care of the issues uncovered in the systemwide evaluation.

The Five Precursors	Kermit's Score	My Score
High levels of stability and quality in both the product and the processes	0.5	0
Excellent machine availability	1.0	0.5
Talented problem solvers, with a deep understanding of variation	0.5	0.5
Mature continuous improvement philosophy	0	0
Strong proven techniques to standardize	0	0
Total	2.5	1.0

TABLE 16-1 The Evaluation of the Five Precursors

The Educational Evaluation

The educational needs were quite extensive and we began immediately to train.

Introductory Training

An introduction to Lean manufacturing was given to everyone by Human Resources as part of a two-hour awareness training, which also included a review of the implementation plan.

Management Training

Management was given training over a one-month period, with each session lasting roughly four hours. Topics included:

- The role of management in Lean
- The House of Lean
- Introduction to OEE

Other Training

Specific skills training was given to those involved, primarily engineers and supervisors, in the areas of:

- The House of Lean
- Introduction to OEE
- OEE calculations for engineers, OEE strategy
- SPC, focus on control charting
- MSA
- TPM
- *Kanban* design and calculation
- Cycle, buffer, and safety stock calculations
- How to level processes and *heijunka* board design

Just How Was the "Rest of the Prescription" Managed

I have summarized The Second Prescription here for your reference. It is an eight-step Prescription for the Lean Project, which includes the following measures:

1. Assess the three fundamental issues to cultural change.
2. Complete a systemwide evaluation of the present manufacturing system (outlined in Chap. 19 and detailed in Chaps. 6 and 7).
3. Perform an educational evaluation of the workforce.
4. Document the current condition of the value stream.
5. Redesign to reduce waste. (This is simply a summary of Chap. 7.)
6. Evaluate and determine the goals for this line.
7. Implement the *kaizen* activities.
8. Following the changes, evaluate the new present state, stress the system, and then return to step 4.

The OEE Initiative

It can be seen that Larana Manufacturing had completed steps 1, 2, and 3. Our evaluations, especially the evaluation of the Five Precursors, told us clearly we were far from being able to get deeply involved in true quantity control. Consequently, we chose to implement an "OEE Initiative" because we had to work on improving the foundational issues of quality control. This OEE Initiative effectively replaced steps 4 and 5. We then set OEE goals and evaluated each line for the appropriate *kaizen* activities. For each line, we developed an eight-week improvement plan, and these plans were executed by the process engineers. To provide support, each week Kermit would review the progress with each engineer, and once a month we had a formal OEE review, which took about 90 minutes. In the formal review, progress was checked, priorities were assessed, and any needed resources were addressed. A Summary Eight-Week Status Report is shown in Table 16-2, which is updated for Week 4 information.

OEE Improvement Plans—Gains ... Line 3				
	First 8-Week Plan			
	Period	Weeks 12–19		
	Week 1 Actual	Week 4 Actual	Goal for End of 8 Weeks	Deviation,* Goal—Wk 4 Actual
Production Information				
Total Units	20,289	20,289	21,100	–811
Total Losses	4,027	2,928	2,932	4
Total Salable Units	16,262	17,362	18,178	–816
Cycle-Time Improvements	5.20	5.20	5.00	–0.20
OEE Performance %	81.17%	85.84%	86.81%	–0.97%
Availability—%	87.38%	87.76%	91.60%	–3.84%
C/T Performance—%	94.74%	99.65%	96.30%	3.35%
Quality—%	98.04%	98.16%	98.63%	–0.47%
OEE Losses/Units				
Availability	2,563	2,483	1,866	–617
C/T Performance	1,067	71	777	706
Quality	398	373	286	–87
Summary Gains		This 8 Weeks	Total to Date	
Total Units/Shift		0	0	
Total Losses/Shift		–1,100	–1,100	
Total Salable Units/Shift		1,100	1,100	

*All positive deviations mean the goals were exceeded.

TABLE 16-2   Typical OEE Summary Status Report

	Line 7	Line 9	Line 3
Total Production	+9.2%	+18.8%	+6.3%
Quality Yield Improvements	+2.0%	+2.6%	+0.0%
Cycle-Time Reductions	5.5%	15.8%	3.9%
Capital Investment	$0	$0	$0

TABLE 16-3   Progress after the First Eight-Week Plan

The Initiative took off, everyone knew their role, and progress was made almost instantly. This goal-setting, implementation of *kaizen* activities, and follow-up were steps 6, 7, and 8 of the prescription. In short order, we had made a great deal of progress. Tables 16-3 and 16-4 shows our results after the completion of the first and third eight week plans.

Comments on the Results

So just how did they improve? Well, immediately upon implementing the OEE initiative, process variation began to show improvements. Not only did quality yield rise, but line cycle times improved. The assigned process engineers had individual work plans and executed them with energy. The engineers grew with the implementation. Problem solving became a way of life with root causes found and eliminated. We developed and defined a structured continuous improvement methodology and it was employed throughout the plant. The processes were showing improvements and the short-term gains were sustained. It was a joy to be a small part of this effort.

This means that:

- With the same staff, the same machinery, and the same raw materials, these three lines could now produce, on average, 24 percent more product. The per-unit cost dropped dramatically. Most of the extra product which they made, they could sell—where they had no incremental demand, they were now able to run the lines fewer hours per week.

- On Line 9, their monthly demand was 1,000,000 units when we started, which could not be met—even when they worked 24/7. Now the demand could be met working only five days per week, with a resultant huge savings in manpower and overhead.

	Line 7	Line 9	Line 3
Total Production	+27.01%	+42.8%	+11.7%
Quality Yield Improvements	+8.1%	+5.73%	+0.63%
Cycle-Time Reductions	16.4%	23.5%	3.9%
Capital Investment	$0	$0	$0

TABLE 16-4   Progress after the Third Eight-Week Plan

- With 5 percent less raw materials, they could produce more product, and quality leakage to the customer was also reduced. Raw materials costs dropped dramatically.

- The financial performance of each line was increased by an average of over 20 percent, with all factors taken into account. Line 9, which, when we started, was about 30 percent in the red, had turned the corner and was now in the black, making money.

The Financial Results Put in Perspective

The financial results are even more impressive when put into perspective. Prior to the initiative, the process engineers were working almost solely on solving quality problems. If they were able to improve the quality yield by 1 or 2 percent on a yearly basis, that was considered good progress. So, figure it out. At about $0.50/unit, making 30,000 units per day, a 2 percent increase meant an incremental gain of over $100,000/yr to the bottom line. That was their definition of "good progress."

We had completely changed the paradigm of achievable progress. Now we were making progress at not 2 percent per year, but 20 percent, and in less than six months. This was ten times as much in less than half the time. *Our rate of improvement was now 20 times larger!*

How Did the Quantity Control Activities Work Out?

Besides working on OEE, during this period we carefully implemented some other Lean techniques on Line 9. For example, on Line 9, which had a particularly problematic process flow, several changes were made.

- Several work stations were modified and two were eliminated by combining with other work stations. This involved process simplification and improved flow.

- *Kanban* were installed to connect the "island processing scheme." This resulted in improved flow, using pull systems with vastly improved inventory controls.

- We slashed the size of the *kanban* containers (transfer lot size). Here, the Lean technique was that the minimum lot size was reduced to accelerate flow, this reduced the manufacturing lead time from 38 to 4.5 hours.

- The final goods inventory was recalculated, set up with cycle, buffer, and safety stocks. A *heijunka* board using production *kanban* was installed to complete the pull system.

- There had been a materials delivery position that was staffed during only part of the shift. We made it a full-time position and added the *kanban* movement to his/her duties. We formalized both a materials delivery route and a schedule.

- In addition, we prepared a present state and a future state VSM for Line 9.

Were They Now Ready for Quantity Control?

Following our third eight-week plan, we completed another assessment on the Five Precursors. These three lines received scores of 12 to 13.5, with no Precursor scoring less than 2. We weren't yet done, but we had already made huge progress. Each line still had a huge upside, just in the precursors alone. Look at the chart in Table 16-5 to see the huge remaining opportunities—and all this without any major capital expenditures.

	Line 7	Line 9	Line 3
Operational Improvements			
Cycle time—Week 12 (secs)	6.10	6.65	5.20
Cycle time—Week 35 (secs)	5.10	5.09	5.00
Achievable cycle time (secs)**	4.40	4.20	4.50
OEE Week 12	79.00%	76.24%	81.17%
OEE Week 35	89.83%	84.31%	88.68%
Achievable OEE**	96.00%	96.00%	96.00%
Production Gains (units/month)			
Monthly salable units—Week 12	812,930	724,559	1,274,778
Monthly salable units—Week 35	1,032,475	1,034,748	1,423,667
Monthly achievable salable units	1,354,579	1,419,083	1,765,970
Gains—First 24 Weeks (Wk 12—Wk 35)	219,544	310,189	148,889
Additional achievable gains**	322,105	384,335	342,303

*Based on 6 days per week
**Achievable without any major capital investments

TABLE 16-5   Gains and Achievable Gains

Now lines 3 and 7 are also ready for quantity control activities. It is time to apply the Four Strategies to Becoming Lean (described in detail in Chap. 6), modify the plan, and have some more fun.

So What Is "the Rest of the Story" Behind This Success?
How did they achieve these huge gains despite the plant's weaknesses?

You have seen the results. To me, these results were not surprising—pleasant for sure, but not surprising. I was not surprised because, in the first case, we had properly addressed the three fundamental issues of cultural change. The plant manager provided excellent leadership, in spite of his inexperience. The staff was and remained highly motivated throughout. Finally, there was a nucleus of problem solvers to work with. With those basics in place, there is always a decent chance of success.

Next, the implementation went well as we followed "The Prescription," as outlined in Chap. 8. There was a good leader and a *sensei*, we made an assessment using the assessment tools from Chap. 19, and utilized the assessment to create our plan. Next, we deployed the plan and turned it into "worker level" goals. This, coupled with some JIT training, fueled the effort and the results speak for themselves.

What about the obstacles and plant weaknesses we mentioned earlier, with all this going against it, just how did the plant achieve all these gains?

The answer to that is simple. Excellent leadership with a good *sensei* working together is the key to success. I cannot say enough about the leadership that Kermit supplied. He showed not only the intelligence and honesty to grasp the situation and see what could be done, he also showed the courage and character required in these stressful situations. He is an impressive young man. Couple that with a good plan, excellent problem solving, and some motivated staff, and that combination will trump all these other so-called problems.

More specifically, if none of the problems stated earlier existed: *If the plant was new and was equipped with the latest technologies and with well-designed process flows, and if the plant had great home office support, with plenty of experienced talented staff, and financial solvency,* it might have succeeded much more quickly and easily. But if the leadership or the *sensei* were mediocre, or the plant personnel lacked motivation, or there was a lousy plan with no good problem solving, such circumstances would doom an initiative to failure. Simply put, those plant weaknesses mentioned earlier are simply not all that relevant.

Do not underestimate the need to have:

- Good leadership with an experience based *sensei*
- Motivated workers
- Good problem solvers

With this formula, you can attack nearly any Lean issue with a strong probability of success. That is the message of Larana Manufacturing and their success.

The Zeta Cell: A Great Example of Applying the Four Strategies to Reduce Waste and Achieve Huge Early Gains

The story of the Zeta Cell is a "back to basics" story. Sometimes when new tools come out, we become so enamored with them that we forget the basic foundation on which these tools were built. The Zeta Cell example shows, in glowing detail, that there are huge early gains to be made, sometimes from the basics alone.

Awareness

One of the better books I have read on Lean manufacturing (besides Ohno and Shingo's works) is *Learning to See* by Mike Rother and John Shook. As I mentioned in an earlier chapter, the key focus of their book is to "learn how to see," to become more aware of how your production processes are performing. They helped us "see" using the performance metrics of:

- Percentage value-added work
- Production lead time

Value Stream Mapping Is Sexy, But ...

Another contribution Rother and Shook made to the Lean movement was popularizing the technique now called *value stream mapping* (*VSM*). Prior to the publication of their book, VSM was known to a small group within Toyota, and only a few outside of Toyota. Within Toyota it was called the material and information flow diagram. VSM is a wonderful tool to use in a Lean initiative. It is also a new and it is certainly, "in vogue," and like many good things, when they hit the market, such items oftentimes are not used quite properly.

There's More to Reducing Waste than Value Stream Mapping

Today, value stream mapping is being treated by Lean practitioners the way a new diet fad is treated by those wishing to lose weight.

Very often, to those seeking weight loss, they put all their faith in the diet and plan to do nothing more than "stick to the diet." Their hope is that in the diet, and in the diet alone, they will reach their goal of permanent weight loss and improved health. If the dieter has the ability to stick to the diet, the results are predictable. Lots of early weight loss. That's not too difficult, but can they sustain the weight loss and improve their health as well? That usually depends on what else they have done. Have they changed their sleeping habits? Have they changed their exercise habits? If the diet is all they did, with certainty the weigh will come back and with that ugly weight will come the emotional baggage of failure and guilt.

In this light, many managers view value stream mapping like the latest diet craze. They hope VSM is the "key" to their success and they think they need to do nothing more to analyze their situation than prepare a few value steam maps, from which they can fully analyze and fully improve their plant. If that is the only analytical tool they use, and they rely on it alone, the results, just like the dieters, are predictable. There will be short-term gains, encouraging gains, but again like the dieter's situation, over time these gains will likely regress and with this will come the emotional baggage of dissatisfaction and disillusionment.

In fact, putting all your eggs in the diet basket, or in the VSM basket, is just another version of the "silver bullet' or the "quick fix" mentality—and history has shown time and again it simply does not work in any field of endeavor.

Should We Ignore Value Stream Mapping?

So does that mean we should ignore value stream mapping? Of course not—no more than the dieter should ignore the diet. Value stream mapping is a powerful tool that allows you to see the production operation from a different perspective, and a powerful perspective—from a distance. It allows you to see how the various parts of the value stream are connected and how these pieces add up; how they increase the lead time and increase the waste accumulation in the value stream. Its major advantages are twofold. First, it focuses your attention on the total value stream and assists you in avoiding the problem of point optimization at the expense of system optimization. Second, it is the most direct and powerful way I have seen to focus on lead time, the key metric of whether a facility is truly Lean.

But just what are its shortcomings as an analytical tool? Value steam mapping is a "macro tool" in that it has an overview aspect and does not get into many of the details that must be understood if you want to make your plant Lean. The biggest weakness is in analyzing cell performance. Cell performance is absolutely crucial to overall performance. So we need other tools to analyze cell performance, and if we do not properly analyze the cell performance, we will end up with good connections between cells that are performing poorly.

I find it is best to prioritize your improvement efforts, inside-out, so to speak. First, start at the cell, make it efficient and effective. Once this is done, connect the cells—this is where the VSM is powerful. There is nothing wrong with improving the cells and improving the cell-to-cell flow, simultaneously. But it is simply incorrect to ignore the cell performance and it is highly inefficient to do it after the value stream mapping effort.

Although that sounds trivial, it is missed by many and with the power of a value stream map driving a project, a lot of progress will be made. In short order, however, the poor performance of the cells will create problems, and these issues will need to be addressed. In the end, the macro work uncovered by the VSM will often need to be redone.

As I have mentioned repeatedly, the healthiest place to start a Lean initiative is to evaluate the "Foundational Issues." Following that, I have found that the place to start is a detailed analysis of the work elements. My mantra is "You gotta know the work!" From there, you can create VSMs, make a spaghetti diagram, create flow, design pull production, do some more analysis, and begin the improvement activities.

> **P**oint of Clarity There is no substitute for "knowing the work."

> **"E**stablishing the flow is the basic condition .... Unless one completely grasps this method of doing work so that things will flow, it is impossible to go right into the *kanban* system when the time comes. **"**
>
> Taiichi Ohno

You Gotta Know the Work!

In their book, *Creating Continuous Flow,* authors Mike Rother and Rick Harris spend some time on value stream mapping. However, the focus of the book is on improving "flow." We often use *Creating Continuous Flow* in our training since it is an excellent book on the subject.

In addition to Ohno's wisdom above, we have found that establishing the flow, as Rother and Harris explain, coupled with line balancing, are two powerful manpower reduction techniques.

Part II of their book is entitled, "What Is the Work?" When we introduce this to the typical manager, usually they just yawn. It seems this is too beneath them to study. And we usually find the same response from the industrial engineers as well. And why not? That's how their boss views it. Ultimately, we find very few firms can really define the work down to the element level, and until they can, their efforts into Lean are inadequate.

An Example of "Not Knowing the Work"

For example, while working to Lean out a very new but large and complex production line at an international tier-one supplier, we encountered this problem. Specifically, they had a detailed present and future VSM, as well as detailed flow with timing studies. These studies had been done on their standard format which was Excel-based. Data could be entered and the output was a series of charts including a line balance chart, a standard work combination table and several other important-sounding documents. The Excel program was filled with lookup tables, dynamic data interchange, data validation, and myriad techniques that, quite frankly, sounded very impressive.

There were two problems, however. First, the work elements, the basic input to the Excel spreadsheet, could not be described by the engineer, the supervisor, or the business unit manager. In fact, no one could. Worse, as I watched the process perform, although you could recognize from the documents that we were watching the correct process, a number of work elements were missing. In addition, the time study of the work was not correct at all. Waiting and work was all mixed up. All in all, the study, although it was presented in an impressive display of charts tables and graphs, was basically useless. Well, "you gotta know the work" ... and they didn't. I find this to be a frequent scenario—that is, much effort on the presentation, but much less on the substance. This always detracts from a facility's ability to attack waste.

They Were Also "Not Very Aware"

In this specific case, the OEE of this line was 61 percent, which was woefully low, all things considered. However, as so often was the case at this facility with its Lean-appearing façade, they had an Information Center and we were able to quickly evaluate only part of the problem. The telltale sign was that, for this process with an OEE of 61 percent; quality losses were 0.95 percent and availability losses were 9.5 percent. What they failed to recognize was that the cycle-time losses were nearly 30 percent! Ouch!

How did we know they failed to recognize the largest problem?

Well, the Continuous Improvement Activities board, at their Information Center, showed a number of projects to improve quality, and two or three to work on availability, but not one single job was focused on cycle time. We were not there to work on that issue but could not help but notice and point it out. They, very promptly, did exactly nothing. For the three months we worked with them, we made good progress on quality and availability, yet the huge losses due to poor cycle-time performance were not addressed.

"To Know the Work": The History of the Zeta Cell

Background Information

An example of not focusing on the work was also seen in the Zeta Cell. It is an odd story in that the company we helped the most was not our client, rather it was the supplier to our client that really got the gains from our efforts.

The background to this is that we were hired for our problem-solving abilities—in this case, to solve a problem with a controller that was produced on the Zeta Cell. The controller was used to guide a robot and would occasionally stick in the "full speed ahead mode," causing the robot to consequently crash into a wall or the production line. This was not only undesirable, it was dangerous.

We analyzed the data using Kepner-Tregoe techniques and found the root cause of the problem. We then assisted them in modifying the production process. We implemented the change and began production with the redesigned process immediately. Meanwhile, we monitored both production and field operations while the company completed the necessary environmental and reliability testing.

The really interesting part of this experience was that during this same time period, this supplier embarked on what they called a "full-blown implementation into Lean." It was at this time that we discovered that this 900-person plant was extremely unprofitable and was up for sale. Their subsequent efforts all looked like a company in crisis. The actions were quick, direct, and their "full-blown implementation into Lean" was done in a very dictatorial fashion.

Previous to our arrival, they had hired a new general manager (GM) and his objective was clear. On one occasion, he told me they had 12 months to make the place profitable or it would be sold. Their costs to produce were just too high and must be brought in line. He fired the current plant manager and replaced him with a man he knew. His "full-blown implementation into Lean" effort was "front-end" focused. He started at the front of the process, the raw materials supply, and worked through the process in the direction of flow.

A huge initial effort was focused on the raw materials warehouse and supplying all materials to the line via *kanbans*, replacing the current practice of kitting. He also started a Lean implementation office with three engineers. They did some basic training regarding the initiative, but there was little if any individual skills training supplied. The three engineers spent the majority of their time working with the raw materials supply, and one thing they did extremely thoroughly was have a Plan For Every Part (PFEP). This is an extremely time-consuming activity, but it is also a good one. It is one I have found best to leave for the later stages of Lean implementation, at least waiting until good flow is established. In addition to having a PFEP, they standardized on a specific work table design and an external U cell and began to change the layouts of all cells to match this. Doing so, they were able to save a great deal of space. Due to their efforts, in-plant supply lines shortened, materials delivery improved, and some clear Lean benefits were achieved.

After six months into the effort for some reason, the GM called me into his office and gave me his six-month update. It was a PowerPoint presentation that described their effort and showed the gains made. Many were paper gains, such as the space savings achieved and reduced lead times for raw materials delivery, and quite frankly the plant looked a lot better. The layout was improved and raw materials flowed more smoothly. He was rather proud that through the materials-handling effort they had been able to shorten delivery times, reduce stock outs, and eliminate three delivery positions. For a three-shift operation that meant a reduction in headcount of nine. The other gain was that on-time delivery had improved from a meager 68 percent to 94 percent, and it was still rising. Considering the effort expended, with the delivery performance excluded, I was not impressed with the results. Since their basic problem was financial in nature, they had made very little progress on improving the operational efficiencies.

Some Specifics on the Zeta Cell

It was about this time that I was working with the production engineer to improve the cell that provided the controller to my client. The line was clearly under-producing and to the trained eye, the line was a "Lean-opportunity-waiting-to-happen." I was buoyed by the energy of their Lean implementation and thought this would be like shooting fish in a barrel.

At this cell, the workers, who were grossly underworked, would leave the cell without warning. Inventory would build up in front of their station and then the operator would return, concentrate on the work for a while, and the inventory would move further down the process. At times, if inventory buildups were too large—that is, they ran out of space—a worker might leave his station to assist in the work-off of the inventory at his colleague's workstation. In short, it was a herky-jerky inefficient operation typical of non-Lean production facilities. No one found it odd. In fact, they were proud of the teamwork and the level of cross-training, which made all this manpower movement possible. Nonetheless, being a Lean practitioner, I proceeded to Lean it out—on paper, that is.

The basic time study and balancing calculations for the Zeta Cell are shown in detail in Appendices B and C of Chap. 7.

Applying the Four Stategies to Reduce Waste

Synchronizing the Supply to the Customer, Externally

First, we did a *takt* calculation. Since they have a 9.5-hour shift with 50 minutes for lunch and breaks, the *takt* was 39 seconds to produce the 800-unit weekly shipment for my client. (As you read on, you will find that the actual cycle time to produce an 800-unit batch was well over 80 seconds!)

Synchronizing Production, Internally

First, we completed a time study and a balancing study, which are shown in Figs. 16-1 and 16-2.

The time study showed we had 157 seconds of work. At a *takt* of 39 seconds, that would be 4.02 operators at 100 percent OEE. No one knew the actual OEE, but they believed it was over 90 percent. With five operators, we would average 31 seconds of

Step No	FC Id	Work Element	Cycle 1	Cycle 2	Cycle 3	Cycle 5	Cycle 6	Cycle 7	Cycle 8	High	Low	Range	Average	Final
1	10	Cut bracket	3	4	3	2	5	11	3	11	2	9	4.4	3
2	20	Assy bushing (3)	11	10	13	12	13	19	12	19	10	9	12.9	12
3	30	Install o-ring and clip	9	6	6	8	7	8	7	9	6	3	7.3	7
4	40	Place in jig, glue	7	8	9	11	10	10	9	11	7	4	9.1	9
5	50	Press in magnets (2)	4	5	6	5	4	7	17	17	4	13	6.9	6
6	60	Insert o-rings, cap, grease	14	12	12	13	19	13	14	19	12	7	13.9	13
7	70	Install support	7	8	7	8	8	9	7	9	7	2	7.7	8
8	80	Install o-ring and clip (2)	6	7	8	9	23	7	8	23	6	17	9.7	8
9	90	Apply epoxy, 3 locations	12	13	15	14	14	14	13	15	12	3	13.6	14
10	100	Install control capacitor	7	8	9	9	8	8	7	9	7	2	8.0	8
11	110	Apply epoxy, topside	7	6	5	9	6	5	5	9	5	4	6.1	6
12	120	Install retainer ring	9	8	9	8	9	7	8	9	7	2	8.3	8
13	130	Install cover cap	6	7	7	8	6	7	7	8	6	2	6.9	7
14	140	Unload/load machine (2)	2	3	3	2	12	3	3	12	2	10	4.0	3
15	150	Apply final sealant (1)	22	14	15	28	14	15	16	28	14	14	17.7	15
16	160	Final test, wrap leads	16	19	17	18	22	17	18	22	16	6	18.1	18
17	200	Package	12	10	28	12	13	11	12	28	10	18	14.0	12
18														
19														
20		Total	154	148	172	176	193	171	166	193	148	45	168.6	157

Process to Monitor: Rayco 43-27 Date: 3/9/2005, 2 shift

Station: Zeta Cell Done by: J. O. Bengineer

Notes 1 Gun required unplugging hence long times, place on PM program
2 Long cycle time was due to dropped parts, attention to details
3 Long cycle times were due to dropped parts, operator needs surgical gloves
4 Hard to do study, so much inventory and lots of movement plus lots of wait times
5 Numerous units dropped on the floor
6 Transportation times not taken
7
8

Figure 16-1 Zeta cell time study.

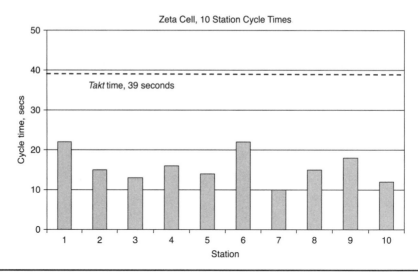

FIGURE 16-2 Zeta cell balancing graph.

work per position. This would be a reasonable place to start and give us a reasonable OEE target of 0.79. Even though one station, on paper, took 35 seconds (see Fig. 16-3), I still felt it would work. Figure 16-3 shows in tabular format and Fig. 16-4, show in graphical format how the work times were allocated with the original ten person layout and now how they are allocated with the newly proposed five person cell. Figure 16-5 then shows the new balancing for the five person cell and compares it to *takt*. In addition

						New Balanced Proposal	
			Zeta Cell, Original Design				
			10 Operators			**5 Operators**	
Step No	FC Id.	Operator No.	Work Element	Final Time	Time Per Operator	New Design, Operator No.	Time Per Operator
1	10	1	Cut bracket	3		1	
2	20	1	Assy bushing (3)	12	22	1	31
3	30	1	Install o-ring and clip	7		1	
4	40	2	Place in jig, glue	9	15	1	
5	50	2	Press in magnets (2)	6		2	
6	60	3	Insert o-rings, cap, grease	13	13	2	35
7	70	4	Install support	8	16	2	
8	80	4	Install o-ring and clip (2)	8		2	
9	90	5	Apply epoxy, 3 locations	14	14	3	
10	100	6	Install control capacitor	8		3	28
11	110	6	Apply epoxy, topside	6	22	3	
12	120	6	Install retainer ring	8		4	
13	130	7	Install cover cap	7	10	4	33
14	140	7	Unload/load machine (2)	3		4	
15	150	8	Apply final sealant (1)	15	15	4	
16	160	9	Final test, wrap leads	18	18	5	30
17	200	10	Package	12	12	5	
			Total	157	157		157

FIGURE 16-3 Zeta cell rebalancing calculations.

Time Stack of Work Elements, Zeta Cell

Time in seconds	Activity	Orig Plan	New Plan	Time in seconds	Activity	Orig Plan	New Plan
80	Apply epoxy 14 secs	Operator 5	Operator 3				
79							
78							
77				157	Package 12 secs	Operator 10	Operator 5
76				156			
75				155			
74				154			
73				153			
72				152			
71				151			
70				150			
69				149			
68				148			
67				147			
66	Install O-ring and clip 8 secs	Operator 4	Operator 2	146	Final test, wrap leads 18 secs	Operator 9	
65				145			
64				144			
63				143			
62				142			
61				141			
60				140			
59				139			
58	Install support 8 secs			138			
57				137			
56				136			
55				135			
54				134			
53				133			
52				132			
51				131			
50	Insert O-ring, grease insert, place in cap. 13 secs	Operator 3		130			
49				129			
48				128			
47				127			
46				126	Apply final sealant 15 secs	Operator 8	
45				125			
44				124			
43				123			
42				122			
41				121			
40				120			
39				119			
38				118			
37	Press in magnets 6 secs			117			
36				116			
35				115			
34				114			
33				113			
32	Place in jig, glue 6 places 9 secs	Operator 2		112	Unload-load machine 3 secs	Operator 7	Operator 4
31				111			
30				110			
29				109			
28				108			
27				107	Install cover cap 7 secs		
26				106			
25				105			
24				104			
23				103			
22	Install O-ring 7 secs		Operator 1	102			
21				101			
20				100	Install capacitor retainer 8 secs		
19				99			
18				98			
17				97			
16				96			
15				95			
14	Assembly bushing to bracket, install cap. 12 seds	Operator 1		94			
13				93	Apply epoxy 6 secs	Operator 6	Operator 3
12				92			
11				91			
10				90			
9				89			
8				88			
7				87			
6				86	Install control capacitor 8 secs		
5				85			
4				84			
3	Cut Bracket 3 secs			83			
2				82			
1				81			

FIGURE 16-4   Time stack of work elements.

to rebalancing the work, I would have liked to perform paper *kaizen,* to further reduce work time, but I knew that going from the current staffing of ten people to five would be a huge cultural change for the group. I thought that would be enough for an initial change. I wanted to keep it simple. Consequently, we did not modify any of the work elements.

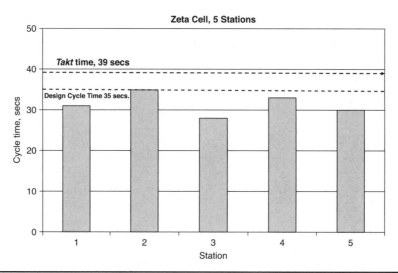

Figure 16-5 Zeta cell balancing graph.

An Effort to Get This Approved

I presented these analyses to management and thought I would be embraced as a Lean hero. Quite the contrary … No one in management wanted to implement the new plan; no one from quality, no one from production, no one from engineering; not one management person. As far as any logical reasons for them to resist the change: They had none. It fit with their Lean initiative, the people were already trained, and we only needed to move one machine about 18″ and re-anchor it and then rearrange the work tables. It was a huge money maker, and it was as simple as it comes—but the unified reaction was that it should not be done now. "Maybe later," they said. I have heard this "maybe later" phrase many times, and it usually means, "not if I have anything to say about it." So I dropped the issue for the time being.

They Had Other Problems, Which Created Opportunity

About that time, we had been investigating some warrantee returns that we thought could be related to the earlier problems. The failure analysis was not clear but it pointed to the root cause possibly being that the units were dropped during production, creating some hidden internal damage, and the hidden damage did not show up until after about 50 hours of operation, hence the warrantee issues.

That was all I needed. With the failure analysis report in hand, I explained that the ten person, large work cell with all the excessive handlings, routinely allowed a lot of dropped product to the floor, which was accurate. They had two options. Scrap all dropped product or close up the cell and eliminate the opportunity to drop the units. Suddenly they were interested in redesigning the cell—so we did.

Creating Flow and Establishing Pull-Demand Systems

We cut the cell size dramatically. Using their new table design, we created a cell using less than 40 percent of the space of the ten-person cell. We moved the press and anchored it. The supervisor was not convinced we could go to one-piece flow, so they set up four-piece space *kanbans* at each work station. Workers were trained to stop producing if the *kanban* location was full. Work instructions were modified to match the work stations, and we were ready to start, which we did the next morning.

We Start Production

Work started and went exceedingly smooth. Quite frankly, since they had been forced into doing this, I expected resistance, but surprisingly none occurred—instead, the whole group was cooperative. With only four pieces at each station, the line filled quickly and production was moving smoothly at about 40 seconds cycle time. Station 4 was clearly the bottleneck. It was extremely awkward for the operator to transfer her part to Station 5. The supervisor recognized the problem, had a small sheet metal slide made, and installed it. By noon, the line was producing at less than *takt*. In fact, by the end of the run, the line was producing smoothly and comfortably at a 28-second cycle time and the 800-unit batch was, for the first time, completed in less than one shift. We had only one major problem. The materials supply and *kanbans* were set up to supply at the older slower rate, so the materials handler had to make many emergency runs to keep up with the cell. Recall that they had started at the supply end with detailed plans for every part, so when we improved the cell, all their part plans needed to be redone.

Quick Early Gains Are Huge

In summary, an 800-unit shipment that had previously taken a ten-person cell over two complete shifts to complete, was now being done in less than one shift with only five people. Hallelujah! (See Table 16-6.) It does not get much better than this when you deal in process improvements. All this without any capital expenditures, and no quality or other downside issues at all! I was pleased. Very pleased.

A large uninteneded consequence resulted from this process improvement. As it turns out, this controller was one of a family of controllers that were all made on this work cell. Plus, the new layout could be used for the entire family. The ten-person cell would normally work 24 hours, 7 days a week to make the demand for the entire family. Now they could produce the demand in less than four days with only five operators. The savings was a net reduction of 30 man days of work per day, forever.

No "Lean Hero of the Month" Here

Wow, think about this. The GM was so proud of his reduction of materials handlers by nine persons, I figured I would be hailed as the Lean hero of the month. Instead, everyone just yawned and didn't even say "thanks." In fact, the lead industrial engineer, who resisted the suggested improvements in the first place, now said, "We had planned on doing that all along." So I guess my entry into the "Lean Hero of the Month Club" would have to wait a while. But the cell was certainly setting new performance standards and was ready for a new time study, and also ready to be rebalanced to achieve further gains.

Metric	Original Case	Leaned Process	% Improvement
First piece Lead Time	4.5 hrs	9 min.	97% reduction
Batch Lead Time	20 hrs	8.5 hrs	58% reduction
Space utilization	425 sq. ft.	160 sq. ft.	62% reduction
Operators per cell	10	5	50% reduction
Labor costs/unit	15 min/unit	3.19 min/unit	79% reduction

TABLE 16-6   Gains Summary, Zeta Cell

Lessons Learned

So just what is the point of the Zeta line experience? It clearly points out that:

- You have to know the work.
- If you want to make materials flow in the basic work cell, you have to deal with the work at the element level.
- "Inside out" is almost always the best way to begin eliminating non-value-added work activities, something that can easily begin during the paper *kaizen* phase.
- There are huge early gains everywhere.

The Case of the QED Motors Company: Another Great Example of Huge Early Gains on an Entire Value Stream

Background

The Prescription—Revisited

The Second Prescription—How to Implement Lean-The Prescription for the Lean Project—is well detailed in Chap. 8. The story of QED Motors shows how all eight steps of the prescription are applied. As is normally the case, the prescription can not always be adhered to in a linear straight-line fashion, but if you follow it through, you will see that all eight steps are well addressed by the team, which improved this process. As a refresher, the Second Prescription is summarized here.

Steps 1–3: Systemwide evaluations

1. Assess the three fundamental issues to cultural change.
2. Complete a systemwide evaluation of the present manufacturing system (outlined in Chap. 19).
3. Perform an educational evaluation of the workforce.

Steps 4–8: Specific value stream evaluations and action items

4. Document the current condition of the value stream.
5. Redesign to reduce waste. (Refer to Chap. 7.)
6. Evaluate and determine the goals for this line.
7. Implement the *kaizen* activities.
8. Following the changes, evaluate the new present state, stress the system, then return to step 4.

Background Information

The QED Motors Company had been making motors for over 30 years. They had a plant in California and had just constructed this new facility in Mexico to take advantage of the low-cost Mexican labor. Their plan had been to run the two plants in parallel for three to six months while the Mexican plant came up to speed, and then shut down the California facility. It had now been over 15 months and the Mexican plant had just that month achieved the design capacity of 3500 motors per month. However, to meet this demand the plant was working seven days versus a business plan of five days, and still had a number of production problems. The largest problems were:

- Low on-time delivery, only 76 percent

- 14 percent scrap rate

- Seven days production lead time.

The new Mexican facility had not demonstrated sufficient capacity or sufficient stability to allow the plant in California to shut down. In addition, at this facility, they had serious management turnover. Since starting up, they were now on their third plant manager, second quality manager, and second purchasing manager.

QED had embarked on a corporate-wide Lean initiative about five years earlier. The effort was being directed from the home office in Minneapolis. It was obvious they had a number of Lean tools in place since *Hoshin* planning matrices were posted and radar charts of all kinds decorated their bulletin boards. But the most encouraging thing was that some of the training had been very effective because most employees understood the basics of Lean.

Nevertheless, this facility was anything but Lean. Due to all of the inventory and recycling on the line, it was impossible to follow the flow even though all work stations were labeled and 5S had obviously been attempted. Scrap was high (with over 100 motors in-process at the rework station), pallets of inventory were all over the place, the CNC lathe had huge inventories in three different locations, and flow was virtually nonexistent. In addition, the process flow path was unnecessarily convoluted, material segregation was a disaster, and rejected parts were mixed with normal production. Their basic design was to produce motors to inventory and have a small stock of motors on hand to account for demand and production variations. Their current inventory was zero. They currently shipped what they could make. This created frequent and disruptive daily changes in the production plan.

We were asked to assist them to further implement their Lean system, which they called the QED Production System (QPS). Luis, the QPS manager, had been there since the startup, he was our contact person. He reported directly to the home office and his job was to guide the plant into QPS maturity. Once that was achieved, he was to take over as plant manager and the plant manager would be promoted to a job in Minneapolis. Luis was very knowledgeable in both the motors manufacturing business and Lean manufacturing, but he had not been dynamic enough to provide the necessary leadership to force this situation.

Specifically, we were asked to:

- Improve line capacity so the plant could provide demand on a five-day basis.

- Make a 50 percent reduction in lead time.

- Implement a make-to-stock, pull system, operating at *takt*.

- Reduce line rejects by 50 percent.

- Increase on-time delivery to over 95 percent.

We were given some specific restrictions. First, no large capital outlays were available. Second, we had 60 days to achieve our gains. Third, and most restrictive, we could not shut down any facilities at all if it caused weekly production to fall below current levels.

The Process Description
The process description is as follows (see Fig. 16-6).

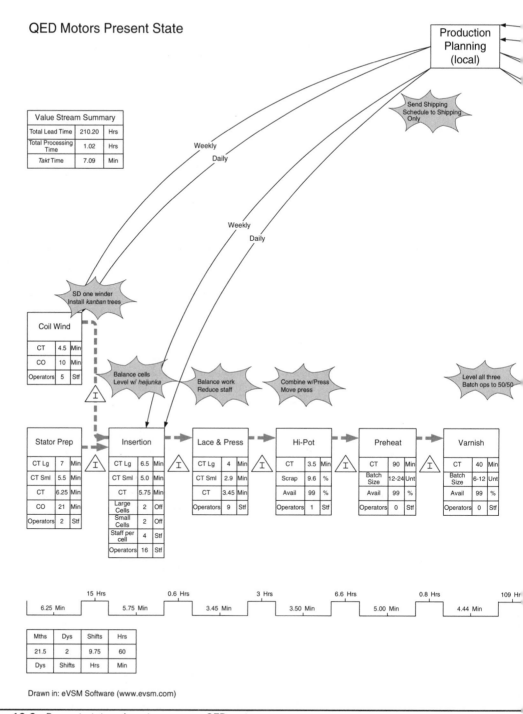

QED Motors Present State

Value Stream Summary

Total Lead Time	210.20	Hrs
Total Processing Time	1.02	Hrs
Takt Time	7.09	Min

Send Shipping Schedule to Shipping Only

Weekly
Daily

Weekly
Daily

SD one winder
Install *kanban* trees

Coil Wind

CT	4.5	Min
CO	10	Min
Operators	5	Stf

Balance cells
Level w/ *heijunka*

Balance work
Reduce staff

Combine w/Press
Move press

Level all three
Batch ops to 50/50

Stator Prep

CT Lg	7	Min
CT Sml	5.5	Min
CT	6.25	Min
CO	21	Min
Operators	2	Stf

Insertion

CT Lg	6.5	Min
CT Sml	5.0	Min
CT	5.75	Min
Large Cells	2	Off
Small Cells	2	Off
Staff per cell	4	Stf
Operators	16	Stf

Lace & Press

CT Lg	4	Min
CT Sml	2.9	Min
CT	3.45	Min
Operators	9	Stf

Hi-Pot

CT	3.5	Min
Scrap	9.6	%
Avail	99	%
Operators	1	Stf

Preheat

CT	90	Min
Batch Size	12-24	Unt
Avail	99	%
Operators	0	Stf

Varnish

CT	40	Min
Batch Size	6-12	Unt
Avail	99	%
Operators	0	Stf

	15 Hrs		0.6 Hrs		3 Hrs		6.6 Hrs		0.8 Hrs		109 Hr
6.25 Min		5.75 Min		3.45 Min		3.50 Min		5.00 Min		4.44 Min	

Mths	Dys	Shifts	Hrs
21.5	2	9.75	60
Dys	Shifts	Hrs	Min

Drawn in: eVSM Software (www.evsm.com)

Figure 16-6   Present state value stream map: QED motors.

Production Planning (local)

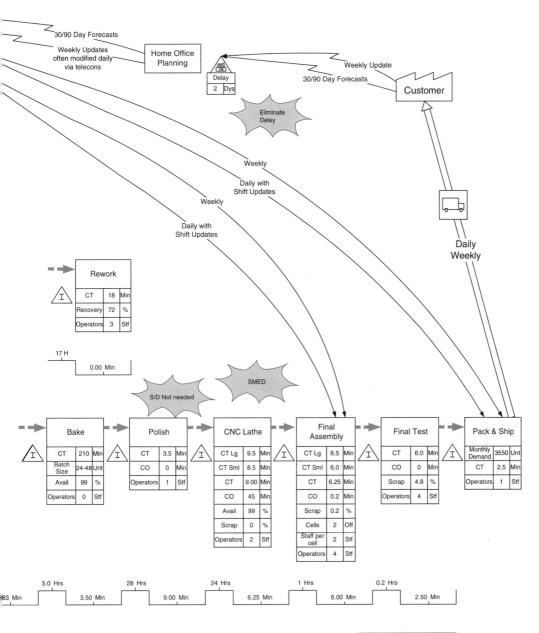

30/90 Day Forecasts

Weekly Updates
often modified daily
via telecons

Home Office Planning

Delay
| 2 | Dys |

Weekly Update

30/90 Day Forecasts

Customer

Eliminate Delay

Weekly

Daily with
Shift Updates

Weekly

Daily with
Shift Updates

Daily
Weekly

Rework
CT	18	Min
Recovery	72	%
Operators	3	Stf

17 H

0.00 Min

S/D Not needed

SMED

Bake
CT	210	Min
Batch Size	24-48	Unt
Avail	99	%
Operators	0	Stf

Polish
CT	3.5	Min
CO	0	Min
Operators	1	Stf

CNC Lathe
CT Lg	9.5	Min
CT Sml	8.5	Min
CT	9.00	Min
CO	45	Min
Avail	99	%
Scrap	0	%
Operators	2	Stf

Final Assembly
CT Lg	6.5	Min
CT Sml	6.0	Min
CT	6.25	Min
CO	0.2	Min
Scrap	0.2	%
Cells	2	Off
Staff per cell	2	Stf
Operators	4	Stf

Final Test
CT	6.0	Min
CO	0	Min
Scrap	4.8	%
Operators	4	Stf

Pack & Ship
Monthly Demand	3550	Unt
CT	2.5	Min
Operators	1	Stf

5.0 Hrs   28 Hrs   24 Hrs   1 Hrs   0.2 Hrs

83 Min   3.50 Min   9.00 Min   6.25 Min   6.00 Min   2.50 Min

Summary		
Total Lead Time	210.20	Hrs
Total Processing Time	1.02	Hrs

Stators were prepared by welding and grinding, and then were placed at the front of four coil insertion cells. The insertion cells had a build schedule and the supervisor hung markers so the stator prep station knew which stator was required. Three models of stators were available: one large model and two smaller ones. Wire was coiled by the coiling machines in a "coiling island" and then mounted on rolling coil trees for transport to the coil insertion cells. The line operators would leave the cell and go get the wire they needed. Forty trees were queued up at the "coiling island." The trees were color-coded for different coil arrangements required on different motors. Four coil insertion cells were available, each staffed with four operators. All four cells were capable of making all ten models.

From insertion, the wound stators went to lacing and press where a pool of nine operators would complete these tasks. The work load varied at this station, due to model mix, so these operators were also used to transport motors in the production line. From lace and press the motors were transported to Hipot, an electrical stress test.

After Hipot, the stators accumulated to form a batch prior to going to the preheat oven. Due to oven size, maximum batches would be 12 to 24 motors, depending upon the large/small ratio. After preheat, they were moved to a batch varnish operation, which varied in batch size from 6 to 12, and after varnish, the motors were moved to the curing oven where batches of 24 to 48 were cured.

Following the cure, stators were first polished and then moved to the CNC lathe were they were trimmed and sent to final assembly, which consisted of two final assembly cells, operating in parallel. Each cell was staffed by two operators. Following this, the motors underwent a series of tests, including visual inspection, and then were passed on to packing.

Some comments on the planning. There was a three month forecast and a monthly plan, both of which were largely ignored at production. A weekly update would come from Minneapolis on Friday, for the following week's production. The planning staff turned this update into a daily production plan and sent it to shipping along with a

Model	Monthly Demand (Avg.)	Pick up Frequency	Stator Size	Wire Coil Pattern
Ia	1130	Daily	Large	A
Ib	720	Daily	Large	B
II	950	3/wk, M,W, F	Small-1	C
III	475	1/wk	Small-2	D
IV	175	1/wk	Small-2	D
V	90	1/wk	Small-2	D
VI	50	1/wk	Small-2	D
VII	20	1/wk	Small-2	E
VIII	20	1/wk	Small-2	E
IX	10	1/wk	Small-2	E
Total	3550			

TABLE 16-7   QED Motors, Model Mix Demand

copy to the insertion cell and also the coil winding island. Several times each week the production schedule would change. Local planning would send updates to all three planning locations on the production line. Many of the required changes were due to internal production or parts concerns. This changing of the schedule created huge variation in the process.

Some More Relevant Information
The production demand of the line was that shown in Table 16-7.

Applying the Second Prescription at QED Motors—How to Implement Lean

The Three Fundamental Issues of Cultural Change

Leadership
Luis and I reviewed the level of leadership, for this short project. I was certain that with both of us working together we could provide the necessary leadership.

Motivation
As for the motivation, the plant was very engaged. In spite of the problems, morale was very high. In most activities, they just didn't really know how to apply the Lean principles in their situation. They knew what to do, just not how to do it. Motivation, which is often a serious obstacle, was a strong asset at this company.

Problem Solvers
Regarding problem solvers, the company did not have enough, but with Luis and I working full time on the effort, that temporarily met the requirements. With that in mind, we put together a plan and felt that inside of three weeks we could increase the production so a five-day workweek was practical, and we could get on-time delivery to reach over 98 percent. We were uncomfortable about making scrap reduction projections. At least that was the plan we submitted to the management team.

The System Assessment

The Commitment Assessment
The commitment assessment was done by Luis and I (behind closed doors) and we concluded that a few problems existed but nothing that we wanted to work on now. We might do this more formally later. We were convinced that for the next two months we had the necessary commitments.

The Five Precursors to Lean
Even though our mission was clear, Luis and I began an assessment of the precursors to Lean. I wanted to get a clear picture of where the facility was on Lean, and I thought it might be an eye opener for Luis. It was. The facility graded out as shown in Table 16-8.

They achieved a score of 7.5 out of a possible 25. Quite frankly, for a company that is five years into a Lean initiative, that is not very impressive, even taking into account the relative age of the facility. What was particularly disappointing was the poor layout,

The Five Precursors	Score
High levels of stability and quality in both the product and the processes	1.5
Excellent machine availability	2.0
Talented problem solvers, with a deep understanding of variation	1.5
Mature continuous improvement philosophy	1.0
Strong proven techniques to standardize	1.5
Total	7.5

TABLE 16-8   The Five Precursors, Summary Evaluation

the absolute lack of flow, and the very low levels of process stability. Luis was extremely disappointed with the condition of the processes but agreed with the evaluation we had jointly compiled. As I hoped, it was an eye-opener for him.

Ten Reasons for Failure

"The Ten Reasons Lean Initiatives Fail" was reviewed briefly and we found there were issues that needed to be addressed later. We did have one issue with outages being caused by poor raw materials supply on several parts, and during the management presentation purchasing committed to working on those immediately.

Process Maturity

As for the specific assessment of process maturity, we did not do one. In this case, we knew that by this point we were going to make major process changes, so we felt it would be best to complete these changes first. However, after the project was completed, Luis had this document turned into a plant standard. On subsequent projects, he later told me, all work stations needed to achieve level 3 status, by all measures, before start of production.

Educational Evaluation

Luis and I performed an evaluation and, since they had done a great deal of training earlier, we didn't really need to do anything at the outset. The entire facility had a reasonable grasp on what Lean was really supposed to be. Their problem was not understanding, it was application. However, as we got into the project, we needed to do some education on quick changeovers, *kanban* calculations, inventory calculations, and *heijunka* board design. When these issues surfaced, we trained only the necessary persons and trained them just in time.

The Implementation of the Four Strategies at QED Motors

Synchronize the Supply to the Customer, Externally

This step has three processes. First is the *takt* calculation. Next, we must level production, and since we have a make-to-stock system, we need to calculate the cycle, safety, and buffer stock inventories.

- The *takt* calculation was: $Takt = (22 - 2.5)^*60/165 = 7.09$ minutes. It was based on two shifts each lasting 11 hours. Each shift included 1.25 hours of meal and rest breaks. Production was 3550 units per month—with 21.5 days/mo = 165 motors/day.

- The processing time for the ten models was relatively straightforward to calculate and it made model-mix leveling easy. All models of small stators take one particular time, while all models of large stators take a longer time. Since large stator motors comprise 52 percent of the total, we decided to level with an initial model mix of 50/50, large and small stators. We were not comfortable designing a perfectly leveled system, rather we would level production based on stator size and work in the specific model planning with the *heijunka* board. For our design, at this point, that was all we needed to do.

- To size the buffer and safety stocks, we did not have good information, so we chose to do it rather arbitrarily. For buffer and safety stock combined, we decided to have two weeks for all weekly pickups and three days for all daily pickups. For cycle stock we used the pick-up volume, as a starting point. Table 16-9 shows the outcome of the plan for the inventories.

Synchronizing the Production, Internally

Defining the Work

This was very difficult, and when we were doing it, I was convinced it would need to be done again because the processes were unstable and cycle times had large variations. Nonetheless, you need to start somewhere, so we elected to use process, not element times. (For example, in the process to Hipot, the motor needed several steps, or work

Model	Monthly Demand	Pickup Frequency	Daily Prod	Weekly Prod	Cycle	Safety and Buffer	Total FG Inv.
Ia	1130	Daily	53		53	159	212
Ib	720	Daily	34		34	102	136
II	950	3/wk M, W, F	44		44	132	176
III	475	1/wk		110	110	220	330
IV	175	1/wk		41	41	82	123
V	90	1/wk		21	21	42	63
VI	50	1/wk		12	12	24	36
VII	20	1/wk		5	5	10	15
VIII	20	1/wk		5	5	10	15
IX	10	1/wk		2	2	4	6
Total	3550						

TABLE 16-9   Inventory Plans

elements, performed. These work elements included burning the leads, straightening the leads, cleaning the leads, connecting to the Hipot machine, cycling the machine, disconnecting from the machine, and passing to the next station.) Since most of the work elements could not really be transferred to another work station, this was not a good tool for us as we tried to internally synchronize the flow. (Recall that this was a key element in balancing the Zeta Cell so efficiently. Unfortunately, it just wouldn't work here).

Table 16-10 shows the results from the time study, with the data segregated into large and small rotors.

The data for the time study was then used to create the line balance chart, Fig. 16-7, showing both the large and small stator cycle times compared to the takt of 7.09 minutes. This was then converted to a balancing chart to show operation with a line that is balanced with 50 percent large stators and 50 percent small stators, shown in Fig. 16-8.

Evaluation of the Balance Chart

As we evaluate this line balance chart, we can see there is lots of waiting and that the process stations are not well balanced. However, the critical problem is the bottleneck of the CNC lathe. Not only is it a bottleneck but the cycle time exceeds *takt*. This shows clearly why a great deal of overtime is required. It is obvious that to meet customer demand we will need to improve the cycle time at the CNC lathe.

Process Step	Large Stator Work Time (Min)	Small Stator Work Time (Min)	Present Staffing	Ave C/T Based on Present Staffing Large/Small (Min)
Weld Stator	14	11	2	7.0/5.5
Coil Insert.	105	65–90	16*	6.5/5.0
Lace&Press	36	26	9	4.0/2.9
Hipot	3.5	3.5	1	3.5/3.5
Preheat	7.5	3.75	1	7.5/3.75
Varnish	6.7	3.3		6.7/3.3
Bake	8.75	4.38		8.75/4.38
Polish	3.5	3.5		4.5/3.5
CNC Lathe	9.5	8.5	2**	9.5/8.5
Final Ass'y	13	12	4***	6.5/6.0
Final Insp.	6	6	1	6.0/6.0
Packaging	2.5	2.5	1	2.5/2.5

* Four cells of four people each
** Lathe personnel operate in parallel
*** Two cells of two people each

TABLE 16-10   Time Study

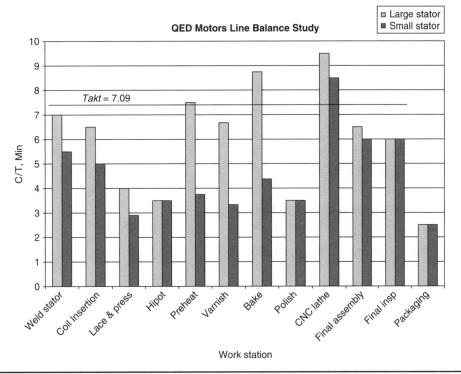

FIGURE 16-7 Line balance chart.

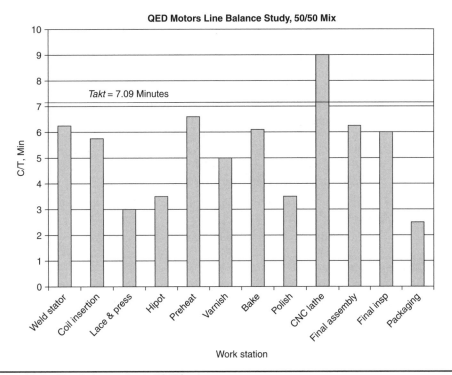

FIGURE 16-8 Line balance chart, 50/50 mix.

	Changeover*	Load	Machine	Unload	Total
Large	3 min	0.5 min	4.5 min	0.5 min	9.5 min
Small	3 min	0.5 min	3.5 min	0.5 min	8.5 min

* Actual changeover (CO) time was 45 minutes, averaging about 15 units per changeover. A unique holding jig was required for eight of the ten models. An average of eight COs occurred each day.

TABLE 16-11   CNC Lathe Cycle-Time Elements

We Reduce the Lathe Cycle Time
A breakdown of the lathe cycle time is shown in Table 16-11.

Kaizen Opportunities in Synchronization
The following were *kaizen* opportunities we could exploit from improvement ideas in both external and internal synchronization.

- It was clear from the balancing chart that the CNC lathe cycle time was the bottleneck, and both models ran at a cycle time exceeding *takt*. Thus, we needed to reduce the average cycle time. After some analysis, we decided to prepare a universal holding jig for small motors and one for large motors. This would reduce the number of changeovers to as few as five per day. Also, using Single Minute Exchange of Dies (SMED), quick changeover technology, we were able to reduce the changeover time to 11 minutes.

- Shut down one winder—this was not required—and go to three operators for the wire coiling cell.

- Model-mix level the production work. The complexity in leveling was in the coil insertion cells. The work to insert coils on a large stator averaged 105 minutes. The work for an average small stator averaged 80 minutes. To level the model mix, we would need two cells of four operators each on the large stators, which would average a motor every 26 minutes/cell (105 min/4 stations per cell). We would also have two cells of three operators each on the small stators and average a motor every 27 minutes/cell (80 min/3 stations per cell). Overall, we would have a cycle time across all four cells of 6.6 minutes. This was below *takt*, yet close enough to avoid overproduction.

- Other opportunities might be possible in staffing the insertion cells and the lacing island.

- Eliminate the polishing step since it was no longer needed.

Creating Flow
It is worth noting that during the making of the present state VSM, all production stopped at coil insertion, and the flow went to zero. Mother nature will always find the hidden flaw. A quick investigation showed there was no wire made with the proper coiling. Yet all 40 transport trees were 100 full. On the floor, we had over four days of coiled wire but none of it matched the demand at the cells. This, of course, was the direct result of sending planning information to multiple points on the value stream. In addition to the planning issues, creating flow was achieved by balancing the flow,

managing the inventories of both finished goods and WIP, as well as reducing transportation steps and distances.

Kaizen Opportunities Related to Flow

The following were *kaizen* opportunities we could exploit that were related to the process flow improvements.

- The process of updating orders currently had a two-day delay at the home office. This was for accounting purposes only. Consequently, we would have the customer copy us on the electronic updates to avoid the two-day delay.

- Lace and press used nine operators, and at a cycle time of less than 4 minutes indicated there was a lot of waiting. The press operation was about 2.5 minutes, regardless of stator size so we decided to transfer the press operation to the subsequent operation of Hipot for better balance and flow. The Hipot/Press station would now have a six minute cycle time, consisting of 3.5 minutes of Hipot and 2.5 minutes to press. Lacing alone would now be 33.5 minutes for large stators and 23.5 for small stators. Consequently we could staff the lacing operation with only five operators and still have an average cycle time of 5.7 minutes, comfortably below *takt*.

- Since many tests and inspections that had been done at final test, were no longer necessary, we could eliminate them and combine final test with packing.

- No motors were allowed to "break the flow" and could only be placed in WIP locations at workstations and supermarkets. Rework was sent directly to and isolated to the rework station.

Establishing the Pull-Demand System

Kaizen Opportunities Related to Pull Systems

The following were *kaizen* opportunities we could exploit that were related to the implementation of pull systems.

- Send daily shipping information from local planning to the storehouse, only.

- The planning function will make minor changes via the *kanban* system and *kanban* replenishment to the *heijunka* board will occur twice per shift.

- Production planning will be by *heijunka* board at the insertion cells only.

- Install a *kanban* system for wire supply using the wire trees as *kanban* carts.

- Establish a storehouse of stators at the Insertion cell so they can withdraw when needed.

- Establish FIFO lanes from insertion all the way to packing. Calculate the maximum inventories and document this on the FSVSM.

- Calculate and establish the finished goods inventory.

Making a Spaghetti Diagram

The spaghetti diagrams, shown in Figs. 16-9 and 16-10, display several process opportunities to reduce transportation and make better use of the floor space. The spaghetti diagrams used in this project are detailed further in Appendix D of Chap. 7.

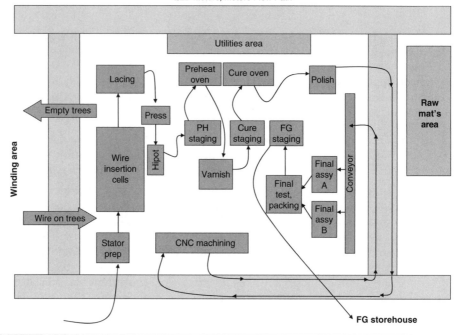

FIGURE 16-9 Spaghetti diagram: QED motors, before.

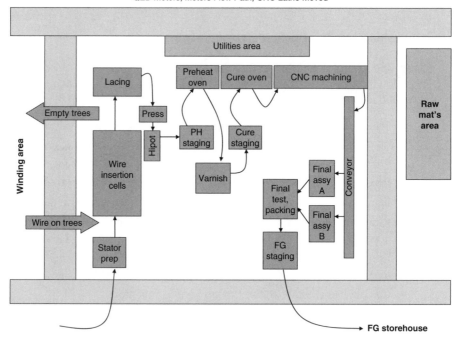

FIGURE 16-10 Spaghetti diagram: QED motors, after.

Kaizen Opportunities Related to the Spaghetti Diagram

The following *kaizen* opportunities were found while studying the spaghetti diagram.

- Move FG Inventory near the aisle on the storehouse side of the process area
- Move the Hipot operation next to the press.
- Move the CNC lathe to the corner near the curing oven.

Document *Kaizen* Activities on a Future State Value Stream Map

In this case, we documented the major *kaizen* activities on a future state value stream map (FSVSM). See Fig. 16-11.

Additional *Kaizen* Activities

In addition to the *kaizen* opportunities mentioned earlier, the following were measures we also took.

- We ran off the huge volume of WIP, almost 700 units. We took special actions to accomplish this. First, we temporarily shut down operation of all stations before the CNC lathe and reassigned those trained to fill in for all workstations from the CNC lathe through packing. Next we sent the extra people to the rework station or assigned them to a *kaizen* project and began rescheduling to cover 24-hour operation, including lunch and breaks, for all stations from CNC lathe through packing.
- We reduced the cycle time at the CNC lathe to achieve 160 units per day by adjusting work procedures at CNC lathe. To do this we transferred work from the CNC operator to the helper, who was lightly loaded, so as to decrease workload and the cycle times of the CNC operator. This reduced the line bottleneck by a half minute.
- We performed SMED analysis on the CNC changeover times to reduce changeover time and further increase production at the lathe. We began immediately producing the universal holding jigs. Next we implemented other SMED activities; changeover time was reduced to 11 minutes. These steps took three days and when complete we were producing 220 units per day at the CNC lathe.
- Final assembly could no longer keep pace, so we increased the number of changeovers at the CNC lathe. This allowed us to decrease the batch size and better level the model mix. At this point, we had all stations running, at least at half rate.
- We leveled the production at the insertion cell using a *heijunka* box with *kanban* and modified the staffing for the small stators to three operators per cell for better balance.
- After two weeks, it was possible to combine final inspection with packing and eliminate another position.
- In addition, over 70 other minor process improvements were executed, mostly by the supervisors, line leaders, and operators in the next four weeks.
- Even though capacity was up, we continued to run seven days per week with two normal 11-hour shifts. This was needed to establish inventories of cycle, buffer, and safety stocks for all ten models.

QED Motors Future State

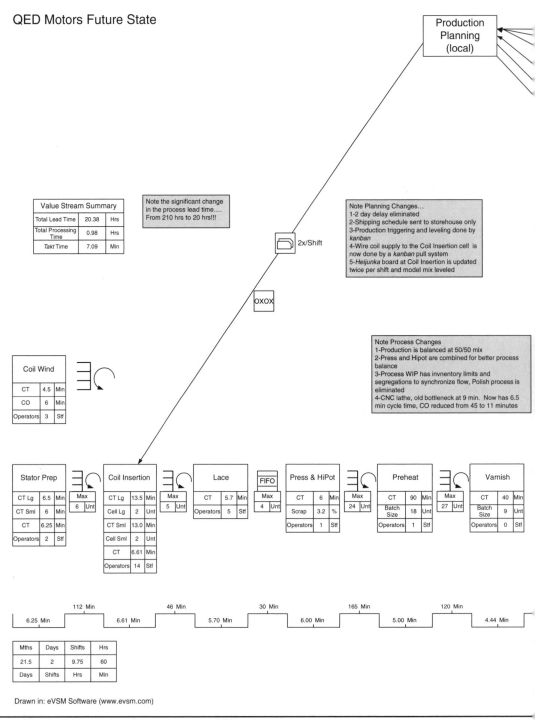

Figure 16-11 Future state value stream map: QED motors.

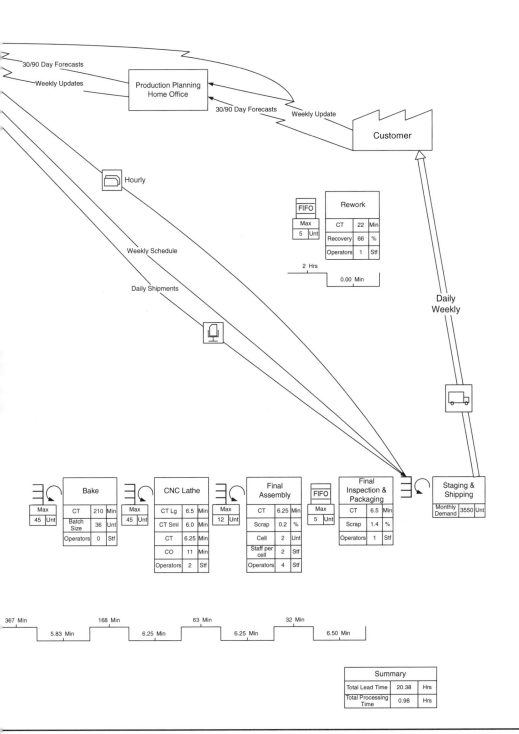

30/90 Day Forecasts

Weekly Updates

Production Planning
Home Office

30/90 Day Forecasts

Weekly Update

Customer

Hourly

Weekly Schedule

Daily Shipments

FIFO		Rework		
Max		CT	22	Min
5	Unt	Recovery	66	%
		Operators	1	Stf

2 Hrs

0.00 Min

Daily
Weekly

	Bake		
Max	CT	210	Min
45 Unt	Batch Size	36	Unt
	Operators	0	Stf

	CNC Lathe		
Max	CT Lg	6.5	Min
45 Unt	CT Sml	6.0	Min
	CT	6.25	Min
	CO	11	Min
	Operators	2	Stf

	Final Assembly		
Max	CT	6.25	Min
12 Unt	Scrap	0.2	%
	Cell	2	Unt
	Staff per cell	2	Stf
	Operators	4	Stf

FIFO		Final Inspection & Packaging		
Max		CT	6.5	Min
5	Unt	Scrap	1.4	%
		Operators	1	Stf

Staging & Shipping		
Monthly Demand	3550	Unt

367 Min

5.83 Min

168 Min

6.25 Min

63 Min

6.25 Min

32 Min

6.50 Min

Summary		
Total Lead Time	20.38	Hrs
Total Processing Time	0.98	Hrs

- It took over three weeks to build up enough inventories so we could take the CNC lathe out of service and move it. We did that over one weekend and the lathe started up on Monday without incident.

- The plant ran seven days per week for two more weeks, and then reverted to five days per week operation.

The Results

Our project had allowed the plant to improve dramatically. Specifically we could now meet demand in a 5 day work week. We were asked to reduce the production lead time by 50 percent and we actually achieved 89 percent reduction using a make–to–stock, pull, production system operating at *takt*, as requested. We were able to reduce defects by almost 60 percent and improve on-time delivery to over 95 percent, actually achieving 99 percent. In short, by focusing on waste reduction, with an aggressive attack on the seven wastes we met or exceeded all our goals and even achieved two significant gains that were not requested. These included major savings in production labor and space reductions. The labor savings alone showed a 44% reduction in direct labor to produce a motor. All of this was achieved by an aggressive waste reduction effort. Except for the minor cost to produce the universal jigs, no capital was expended. This was a massive success and the results are summarized in Table 16-12.

Some comments on the results

- The gains were huge and early, all achieved in less than 2 months.

- Of particular note were the quality improvements. Luis was very worried because we had no specific line items to work on any of the defect modes. As it turned out, by reducing inventory and implementing standard work we accomplished quality improvements.

- The facility still needed to work on a number of projects. First, and most importantly, although we cut the defects by 60 percent, line rejects are almost 6 percent. This is not acceptable by today's standards, thus they still need to work on this.

	Initial Situation	After First Pass of Lean	Improvement
Production Rate	125 u/day	165 u/day	+32%
On-time Delivery	76%	99%	+23%
Lead Time	92 hrs (7.5 days)	16 hrs (0.8 days)	−89%
Line Rejects	14.4%	5.8%	−59.7%
Operators per Shift*	50	43	−17.5%
Labor Consumed (Mh/motor)	9.33	5.21	−44%

* In addition to the staffing on the FSVSM, there were three operators in rework, four materials handlers, and three in material prep. None of this was changed as yet.

TABLE 16-12 QED Motors Results Following the Lean Applications

- They were now able to shut down the California plant and did so.

- One weakness in this effort was that we did not implement any kind of a *jidoka* concept. We were not asked to but I was disappointed we were not able to weave it into the project.

What Is Next for QED Motors?

As is always the case, these improvements brought about a new present state, which will need to be fully evaluated, including applying the Four Strategies to Becoming Lean, starting with the *takt* equation and ending with a spaghetti diagram, as well as a future state value stream map with *kaizen* activities noted.

The company's *kaizen* activities will clearly depend upon their specific goals, but we made a starter list based on what we had learned to date. Possible measures included:

- Implement a reasonable *jidoka* concept.

- A new floor layout to take advantage of the space savings achieved.

- Review the replenishment time to see if cycle stocks could be further reduced. Also review production and demand variations to see if buffer and safety stocks can be further reduced.

- Since the major impediment to lead time reductions was the batch-operated preheat, varnish, and bake operation, they needed to consider converting this to a continuous flow operation. This would shorten lead time significantly, but would require new capital equipment.

- The majority of the defects found at Hipot were due to the manual coil insertion operation. Automatic winders should be explored, which of course would require capital but would substantially reduce lead time and quality dropout. We did some preliminary economic analyses and it appeared to have a rate of return on invested capital exceeding 80 percent, based on labor, rework, and waste savings.

- QED Motors had made significant progress in a very short period of time and they can be proud of their accomplishments. Now they must focus on their future improvement ideas and continue on the journey to becoming Lean. It is a never-ending cycle of assessment, analysis and improvements through the continuous reduction of waste on the way to making their facility a better money-making machine with a more secure work environment.

CHAPTER 17

The Precursors to Lean Not Handled Well

The story of the ABC Widgets Co. is an example of a company that wanted to look Lean more than they wanted to *be* Lean. Internal conflicts, along with too few onsite resources, created a large resistance to becoming Lean. In spite of these deficiencies, we assisted them in making large improvements. However, they are an example of "How *Not* to Implement Lean," in that after four years into their Lean initiative, basic and large foundational problems still persisted. They are a story of the "Precursors to Lean … not handled well."

Background to the ABC Widgets Story

We were called in by Miguel, the reliability manager. Our task was defined as problem-solving assistance, not Lean implementation. The situation was this: a mature production line, one of four within the plant, previously had 100 percent on-time delivery for two years, but was now having serious problems: On–time delivery had dropped to 74 percent and remained there for over five months, despite repeated efforts to improve the situation.

Also, five months earlier, the daily production demand had increased from 30 units/day to 36 units/day, and the customer, in keeping with their own Lean initiative, had begun daily pickups. Previously, customer demand was 150 units per week and would be picked up in a single weekly shipment.

The reliability manager was sure one of the major contributors to the on-time problem was a welding machine with low availability, but he felt powerless to correct it. So, he sought our assistance.

In addition, two months earlier, the plant received a visit from the home office, and a Lean expert had assisted them in problem-solving activities. After one week of analysis and meetings, the plant was given a list of 22 projects they were to work on, but after nearly 60 days they had not been able to improve the delivery situation.

Our Challenge

The problem was stated as: "We have low on-time delivery. It is 74 percent and needs to be greater than 99 percent." Thus, we were asked to:

- Survey the situation.
- Make recommendations.

- Sell the recommendations consistent with their Lean initiative.

- Assist in the implementation.

Our constraints were many, but for the sake of brevity I will distill them to the critical few.

> **P**oint of Clarity   When discussing variation, the standard deviation has no meaning unless the data show stability.

- There was really no capital available to solve this problem.

- We could not increase inventories. (Miguel noted that they were presently holding 30 units of safety stock to handle the internal supply variations. This volume was calculated to be three sigma of inventory, but they were still missing deliveries.)

Some More Relevant Information

Of course, a little more information is necessary to appreciate this particular situation. This plant was a Maquiladora and it had only a skeleton of technical personnel onsite, with little or no design responsibilities. Like many Maquiladoras, which are the Mexican half of the Twin Plant concept, taking advantage of low cost Mexican labor, their task was to meet the production schedule at minimum costs. Purchasing was done centrally, while planning was local. Four years earlier, the ABC Widgets Co. had embarked on a Lean initiative, but most Lean-specific skills were centrally located at the home office. Both the metrics of performance and the actual performance results of the plant were much the same now as they were before the implementation of the Lean initiative. They had made great strides in one area: inventories. Both raw materials and finished goods inventories had been dramatically reduced. The entire plant took great pride in this because they were among the leanest plants in the corporation by this measure. Raw materials had been improved from 12 to 35 turns and finished goods for this line improved from 24 to nearly 70 turns. No information was available about WIP.

Another issue was influential in this problem. The reliability manager, Miguel, was the son-in-law of the general manager of their North American manufacturing division. Miguel had graduated three years earlier from an engineering university and was hired 18 months before I arrived to assist in the Lean initiative. "Water cooler rumors" of nepotism had arisen and several managers questioned his capabilities. Since his arrival, machinery reliability had improved, and availability associated with machinery reliability had risen over 7 percent. These gains were clearly seen in Overall Equipment Effectiveness (OEE) on several production lines. To make things more problematic, Miguel had apparently created a serious rift with the production personnel when, earlier in the year, he had initiated autonomous maintenance as part of their TPM efforts. It did not go well, and that effort was aborted after just a few months due to the friction it caused. Obvious tension remained between the Reliability and Production departments.

We Analyze the Data

Our efforts began with a look at the data associated with this line for the last 30 normal production days. This is shown in Table 17-1.

	Daily Prod	Dev. from Goal[1]	Avg. for the Week[2]	SD for the Week[3]	Total Weekly Prod'n	Weekend Run[4]
			Base Case			
1	35	−1				
2	19	−17				
3	38	2				
4	8	−28				
5	36	0	27.20	13.14	136	44
6	22	−14				
7	36	0				
8	18	−18				
9	35	−1				
10	29	−7	28.00	7.91	140	40
11	38	2				
12	12	−24				
13	32	−4				
14	28	−8				
15	27	−9	27.40	9.63	137	43
16	29	−7				
17	39	3				
18	40	4				
19	35	−1				
20	31	−5	34.80	4.82	174	6
21	37	1				
22	0	−36				
23	14	−22				
24	29	−7				
25	37	1	23.40	16.10	117	63
26	40	4				
27	18	−18				
28	17	−19				
29	27	−9				
30	40	4	28.40	11.28	142	38
		Avg.	28.20	10.66		

---

[1] Deviation from the planning goal
[2] Average daily production for the five-day week
[3] The daily variation measured as the standard deviation (SD)
[4] The production over the weekend, needed to make the weekly demand

TABLE 17-1 Production Data, Base Case

Point of Clarity   Prior to implementing a Lean initiative, processes must show statistical stability. This process does not!

Even though these data had huge variation, we were assured that these production data were typical. In this six-week period:

- Six daily shipments had been missed.
- Production was well below the 36-unit target.
- The production rate had a huge day-to-day variation.
- Weekend work (both Saturday and Sunday) was required to replenish inventory.

We Pareto-ize the Days of Low Production, We Find Two Problems

We reviewed production data to find the reasons for low line production. In this 30-day period, there were nine days where actual production was ten or more units below demand. (Each tray had 25 production items, with four trays per box, so a box contained 100 items. This was their production jargon: boxes, which they called "production units" or just "units" for short). In all nine cases, either the automatic welding machine failed or the sensor, a high-cost component, had a stock out, or both occurred. At any rate, their large problems were not quality problems, but availability problems caused by these two items.

We Investigate Further

With this information in hand, we wanted to have another discussion with Miguel to gain more insight. First, the welding machine availability was number one on our list of items to correct. Unfortunately, the home office facilitator had characterized the problem as inadequate maintenance: specifically, the need to train a replacement for Jorge, the welding technician, who had retired four months earlier. The plant was not allowed to replace Jorge because they had been asked to reduce manpower. Miguel told us the problem had nothing to do with Jorge, it was a capacity problem with the welder. It would simply overheat at the new rate and the electrode and holder would fail, requiring a shutdown to replace the parts and several hours to complete the setup, which included alignment and testing. Work was underway to implement Single Minute Exchange of Dies (SMED), quick changeover technology on the machine startup, but a practical solution was months away; however, Miguel did not believe that starting up faster was the issue. We did some more investigating and found out he was right. The problem was poor reliability.

"Let's not work hard to get good at something which should not be done at all. **"**
J. Keating

As for the sensor stock outs, this sensor was a high-cost component comprising 44 percent of the total raw material cost of the product. The home office had implemented strict inventory guidelines and aggressively reduced the inventory levels of all components, including the sensor. After the ramp-up, the supplier could meet demand, but only by working overtime. Since these issues were managed by the central purchasing group, Miguel thought it would be futile to attempt to increase inventory levels, although that would surely solve the problem of stock outs.

We inquired about the line capacity and the recent increase from 30 to 36 units per day. The nameplate bottleneck on the line would limit the line to 40 units in a 24-hour work day. Line OEE would need to be at 90 percent, and these levels had not been previously achieved.

We asked for the background on the recent increase to 36 units per day. We were told the customer had demanded a 15 percent price reduction over two years and would then give the plant additional demand. According to Miguel, the line was very profitable and management jumped at the chance with very little review, not even an update of the Process Failure Mode Effects Analysis (PFMEA). Management was very confident the line would perform. Prior to the ramp-up, the customer returns were less than 500 PPM, on-time delivery was practically 100 percent, and the line name plate capacity was adequate. It appeared to be a win-win situation, until they could not meet demand. That problem cropped up almost immediately. In addition, with all the weekend overtime and expediting, profits had eroded to practically zero and Miguel was convinced they were not on the correct path to return to profitability.

We Investigate More, and Fix the Welding Machine, the First Problem

With this background, we did some digging and discovered the following. The welding machine had numerous previous problems, but those problems were masked by excess capacity in the line and the inventory situation, specifically the weekly pickups. Earlier, they could have an off day or two and still be able to make up volume with the excess capacity and even a little overtime. Also, it was only recently that they achieved the low inventory levels. At higher inventory levels, this also helped mask the problem of the welder's poor reliability. We asked the technician in charge of solving this problem to bring in the manufacturer's rep for the welder. The welder's rep readily confirmed that the machine was probably capable of 30 units per day, but would surely overheat at 36 units per day. He suggested a minor upgrade to the cooling system, and for $1500 he would expect greater than 99 percent reliability at our needed cycle times. The equipment was ordered and installed. Table 17-2 shows the next 25 days of production.

In this five-week period, significant progress had been made. The welder did not fail once. In addition:

- Only one shipment was missed.
- Production had improved by over three units per day.
- The production variation shrank from 10.7 to 7.6 units.
- Weekend work was significantly reduced.

The Second Problem, Sensor Stock Outs Is Addressed

The second problem, the sensor stock outs, proved to be a little more slippery. The supplier was contacted and he was doing all he could but was severely stretched. His production facility was at capacity and earlier he had informed the buyer, who did not pass on this information. They had plans to increase capacity, but any future increases were six months away. It seemed the only viable short-term solution was to increase inventory levels to account for the supplier's variation. No one wanted to hear that.

Point of Clarity The only purpose of inventory is to protect sales, not production, but sales. So where should the inventory be located? Inventory is a waste, so minimize the inventory cost, be it in raw materials or finished goods.

	Base Case—Welding Machine Now Reliable					
	Daily Prod	**Dev from Goal**	**Avg. for the Week**	**SD for the Week**	**Total Weekly Prod'n**	**Weekend Run**
1	36	0				
2	32	–4				
3	39	3				
4	21	–15				
5	37	1	33.00	7.18	165	15
6	20	–16				
7	36	0				
8	35	–1				
9	33	–3				
10	17	–19	28.20	8.98	141	39
11	40	4				
12	36	0				
13	33	–3				
14	14	–22				
15	30	–6	30.60	9.99	153	27
16	33	–3				
17	36	0				
18	37	1				
19	16	–20				
20	34	–2	31.20	8.64	156	24
21	39	3				
22	33	–3				
23	34	–2				
24	35	–1				
25	36	0	36.00	2.30	177	3
		Avg.	31.68	7.62		

TABLE 17-2   Production after Welding Machine Upgrade

Nonetheless, we made an XL model from our data, which showed that although sensor inventory costs would need to increase by 25 percent, finished goods safey stock inventory could be reduced by 60 percent. It turned out that for each $1.00 of sensor inventory that was added, $8.00 of finished goods inventory could be reduced. This was a clear winner in terms of total money tied up in inventory.

Point of Clarity The system optimum is not always the sum of the local optima!

The computer model showed a new financial optimum. This, coupled with our prior success, got us an audience with the general manager (GM) of the North American manufacturing division. We met the GM during his monthly visit and convinced him that although sensor inventory costs would go up, we could reduce finished products inventory by a greater dollar volume so that the overall effect would be a significant savings in money tied up as inventory. For finished goods, we now had a safety and buffer stock inventory of 12 units, which was less than one day of production. We got the go-ahead and the supplier was able to pull some strings and use capacity from another plant in order to supply some additional product to create the inventory. When the new sensor inventories had been established, sensor stock outs stopped. The production now looked like that shown in Table 17-3.

	Daily Prod	Dev. from Goal	Avg. for the Week	SD for the Week	Total Weekly Prod'n	Weekend Run
	Base Case—Welding Machine Reliable, Sensor Supply Resolved					
1	35	−1				
2	36	0				
3	38	2				
4	36	0				
5	40	4	37.00	2.00	185	−5
6	34	−2				
7	36	0				
8	38	2				
9	35	−1				
10	32	−4	35.00	2.24	175	5
11	38	2				
12	39	3				
13	36	0				
14	28	−8				
15	37	1	35.60	4.39	178	2
16	31	−5				
17	40	4				
18	36	0				
19	35	−1				
20	34	−2	35.20	3.27	176	4

TABLE 17-3 Production after Welding Machine and Sensor Countermeasures

Base Case—Welding Machine Reliable, Sensor Supply Resolved						
	Daily Prod	Dev. from Goal	Avg. for the Week	SD for the Week	Total Weekly Prod'n	Weekend Run
21	37	1				
22	35	−1				
23	36	0				
24	32	−4				
25	40	4	36.00	2.92	180	0
26	38	2				
27	35	−1				
28	37	1				
29	30	−6				
30	35	−1	35.00	3.08	175	5
		Avg.	35.63	2.89		

TABLE 17-3 Production after Welding Machine and Sensor Countermeasures (*Continued*)

In this six-week period, additional progress had been made, specifically:

- Zero shipments had been missed.
- Production had improved by an additional four units per day.
- The production variation shrank from 7.6 to 2.9 units.
- Weekend work was reduced to practically zero.

Summary of Results

The summary of results is shown in Table 17-4.

Demand could now be met, overtime practically ceased, there was no need to airship product, labor cost was reduced, and the product was once again profitable. As a result, almost everyone was happy.

	Prod Rate	Std. Dev. of Rate	Missed Shipments	Weekend Production
Base Case	28.20	10.66	26%	196
Fix Welding Machine	31.68	7.62	4%	108
Eliminate Stock Outs	35.63	2.89	0%	11

TABLE 17-4 Summary Results

How Did the Management Team from ABC Widgets Handle the Fundamentals of Cultural Change?

Let's look at the actions of the management of ABC Widgets in relation to the implementation of any cultural change initiative (see Chap. 6). How did plant management handle the three fundamental issues?

Did They Have the Necessary Leadership?

The leadership was clearly deficient in several areas. First, when they ramped up, they had nothing that resembled a plan. Recall that they did scarcely nothing except hope they could make it work. Next, once they got in trouble, they did not have much of a plan for that either. They counted on the home office for help that did not materialize. Third, when it become obvious they could not solve the problem themselves, management resisted outside involvement. Finally, as you hear the entire story, a grand deficiency in leadership exists at the corporate level. Lean cannot be managed from afar. Lean leadership and the problem solvers—need to be on site. There is no substitute for this.

Their Motivation

So did they have the motivation to make it work? At one level, the answer to this was "yes," since they were actively trying to solve the problem—that is, the problem of low production, once they got into it. But at the larger level, their motivation was grossly lacking. Motivation is measured by actions. And just what were their actions to make the plant Lean? Superficial and clearly inadequate. Several years into the initiative they discovered basic, foundational problems that should have been found and resolved during the first wave of activities. Yes, problems need to be both uncovered and resolved. They did not find them because they did not look, nor did they listen to those who knew about the problems. Just for practice, based on these data, do a commitment evaluation on their management. (Turn to Chap. 19 and see how ABC Widgets scores on the Five-Part test.)

Their Problem Solvers

So, did they have the necessary problem solvers in place? Clearly not. Regarding this rather straightforward production problem, they could not solve the problems themselves. For example, the "help" from the home office was not helpful, and finally, to resolve the issues, they needed outside assistance.

Think Back to an Earlier Time

For just a moment, think back to the major problems they had prior to the ramp-up. They had supply and process capability problems of which they were unaware. They had inadequate machine and line availability. It was only the excess line capacity and excess inventory that masked these problems. The problems were there from the beginning! They simply failed to "see" them and take them on.

A Lean plant will see these as opportunities and attack them to reduce space needs, manpower needs, lead-time issues, and inventories, to name just a few. All of this could have been done earlier, but it was not. It wasn't even recognized.

The Real Message

What was the message of the ABC Widgets story? The lessons are multiple, but can best be described by a call we received about three months after our work with ABC Widgets. The caller was Miguel. We had spoken with him several times since our work together. He would occasionally call for advice or sometimes just to chat. This call was particularly revealing in that he told us he was taking a new position with a competitor of ABC Widgets. He was going to be in charge of their Lean implementation. He was changing companies because, as he put it:

> "I was getting so frustrated with our efforts. I could easily see the opportunities but we were not staffed nor were we structured to capture them. I recall some advice you had given me earlier when you said, 'You cannot have a just in time (JIT) materials system without a JIT support system, including JIT maintenance and JIT problem solving.' I began to see the problems caused by our centralized approach, especially the problem-solving issues. When you were at the plant, then and only then did we have sufficient problem-solving support and we made great headway. Once you left, progress ceased. In addition, it became obvious how we had completely circumvented Lean principles when we increased the line capacity. There were no process analyses, no FMEAs. Only a superficial economic analysis and the hope that things would go well. It was at this point that I decided our efforts were more at *looking* Lean than *being* Lean. I discussed all these issues with our management and there was no interest in changing these areas, so I decided to move on."

Other issues of note here, which are very common in many plants and are always destructive, are the issues of favoritism and politics. These have no place in dispassionate problem solving and Lean analyses. Lean decisions need to be data driven and fact-based. Here, the relationship of the reliability manager, Miguel, and the general manager, Juan Pablo, his father-in-law, created some problems. We found Miguel to be bright and unbiased. His decisions were fact-based and good. Unfortunately for him, many others were biased against him since they thought he got his current job because he was the "son-in-law," rather than because he was a good engineer and manager. It was this bias that was largely the reason the implementation of autonomous maintenance had failed. We found that many decisions here were not fact-based. The entire plant was short on problem solvers, and existing personnel were neither trained in, nor accountable for, problem solving. Yet there was no shortage of problems. This is not a good combination!

> **P**oint of Clarity   A JIT manufacturing system needs JIT problem solving and JIT decision making. Hence, central controls are incompatible with Lean!

Chapter Summary

The problem solving worked and we helped the facility fix their problems. We increased capacity, improved reliability, reduced labor costs, and reduced inventory as we let them return to excellent on-time delivery. However, this was not the most revealing part of this story, which is a sad lesson we must all learn.

Earlier, I had characterized this as a "not-so-Lean-but-we-want-to-appear-Lean" plant. I hope it is obvious now what I meant. Superficially, the plant had a very Lean appearance. 5S was mature and cells were set up neatly with visual information centers. Centralized materials handling was in place and, quite frankly, from a visual standpoint the plant had the appearance of being Lean. However, as can now be seen, the plant was anything but Lean. When we helped them solve these problems, it uncovered issues that should not have been lingering in a mature line after four years of Lean efforts.

Still, the killer issue was that they actually *thought* they were Lean.

Another serious issue should be remembered as well. The entire plant, especially management, refused to "see" many of the problems. During our work there, it was obvious that many of the issues had earlier been "seen" by a number of staff, but there was a marked reluctance to bring them up, as if by ignoring the problem, it would politely stay in the closet. Unfortunately, these problems have a way of surfacing and resurfacing. In this case, the problems found the light of day when an opportunity arose to increase capacity and make money.

Like ABC Widgets management, we would all be well advised to listen to the wisdom of those from the past. It is the wisdom quoted earlier in the preface:

> **"I**n the choice between changing one's mind and proving there's no reason to do so, most people get busy on the proof. **"**
> John Kenneth Galbraith

> **"O**pportunity is missed by most people because it is dressed in overalls and looks like work. **"**
> Thomas Edison

The lesson of ABC Widgets is that there are no shortcuts to becoming Lean. We must

- Address the fundamental issues to any cultural change initiative.
- Take care of the foundational issues of quality control.
- Implement quantity control measures.

This must be done to make our facility a securer and better money-making machine.

CHAPTER 18

An Experiment in Variation, Dependent Events, and Inventory

Background

This is a simple experiment that can be done at your desk or in a small group. All that is needed are some dice and a simple form (shown later in this chapter). The experiment is a factory simulation using pull production while studying the effects of variation and dependent events on factory performance.

This is a phenomenon that is understood by only a few.

In short, when we have variation, as we do in any process, coupled with dependent events, as we do in a multistep process, then the process will not produce to the average rate of the processing steps, unless we have inventory between the dependent steps. Plus, as the variation is increased, the inventory levels must increase to maintain production. In addition, as the number of sequential steps are increased, the inventory increases by an exponential factor. It is for these reasons that most *factories cannot produce at the nameplate average rate of the equipment, unless inventory levels are extremely large or variation is reduced to very low levels.*

Read that sentence again.

It gives tremendous insight into why, *when projects or production schedules are planned, halfway through the project, overtime is needed, and in the end more overtime is needed, and yet we still often need to pay to expedite the shipment.*

This experiment explains this phenomenon and more.

Variation and dependent events are everywhere in a factory. Take a simple cell, for example. Let's say we have a six-station cell and all work stations have 60 seconds of work, which is also *takt*. Also, there is one piece at the workstation and there is no inventory between stations—true one-piece-flow. When station 1 finishes a piece, so do stations 2 thru 6, and in unison, all six pieces of in-process work are simultaneously pulled to the next work station every 60 seconds—the perfect synchronization of process flow: the ideal state.

But, for the moment, let's imagine that the cycle time for station 4, although it averages 60 seconds, varies from 50 to 70 seconds. When station 4 performs at 50 seconds, it finishes its process, and then station 4, has a ten-second wait time before its product is pulled by station 5. Ten seconds of waiting time elapse, which is a waste for station 4, *but this is not a production rate problem,* the cell will still produce to *takt.* It is just that the operator at station 4 will sit around a while. On the other hand, when station 4 takes 70 seconds to produce its work, that subassembly is held up and station 5 is starved for work for ten seconds. This delay passes through all the work stations of the cell in a wave and that piece is produced on a 70-second cycle time.

So let's recap… If the station that varies—in this case, station 4—operates faster than *takt,* station 4 must wait for the subsequent station to pull the production; however, when station 4 just happened to operate slower than *takt,* station 4 would slow down the whole cell on that cycle and there is no recovery with a resultant loss of production rate.

So even though the station may have a 60-second cycle time *on average,* any time the cycle time is above average, the production rate drops. This concept is known as the effect of variation and dependent events. (The dependency is that the "next step" depends on the "prior step" for supply.)

So the solution is, guess what …? You got it! Add some inventory. We will need to add inventory both before and after station 4, the one with the variation. We need the inventory in front of station 4 so when it produces faster than *takt,* say at 50 seconds, there is raw material available to keep it producing. We also need the inventory after station 4, so when it is operating slower than *takt,* say 70 seconds, there is raw material to supply station 5. Then, station 4 can have the variation *and* maintain production at *takt* on average.

This effect of variation and dependent events is not well understood and is a problem, always. Two solutions can be employed: Either totally remove the variation, or totally remove the dependency. To totally remove the variation is an impossibility. Recall that the definition of variation is, "the inevitable differences in the outputs of a system." Since it is inevitable, total removal is an impossibility. Okay, so let's totally remove the dependency. This means tons of inventory, the exact thing we are trying to eliminate in a lean solution.

So, guess what? The solution is to find the happy medium and it is best done by first reducing the variation to a minimum so inventory can then be reduced accordingly.

That's enough background for now; do the experiment and see firsthand—under controlled conditions—exactly how this phenomenon plays out.

The Experiment

To do the factory simulation experiment, get 12 ordinary dice, some students; four teams would be ideal. If you only have enough for two or three teams, do that, the experiment it totally flexible. We will use dice to get random numbers, and we will vary the number of dice for each team to modify the amount of variation for each team. Distribute the dice as shown in Table 18-1. A little math will show that the average values, independent of the number of dice used will approximate 21. For example, the average for any die is 3.5, and in each case the multiplier times the number of dice is six, so each team will average 21. Hence, the rate of the production process simulations for all four

Team No.	No. of People	No. of Dice	Multiplier
1	1	1	6
2	2	2	3
3	2	3	2
4	3	6	1

TABLE 18-1   Team Distribution of Dice

teams will average 21, but their factory simulations will perform differently because ... well, let's do the experiment.

Creating the Data

1. Roll the dice (or die) and count the total spots.

2. Multiply the total by the multiplier for your team.

3. Enter this number in the oval on the plant simulation spreadsheet, starting with Cycle 1, station 1.

4. Repeat steps 1 thru 3 for Cycle 1, station 2, and so on, through Cycle 1, station 8 ...

5. Do this for 20 cycles at least, which means several copies of the spreadsheet will be needed.

6. To simulate the process, fill in the rectangles, described in this chapter in the section entitled, "Processing the Data."

7. Calculate the production.

8. Sum the totals, as in the summary data table shown in Table 18-2.

Processing the Data

1. Each horizontal row of information is a cycle for this process, and each cycle of production goes through eight processing steps.

   The oval (see Figs. 18-1 and 18-2) represents the instantaneous capacity of that work station based on station capacity alone. Since each die has the potential of numbers 1 through 6, for Team 1 since its multiplier is six, it can be seen that the instantaneous capacity for each station is a value from 6 to 36. And in the high variability case of 1 die, the only possible values are 6, 12, 18, 24, 30, and 36.

2. The rectangle directly below the oval, Rectangle 1, is the amount of material (think of them as kits) that is available for the next station in the cycle. We assume the warehouse has infinite material, so there is no material constraint at the first station. The following then becomes the two possible constraints on the system:

   a. Instantaneous Station capacity given by the dice data

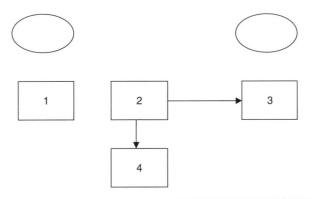

Figure 18-1 The data format for the experiment.

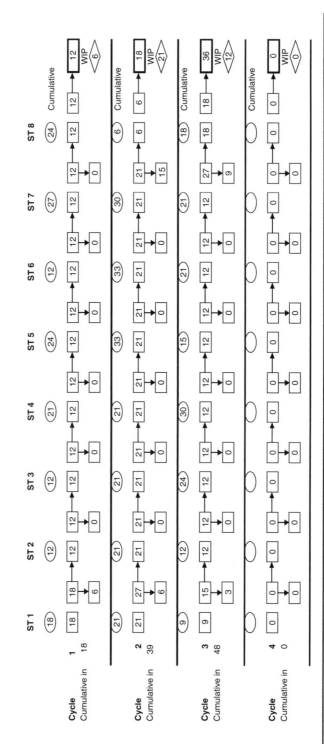

Figure 18-2 Experiment spreadsheet.

    b. Material availability, which is the kits processed and the kits remaining in WIP (the volume in Rectangle 2)

3. Next to Rectangle 1 is Rectangle 2. It represents the total available material for the subsequent processing station.

4. Rectangle 3 is then the actual production for that station based on the system constraint. Rectangle 4 then represents the amount of material left by that station in that cycle when the instantaneous station capacity (the oval) is the constraint. This amount of material, the WIP, is then available for the subsequent cycle.

    a. To make this material available for the subsequent cycle, write this value in Rectangle 4 and add it to the value in Rectangle 1 in the subsequent cycle.

    b. Place this value in Rectangle 2. This now represents the total amount of material for this station during the next cycle of production. It is the sum of the material not used in the previous cycle, plus the amount produced by the prior station, in that cycle.

5. At the end of the cycle are three rectangles and a triangle. The first two rectangles are the net production from that cycle. They are the same number. The last rectangle is the accumulated production from that cycle, plus all previous cycles.

6. The triangle is the total WIP that was accumulated in that cycle. It is the sum of all the Rectangle 4s in that cycle.

7. A simple check of your logic and math is to add up the cumulative production, plus the WIP for that cycle. This total should be equal to all the materials withdrawn (station 1 accumulated totals), up to and including that cycle.

8. A simpler approach is to write me at www.qc-ep.com and I will send the spreadsheet on a CD. Shipping and handling will be the only charges.

9. For example, I have attached the first three cycles of a spreadsheet.

    a. Review that in Cycle 1, station 1, 18 units are withdrawn from the storehouse. Hence, station 1 does its work and they are processed and transferred to station 2.

    b. However, station 2's pull capacity is only 12, so six units remain in WIP for the next cycle, 12 units are passed to station 3, and so on.

    c. At station 4, the instantaneous capacity is 21 units; however, there are only enough raw materials for 12 units, hence only 12 are processed and passed on. It goes this way thru station 8.

    d. Consequently, 12 units are produced, six remain in WIP, which checks since we started with 18.

    e. In Cycle 2, station 1 has the capacity to produce 21, so 21 units are withdrawn from the storehouse, but six units were left in WIP from the first cycle, so now station 2 has 27 units of material at its disposal. Station 2 has enough raw materials to produce 27 units, but its limit is the instantaneous station capacity of 21, so 21 units are pulled, processed, and sent to station 3. The six units not pulled by station 2 now remain for the next cycle.

    f. In Cycle 2, since the capacity of stations 3 thru 7 all exceed the raw materials availability, they all process and pass on 21 units until we get to station 8.

g. At station 8, we have enough material for 21 units, but the machine capacity is only six; hence, only six units are produced, leaving 15 units in WIP available for station 8 for the next cycle.

h. Six units are produced and added to the 12, so we have accumulated 18 units of finished goods. Our WIP totals 21. The sum of finished goods plus WIP is 39, which agrees with the total raw materials input of 39. It checks.

i. Now look at Cycle 3. Station 1 which only produces nine units, but since it had an inventory of six from the previous cycle, it was able to produce 12 units. Had there been no inventory, the rate would have been limited to nine.

j. In Cycle 3, the capacity and materials availability limit the production at stations 2, 3, 4, 5, 6, and 7 to 12 units. However, because of the inventory left over from Cycle 2, now station 8 can produce to its capacity and make 18 units. Here, the inventory has "uncoupled" the process and allowed it to produce more than the previous processing step.

The Results

We have done this experiment in many training sessions, and the results are always very similar. I have included results from one of the sessions in Table 18-2.

As expected, there is some variation in the results. Nevertheless, we will always find the following:

- Variation exists in the instantaneous capacity of the work stations.

- As the variation is reduced, WIP will decrease and both the total production and percentage of RM Turned into Sales will always increase.

- As the variation increases, WIP inventory increases and the process steps become uncoupled, creating what we call "island like production." Only then will production increase, but at the expense of huge inventory accumulations which reduce flow dramatically and consequently increase lead time significantly.

- This all occurs because we have variation and dependent events! This is a demonstration that:

  - If your system has variation, the system will need inventory to sustain production.

  - The more variation the system has, the more inventory the system will need to sustain with any given level of production.

Team	No. of Dice	Total Production Yield, Units	Total WIP, Units	% RM Turned into Sales	Avg. Prod Rate (Dice Average)	Std. Dev. of Capacity (Dice Std. Dev.)
1	1	246	168	59.4%	20.59	10.76
2	2	294	162	64.5%	20.79	6.69
3	3	328	98	77.0%	20.29	5.64
4	6	350	36	90.7%	21.18	4.01

TABLE 18-2 Typical Results from the Experiment

CHAPTER 19

Assessment Tools

Based on our experiences, which cover over 35 years of implementation efforts, I have catalogued a number of problems we have encountered in assisting various companies in their Lean initiative efforts. I have documented the key issues in four attachments:

- The Five Tests of Management Commitment to Lean Manufacturing
- The Ten Most Common Reasons Lean Initiatives Fail (in Part or Totally)
- The Five Precursors to Implementing a Lean Initiative
- Process Maturity

My hope is that facilities wishing to implement a Lean initiative will read, digest, and be guided by these documents. To this end, I hope our experiences will be helpful to others.

The Five Tests of Management Commitment to Lean Manufacturing

1. Are you actively studying about, and working at, making your facility Leaner and, hence, more flexible, more responsive, and more competitive? (We must be intellectually engaged. All must continue to learn. No spectators allowed!)

2. Are you willing to listen to critiques of your facility and then understand and change those areas in your facility that are not Lean? (We must be intellectually open.)

3. Do you honestly and accurately assess your responsiveness and competitiveness on a global basis? (We must be intellectually honest.)

4. Are you totally engaged in the Lean transition with your:
   - Time
   - Presence
   - Management attention
   - Support (including manpower, capital, and emotional support)

   (We must be engaged at the behavioral level; we must be doing it. We must be on the floor, observing and talking to people, and imagining how to do it better. Lean implementation is not a spectator sport.)

5. Are you willing to ask, answer, and act on the question: "How can I make this facility more flexible, more responsive, and more competitive? (We must be inquisitive, and be willing to listen to all personnel, including peers, superiors, and subordinates alike, no matter how painful it may be, and then be willing and able to make the needed changes.)

A "Yes" to all five questions means you have passed the commitment tests. Any "No" answer means there is an opportunity for management improvement.

The Ten Most Common Reasons Lean Initiatives Fail (in Part or Totally)

1. The facility really does not understand what a Lean initiative is—and more specifically, they do not understand the concepts of quantity control.

2. Management has not developed a rational, experience-based, sequential plan for implementation.

3. The company does not have facilitywide goals that properly promote Lean.

4. Management is not really committed to making it successful—in other words, they cannot pass the five questions in the Commitment Test.

5. Unreliable raw materials suppliers.

6. Inadequate understanding of process management, especially process stability, process capability, process bottlenecks, and machinery availability.

7. They underestimate the effort required.

8. There is an inadequate understanding of variation and its effects.

9. The planning model for production, usually MRP or something like it, is not properly managed.

10. The five precursors to Lean are not properly addressed!

The Five Precursors to Implementing a Lean Initiative

There are five major precursors to implementing a Lean initiative. These precursors are foundational issues that must be in place to make a Lean initiative successful. If they are not in place at the kickoff of the Lean initiative, it can still proceed; however, these precursors must be recognized and they must be built into the Lean initiative plan in the proper sequence (See Table 19-1).

1. High levels of stability and quality in both the product and the processes

2. Excellent machine availability

3. Talented problem solvers, with a deep understanding of variation

4. Mature continuous improvement philosophy

5. Strong proven techniques to standardize

Needed Trait	0	1	2	3	4	5
1 - High levels of stability and quality in both the product and the processes	Does not meet Level 1 criteria	"Critical" product characteristics identified; the stability and capability are known for these; many unknown losses	All key product and process characteristics are identified; most are stable; some meet min. Cpk	All key product and process characteristics are identified; are stable and meet min. Cpk requirements	A shift is in place to eliminate the need to monitor processes and build in process robustness; the need for SPC and numerical techniques is reduced	Significantly reduced need to monitor the process; robustness is built into the process
2 - Excellent machine and line availability	Does not meet Level 1 criteria	Mach availability is unknown; stock outs unknown; most maintenance is reactionary; line losses are known; >8%, not stratified	Avail. known; total losses <8%; stock outs <1%; no formal plan to improve; planned and predictive maintenance done	Total losses <3%; active plan to improve; first 4 elements of TPM in full use	Losses <2%; Level 3 plus the 5th pillar of TPM; early management and MP design	Losses <1%; Level 4 plus actively improving design of production equipment with engineering and suppliers
3 - Talented problem solvers, with a deep understanding of variation	Does not meet Level 1 criteria	Firefighting only; done by supers and engrs. only; little training; few understand variation effects	Mostly firefighting, engrs., super do PS; trained in SPC, 5 Whys, 6 Sigma, and so on	Some firefighting; all do PS; training given to all; all given training in variation reduction	Little or no firefighting; all are involved; all are trained in PS and variation reduction	No firefighting; total involvement including customers and suppliers; joint training; Lean implemented through suppliers and customers

TABLE 19-1   The Five Precursors to Lean—An Evaluation Matrix

Needed Trait	0	1	2	3	4	5
4 - Mature continuous improvement philosophy	No policy exists	Policy, but no methods to implement	Policy and methods in place; done by engineers, supers, and mgrs.	Level 3, but all are involved; CI is recorded and posted	Everyone is involved at the facility; use policy deployment to get all involved in CI planning	All are involved, including customers and suppliers
5 - Strong proven techniques to standardize	No policy; problems repeat	Policy in place; many problems repeat quickly; no clearly defined methods	Policy and methods are documented; problems repeat but after a significant time span	Policy and methods are documented; some problems repeat	Problems do not repeat	Everyone is involved, including customers and suppliers

TABLE 19-1 The Five Precursors to Lean—An Evaluation Matrix (*Continued*)

Process Maturity

Frequently, I receive requests from manufacturing firms to assist them in a Lean initiative. Typically, they have done some background research on Lean manufacturing and want guidance in making their operations Lean. Unfortunately, the most common thing I find is that they do not have the foundational work elements in place to begin a Lean initiative. More importantly, they do not recognize this and think that, independent of the current state of the process, a Lean initiative can be implemented and techniques such as *kanban* can then be readily applied. Their common perception is that they only need to complete some training and they will be on the road to a Lean enterprise. Nothing could be further from the truth. First, a number of precursors must be taken care of. For example, they must have a reasonably stable materials supply, and they must have good machine availability, to name a few. Most importantly, they must have in place a process that already produces at high-quality levels; processes that exhibit both process stability as well as adequate capability. This item is of particular importance to the success of many Lean techniques.

It is to those groups who wish to undertake a Lean initiative that this document is written. In this format, a specific order is implied. In practice, I have found that those firms that deviate greatly from this basic format spend much more time and much more of their resources in reaching the goal of becoming Lean. Those processes that produce to this Lean standard I refer to as "mature."

Process maturity differs from process "goodness." Process or product "goodness" usually means the product meets the needs of the customer. In this regard, "goodness" is usually measured by the process capability indices of Cp and Cpk for a few key product and process characteristics. If these two indices meet minimum standards, for example, they are greater than 1.33, then the customer is satisfied, declares the process to be "good enough" and the supplier is allowed to proceed with greater independence. If, however, the indices fall short of the standard, then the supplier is required to do extra processing, which is usually some form or containment or extra inspection coupled with a specific action plan to achieve the desired levels of Cp and Cpk.

Process maturity goes beyond the measure of "good enough" so that a product is not only good, but is produced with a minimum amount of waste. These processes have other characteristics, such as minimum inventories and short production lead times, to name a few. These processes are now widely referred to as Lean, and in this document, the process that produces a Lean product is a mature process.

This treatment addresses levels of process maturity for a typical manufacturing process and does not address some topics, which are outside the scope of this treatment. Other aspects of the process not addressed herein, might include ergonomics, environmental issues, and safety, to name but a few.

A process that is mature has five characteristics, which are:

1. Documentation
2. Flow, that is, a specific process routing
3. Quality understanding and performance
4. Inventory understanding and control
5. Leanness by all 20 measures; exhibit advanced levels of continuous improvements, with Lean goals driving the process

In short, process Levels 1 through 3 work to achieve a process that has met all *quality* standards of the customer. Levels 4 through 5 address the issues of *quantity* control.

Generally, the development of a process should follow a natural pattern as outlined here and in this order. Although this is not true for all work stations or even all complex process flows, it is a good general guideline and will serve you well in describing the level of process maturity. The levels are labeled as Level 1 through Level 5 and are characterized by the following:

1. Level 1

   a. A Level 1 process has a set of documentation that allows a firm to design and build a product, as well as a means to assess quality and delivery capabilities. Documentation usually includes:

      i. A complete up-to-date drawing of the part or assembly, including any part or subassembly requirements. All construction goals are clearly understood, with product and process critical characteristics defined and agreed upon

      ii. A test specification, plus all operational definitions to determine if a product is accepted or rejected

      iii. Packaging specifications

      iv. A plot plan

      v. A flow chart

      vi. A PFMEA

      vii. A Part Number Control Plan

      viii. Appropriate work instructions, including all instruction for rework

      ix. Demand rates or projections good enough for production scheduling

   b. Nearly all of the 20 Lean techniques should be designed into the process. For example, cells, multiskilled workers, *kanban*, *takt* time, leveling, and standard work, to name a few, should be built into the design.

   c. Since a Level 1 process is basically a preproduction condition, successful attainment would include such items as completion of PPAP and run-at-rate to meet customer requirements.

2. Level 2

   a. A Level 2 process has all the characteristics of a Level 1 process, plus the production process flows (has an actual process routing) in full accordance with the Level 1 documentation. Specifically:

      i. Raw materials enter the process only as noted; products leave only as documented.

      ii. No other intermediate assemblies, and so on, enter or leave the process unless documented on the flow.

      iii. All work in process flows just as in the flow chart. All good product follows the process flow diagram. All scrap and potential reworked product are properly segregated from the normal process flow. All scrapped and reworked products follow the prescribed process flow for scrap and rework.

    iv. All critical processes or process steps have completed and acceptable gauge studies.

    v. All process steps are statistically stable.

    vi. Work instructions are in place and adequately describe the work. Work instructions are scrupulously followed.

b. Again, with a Level 2 process, many Lean techniques have been incorporated into the design but the real focus of this step it to institute continuous flow of the product through the process, and to make the process statistically stable. Other Lean techniques that are focused on usually include:

    i. Flow

    ii. *Jidoka*

    iii. 5S

    iv. *Takt* time

    v. Usually only minor *kanban* implementation, such as carts, and so on

    vi. Balanced operations

    vii. Transparency

    viii. Standard work

    ix. *Kaizen*

    x. 5 Whys

c. Key process and product variables monitored include production rate, stratified defects, machine availability, First Time Yield, and OEE.

3. Level 3

a. A Level 3 process has all the characteristics of a Level 2 process, plus the quality levels are understood and are acceptable to the needs of the customer—in other words, all process steps have adequate Cpks. A continuous quality improvement effort is in place. Specifically:

    i. A modified Part Number Control Plan that addresses the quality issues found.

    ii. Specific measures of internal quality, with long-term graphs available on the shop floor.

    iii. Specific measures of external quality, with long-term graphs available on the shop floor.

    iv. A documented plan of continuous improvement for quality, for both internal and external quality measures.

    v. Continuous improvement quality goals in place, for both internal and external quality measures.

b. Lean techniques that can now mature at this phase, include:

    i. More *kanbans*

    ii. Minimum lot sizes

    iii. SMED/OTS

        iv. Store/buffer/safety stocks

        v. *Poka-yoke*

  c. Quality measures are fully mature at this point and used as process improvement drivers.

4. Level 4

  a. A Level 4 process has all the characteristics of a Level 3 process, plus all inventory levels are controlled and minimized.

  b. Lean techniques at this level include:

      i. *Kanban* is further improved. *Kanbans* control all in-plant materials flow.

      ii. SMED/OTS become critical for further lot size reductions.

      iii. Standard inventory is fully reviewed.

      iv. Autonomation is more fully developed.

      v. JIT must be fully embraced with a "JIT support system," including the culture change to:

        1. JIT material supply and product production

        2. JIT problem solving

        3. JIT maintenance

      vi. At Level 4, frequently major improvements are made in the flow, and a cell redesign is often beneficial, with a critical look at staffing, flow, rebalancing, and even the basic layout.

      vii. Also at Level 4, it is common to prepare a meaningful future state value steam analysis map to guide future projects.

  c. Key process measures include the previously mentioned ones, especially manufacturing lead time plus inventory turns. Also, it is common to address value added (VA) steps as a percentage of total process steps, and also to see VA time as a percentage of manufacturing lead time as process improvement focusing tools.

5. Level 5

  a. A Level 5 process has all the characteristics of a Level 4 process, plus all 20 Lean techniques are in full maturity.

  b. Efforts to improve VA steps and VA time are in place and the supply value stream becomes a key focal point, whereby suppliers and customers are included in improving the overall process. A dynamic future state value stream map is a fully functional tool that is guiding process improvements. This will, by necessity, take you outside the bounds of the plant.

CHAPTER 20

A House of Lean

Figure 20-1 shows a "House of Lean," which was developed for use with a specific client based on their individual needs. This is a good tool that gives a graphic description of your Lean initiative to all employees. It is a way to capture the entire program in a pictorial format. I have found that it often makes the overall effort more easily visualized and hence understood. Most of the techniques listed are common but some unique items are listed, too. For example, Process Simplification was a key component for this company. They needed to make process changes to eliminate unneeded steps and also to change some processes that had poor availability due to using outdated technology. These items needed to be taken care of first, or they would have created unnecessary future work. Your *sensei* should be able to identify these issues and help you with the order of activities.

The House of Lean is a metaphor that is designed to show how the various topics discussed in Lean manufacturing fit together and interact. As with all metaphors, they have limits. Nevertheless, it is a good tool to use in teaching:

- Objectives
- Strategy, tactics, and skills
- Foundational elements

Which ones comprise *your* Lean manufacturing system?

When we discuss strategies, we speak at the conceptual level, tactics are normally small group activities that are required to achieve the strategies, while skills are the individual behaviors that must be executed to accomplish the tactics.

For example, it is clear from Ohno's writings that the two key strategies of the Toyota Production System are just in time (JIT) and *jidoka*. One of the tactics of JIT is *kanban*. To execute *kanban*, we need a variety of skills, such as the skill of making the *kanban*, sizing the *kanban* volumes, planning the circulation of the *kanban*, and so on.

My advice is to thoroughly evaluate your needs and create your own House of Lean to explain them.

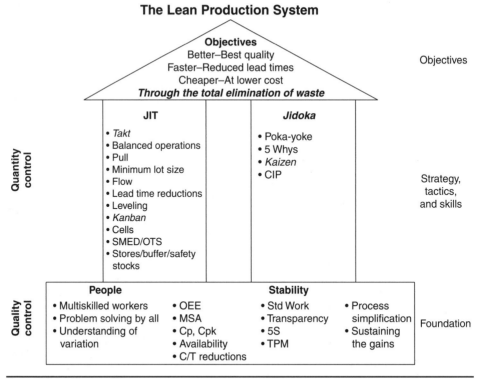

FIGURE 20-1 House of lean.

Glossary

5 Whys   A problem solving method which employs the technique to continue to ask "Why?" to explore the cause and effect relationship, in an attempt to find the root cause. It requires a great deal of experience and a strong knowledge base to do well.

5S   A tool in the Lean Toolbox which focuses on workplace readiness. The 5 S's stand for Sort, Set to order, Shine, Standardize, and Sustain.

Andon   A warning device, normally a light, to signal an abnormality, it is a part of the system of transparency.

Autonomation   Ohno's word for *Jidoka*. It literally means, automation with a human touch designed to supply 100% inspection, sort and then initiate problem solving.

Availability   The concept that production facilities are capable of producing when they are scheduled to be producing, or when needed.

Balancing   Synchronizing operations, generally making sure that each step has the same process cycle times.

Balancing chart   Normally a bar chart which shows process steps on the X- axis and cycle times on the Y- axis.

Bottleneck   Any activity or process step which limits production.

BRP   Business Resource Planning, another version of MRPII.

Buffer stocks   A type of inventory which is held to account for external demand variations.

Buffers   An excess resource in the production system, usually designed to compensate for some type of variation so demand can be met. There are three types of buffers: capacity, time, and inventory.

Catchball   A process used in *Hoshin Kanri* planning where the boss decides what to do, the subordinate decides how to do it and an interchange is effected to make sure the goals and the means to achieve the goals are understood and doable.

Cells   A manufacturing equipment layout where people and machines are in close proximity to reduce transportation and WIP inventories. Cells are designed to achieve one piece flow.

Cells, U & C shaped   Cells' layouts are normally in a U or C shape so the incoming raw materials and outgoing products are near one another. It aids in materials and information handling.

Changeover   Converting a machine, or process to make a different model or different product.

Changeover time   The time it takes from the last good part prior to the changeover to the first good part after the changeover.

CIP   Continuous Improvement Process.

Constraint   Another word for bottleneck, see bottleneck.

Continuous improvement process   A series of sequential steps to forever analyze a product or a process and continue to increase the value added portion.

Control chart   A statistical tool invented by Walter Shewhart to evaluate the statistical properties of a process. Control charts will allow you to characterize both the variation in your process and if you are producing to the target specification or not.

Correlation and regression   A technique used to study the relationship of cause and effect and the impact that variation has on this relationship.

Cp, Cpk   Process capability indices.

Culture   The combined thoughts, actions, beliefs, artifacts and language of a group of people. It is "How we do things around here".

Current state, VSM   A current state value stream map, sometimes called an Information and Materials Flow Diagram, see also PSVSM.

Customer   Your client; they are usually defined by four characteristics; they are courted to consume your product; they pay for your product or service; they pick up and use your product or service; and if they are dissatisfied with your product or service they can cause you immediate discomfort, that is they can complain and get action. Your external customer is the entity which pays you, however, the customer is also the next step in the process and the needs of the internal customers must be met, just as the needs of the external customers must be met.

Defects   Things gone wrong with your products; quality characteristics which are not met.

Deming   W. Edwards Deming, the great statistician and quality guru, creator of Deming's 14 Obligations of Management and author of Out of Crisis.

DOE   Designs of experiments, an advanced statistical tool used for in-process understanding and optimization.

Downtime   Time that a process or machine is not running.

Effectiveness   The ability to achieve a goal.

Efficiency   Achieving a goal using minimum resources.

ERP   Enterprise resource planning, another version of MRPII.

Excess processing   One of the seven wastes. Performing work on a product beyond what the customer considers value, beyond what they are willing to pay for.

FIFO   Acronym for First in, First out.

FIFO lanes   Processing lanes of goods where FIFO materials handling is practiced.

Finished goods inventory   The completed production of a product held to assure supply to a customer in a make-to-stock supply system. It has three components: stores inventory, buffer stocks, and safety stocks.

First piece lead time   The time it takes to produce the first piece of a batch. A key factor in quality responsiveness. The objective is to reduce this to a minimum.

Flow   The concept that once started a product continues to move with value added work being performed, during the entire manufacturing process.

Flow line   A linear arrangement of processing close coupled equipment, as distinguished from a U cell, for example.

FMEA   Acronym for Failure Mode Effects Analysis, a quality tool designed to sort out quality problems, preproduction and implement countermeasures.

FSVSM   Future state value stream map, a map showing the information and materials flow for a product depicting some hoped for, future state.

Future state, VSM   Value stream map with future conditions designed into it, see Present State VSM.

Hawthorne Effect   The positive effect that is achieved in improved performance when attention is paid to people.

Heijunka   Japanese word for leveling, specifically leveling production which means stabilizing the rate in a narrow band; no large ups or downs in rate. A *heijunka* board is a planning board used to level production and become part of the visual system to evaluate production status.

***Hoshin Kanri* planning**   A strategic planning method developed in Japan, it means policy deployment and is one of the very few tangible top management tools in the Lean toolbox.

Inventory   One of the seven wastes. It includes, finished goods which have not been picked up by the customer and all the materials in the system which you intend to convert to finished goods, including raw materials and WIP. All inventory is waste, although some is necessary considering the present conditions.

Inventory turns   A measure of the rate at which inventory turns over each year. Twelve inventory turns mean that you have 30 days of inventory on hand.

Ishikawa   A Japanese quality guru who wrote extensively about quality, the creator of the *Ishikawa* diagram or sometimes called a Fishbone or Cause-Effect analysis.

Island production   A type of production where the work stations are set up far apart, typically with lots of inventory in front of and behind the island. Generally an inferior method of production to the cellular manufacturing.

Jidoka   One pillar of the TPS. *Jidoka* is a method to prevent bad material from advancing in the production system and to find system weaknesses and fix them.

JIT   Just in time, the other pillar of the TPS. The concept is to avoid waste by supplying exactly the right quantity of materials to exactly the right location at exactly the right time. It is quantity control.

JUSE   Japanese Union of Scientists and Engineers.

Just in time   See JIT.

Kaikaku   Radical change, it is super fast, super large *kaizen*.

Kaizen   The concept of making continual product and process improvements, usually small and typically done by the entire workforce.

Kanban   *kanban* means card, it is the method the JIT pillar uses to minimize inventory and follow pull-demand system rules to reduce wastes.

Lead time   The elapsed time it takes, from start to end, to produce a product.

Lead time, first piece   The elapsed time it takes for one piece to completely flow through the production process.

Lean   Short form of Lean Manufacturing System. The generic name given to the Toyota Production System. Through use and misuse it has come to have many applications, some good, some bad, but the popular concept is to be able to make more product, using less resources. This addresses many of the technical aspect of Lean. However, to get the full definition of Lean refer to Chap. 2.

Leanspeak   The unique language used in Lean manufacturing using common words like waste, pull and flow, in a different sense than the typical definition, plus the use of unique lean terms such as catchball and autonomation.

Leveling, model mix   Avoiding the batch production of models of a given product.

Leveling, production   Avoiding the unnecessary changes in production rates.

Make-to-order   A production system with no finished goods inventory. Production does not start until a specific order is received and they are shipped directly. A system with no finished goods inventory except those materials awaiting pick up.

Make-to-stock   A production system with finished goods inventory and all production is sent to a storehouse for holding prior to shipment.

Mass production systems   Production systems designed to produce in large volumes using large batch philosophy in an attempt to be cost effective, which it seldom is. Characterized by push production systems using large batches, long production runs, large inventories, and island production. It is characterized by inflexible, nonresponsive systems with long lead times and a very low percentage of value added work; as compared to Lean.

MassProd   The abbreviation for Mass Production Systems.

Metric   The measurements used to evaluate plant performance. For example, OEE, On time delivery, and Lead time are all metrics.

Minimum lot size   An attempt to shrink lot sizes to reduce inventory, specifically WIP inventory, and to improve flow and reduce lead time for the production lot AND the first piece.

Movement   One of the seven wastes, movement of people.

MRPII   Manufacturing Resource Planning Two, is a planning program designed to integrate business needs down to the production floor and among other things, create a meaningful production plan. In short, it is too slow for production floor planning and is uniformly overused in a lean facility but has other necessary functions. For the purposes of Lean, it can be defined as an inadequate planning tool for hourly, daily and sometimes even weekly production plans.

MSA   Measurement System Analysis, a statistical method to determine the usefulness of the measurement system for both products and processes. Its chief benefit is the ability to find and classify variation in the measurement system.

NVA   Non-value-added, the opposite of VA.

OEE   Overall Equipment Effectiveness. A means to numerically describe production effectiveness, the ability to produce good product. Within OEE are characterized the three key production losses: quality, availability, and cycle time losses.

Ohno   Taiichi Ohno, long time Chief Engineer for Toyota and accepted architect of the TPS.

One piece flow   This concept starts at the customer, whereby the customer purchases a single piece and the manufacturing system should replenish only that piece. Hence the Lean system strives to make just one piece at a time, this is true one piece flow.

Optimum, local   An optimum condition for some local situation.

Optimum, system   An optimum condition for the overall system, and it must not be subordinated to any local optima.

OTED   One Touch Exchange of Dies, see OTS.

OTS   One Touch Setup, see OTED.

Overproduction   The largest of the seven wastes, it includes all excess production and production made too soon.

Pacemaker step   The step of the process which determines the process rate and the process model mix. It is the step where scheduling will send production orders.

Paradigm   Your mental image of a concept, often developed unconsciously but paradigms often shape how you act.

PDCA   Plan-Do-Check-Act. This is the iterative process improvement cycle which is inherent within the *kaizen* improvement process.

PFEP   A Lean acronym, Plan for every part, used in *Kanban* design, for example.

PFMEA   Process failure mode effects analysis, a structured process to determine, before final design, which aspects of a process need additional controls so the production process will be more safe, stable and have a higher yield.

Pitch   A time interval equal to *takt* time multiplied by pack out size, normally the minimum quantity released from the pacesetter and the practical extent to leveling, considering the current packaging.

Poka-yoke   Error proofing. For example most cars have thousands of *poka yokes*. While filing your gas tank there could be several such, for example a device to connect your gas cap to the car so you do not lose it; an automatic shutoff on the gas pump; a ratcheting device to prevent over tightening of the gas cap; and a warning light on the door to warn you if it is not closed properly.

PPAP   Production Part Approval Process, a method developed by the Automotive Industry Action Group, to standardize the process of obtaining customer approval of a product prior to mass production.

Process   A sequential series of steps which are designed to produce a product or a service.

Process cycle time   The time it takes to complete the work in a process or a process step. Generally lead time is the term used instead of cycle time when we speak of the entire production process.

Product family   A group of products which have the same basic complement of parts and are produced using the same basic production process.

PSVSM   Present State Value Stream Map, a map showing the information and materials flow for a product using present state condition.

Pull   The Lean production supply concept; production should only occur when the customer removes a product, the opposite of a push system.

Push   Production is determined by schedules, resource rates, and goals which are generally designed to create an optimum condition at the production source, but it ignores the system optimum. Production will continue, regardless of usage, until the planning system is tweaked to modify the release of jobs.

QFD   Quality Function Deployment; a technique to connect customer needs to process parameters.

Replenishment   To restock. However, in Lean the replenishment concept is JIT.

Safety stock   A type of stock, the volume of which is statistically determined. It is designed to take care of internal variations in a make-to-stock-system.

SAP   Another version of MRPII.

Sensei   A teacher, literally one who has gone before, hence the concept of wise and experienced.

Shingo   *Shigeo Shingo*, one of the architects of the TPS. Credited with much of the technology of SMED and *Poka yokes*, wrote extensively on these two topics.

Skills   Individual behaviors necessary to execute work.

SKU   Another term for a unique part number.

SMED   Single Minute Exchange of Dies, the quick changeover methodology, largely developed by *Shingo*, and absolutely necessary in most plants to avoid large batch production.

SPC   Statistical Process Control, a series of technical tools often equated to *Ishikawa*'s Seven Tool, but more and more equated to just control charting. See control chart.

Standard Inventory   The inventory designed to be at any given work station, documented on the Work Instructions and Standard Work also.

Standard work   Not standardized operations, Standard Work is a document written for the manager and the engineer, not the line worker. It contains three elements: the work sequence, the standard inventory and the cycle time. It is part of the system of visual management, transparency system.

Statistical stability   A technical term, developed by Walter Shewhart and operationally it means the process is in statistical control when placed on a control chart. In lay terms, it means the system is predictable.

Stores inventory   The inventory of finished goods which is built up between customer pickups.

Strategies   Concepts of how you intend to attack a problem or situation; supported by tactics, which are in turn supported by skills; usually expressed in the form of a plan.

Supermarket   A controlled volume of inventory to be replenished by the upstream process, also called stores.

Supplier   Those entities which provide resources, usually, raw materials to a process. We have external suppliers and our own employees are internal suppliers.

SWCT   Standard Work Combination Table.

Synchronized production   The concept that all process steps take the same cycle time. So in theory, in a cell, all parts are completed simultaneously and consequently are moved to the next step simultaneously. A concept to be achieved, rather than a reality.

Synchronized supply   The concept of supplying the product to your customer not only in the volume he desires on the delivery date he desires but also producing it at the rate he consumes it even if he has periodic pickups. This concept provides maximum flexibility and responsiveness. This is the manifestation of the concept of leveling and a key batch destruction strategy.

Tactics   Small groups of people acting together to comply with the strategy.

Takt   German word for rhythm. In Lean manufacturing the formula is, the available work time divided by the customer demand, over a time interval such as a month, week, or day. It is the "normalized" rate of supply to the customer. It is normalized to your production schedule.

TPM   Total Productive Maintenance, (not Preventive but Productive); a methodology to eliminate the 6 maintenance losses.

TPS   Toyota Production System.

Transparency   A concept for management which allows you to "see" what is happening in production without using computers, charts tables, or graphs. See visual management.

Transportation   One of the seven wastes, movement of inventory, WIP and finished goods, including all activities necessary to achieve the transportation including packaging.

TWO DIME   A mnemonic for the seven wastes: transportation, waiting, overproduction, defective parts, inventory, movement, and excess processing.

Uptime   The time that a process or a machine is running.

VA   Value-added. In Leanspeak, it refers to something the customer is willing to pay for.

Value   What the customer is willing to pay for.

Value added work   Those work steps which add value to the product; processing which augments the form, fit or funciton of a product.

Value stream   The process flow which applies value to the raw materials. The value stream culminates in a product for the customer.

Value stream mapping   A technique to graphically describe the value stream so a system review of lead time and value added time can be made. A key tool in the battle of waste reduction.

Visual management   The placing of tools, materials, and information in plain view using simple tools so the status of the process or product can be understood at a glance. Transparency.

VSM   Value Stream Mapping.

Waiting   One of the seven wastes, it is the waiting of people for any reason including waiting for information, parts, or machines.

Waste   Things the customers is not willing to pay for; the focus of the TPS; "the absolute elimination of waste," T. Ohno.

WIP   Work In Process. All materials in the production process once they are withdrawn from the storehouse until they are stored as finished goods. One of the three basic forms of inventory which are raw materials, WIP, and finished goods.

Work element stack   A graphic tool to show how work elements combine so line balancing can be achieved.

Yokoten   The concept of sharing process improvement ideas with others and applying these in other applications beyond the original concept.

Bibliography

Akao, Y. (1991). *Hoshin Kanri: Policy Deployment for Successful TQM*, Productivity Press, Cambridge, MA.

Bhote, Keki R. (1988). *World Class Quality*, AMA, New York, NY.

Deming, W. E. (1982). *Out of the Crisis*, MIT CAES, Cambridge, MA.

Kemp, S. (2004). *Project Management DeMYSTIFIED*, McGraw-Hill, New York, NY.

Goldratt, E. M. and Cox, J. (1986). *The Goal*, North River Press, Croton-on-Hudson, NY.

Hall, R. W. (1983). *Zero Inventories*, Dow Jones-Irwin, Homewood, IL.

Harry M. J. and Lawson, J. R. (1990). *Six Sigma Producibility Analysis and Process Character Publishing*, Addison-Welsley Publishing Co., Reading, MA.

Hopp, W. J. and Spearman, M. L. (2008). *Factory Physics*, McGraw-Hill, New York, NY.

Jackson, T. L. (1996). *implementing A Lean Management System*, Productivity Press. New York, NY.

Kepner, C. H. and Tregoe, B. B. (1981). *The New Rational Manager*, Princeton Research Press, Princeton, NJ.

King, B. (1989). *Hoshin Planning: The Developmental Approach*, GOAL/QPC, Metheun, MA.

Ohno, T. (1988). *Toyota Production System Beyond Large-Scale Production*, Productivity Press, Portland, OR.

Peck, M. S. (1987). *The Different Drum*, Simon and Schuster, New York, NY.

Rother, M. and Harris, R. (2001). *Creating Continuous Flow,* Lean Enterprise Institute, Brookline, MA.

Rother, M. and Shook, J. (1999). *Learning to See*, Lean Enterprise Institute, Brookline, MA.

Scherkenbach, W. (1986). *The Deming Route to Quality and Productivity*, CEE Press Books, Washington, DC.

Scholtes, P. R. (1988). *The Team Handbook*, Joiner Associates, Madison, WI.

Schonberger, R. (1986). *World Class Manufacturing*, The Free Press, New York, NY.

Shingo, S. (1985). *A Revolution in Manufacturing: The SMED System*, Productivity Press, Cambridge, MA.

Shingo, S. (1989). *A Study of the Toyota Production System from an Industrial Engineering Viewpoint*, Productivity Press, Portland, OR.

Small, B. B (ed.) (1956). *Statistical Quality Control Handbook*, AT&T, Indianapolis, IN.

Womack, J. P., Jones, D. T., and Roos, D. (1990). *The Machine That Changed the World: The Story of Lean Production*, Rawson Associates, New York, NY.

Index

7/10 -0

11|13 H